THE ASTRONOMY BOOK

THE
ASTRONOMY
BOOK

DK LONDON

SENIOR EDITOR
Victoria Heyworth-Dunne

US EDITOR
Margaret Parrish

SENIOR ART EDITORS
Gillian Andrews, Nicola Rodway

MANAGING EDITOR
Gareth Jones

SENIOR MANAGING ART EDITOR
Lee Griffiths

ART DIRECTOR
Karen Self

ASSOCIATE PUBLISHING
DIRECTOR
Liz Wheeler

PUBLISHING DIRECTOR
Jonathan Metcalf

SENIOR JACKET DESIGNER
Mark Cavanagh

JACKET EDITOR
Claire Gell

JACKETS DESIGN
DEVELOPMENT MANAGER
Sophia MTT

PRE-PRODUCTION PRODUCER
Jacqueline Street-Elkayam

SENIOR PRODUCER
Mandy Inness

DK DELHI

JACKET DESIGNER
Suhita Dharamjit

EDITORIAL COORDINATOR
Priyanka Sharma

SENIOR DTP DESIGNER
Harish Aggarwal

MANAGING JACKETS EDITOR
Saloni Singh

produced for DK by
TALL TREE LTD.

EDITORS
Rob Colson, David John

DESIGN
Ben Ruocco

ILLUSTRATIONS
James Graham

original styling by
STUDIO 8

First American Edition, 2017
Published in the United States by
DK Publishing, 345 Hudson Street,
New York, New York 10014

Copyright © 2017
Dorling Kindersley Limited
DK, a Division of Penguin
Random House LLC
17 18 19 20 21 10 9 8 7 6 5 4 3
003—283974—Sep/2017

Published in Great Britain by
Dorling Kindersley Limited.

A catalog record for this book is available
from the Library of Congress.

ISBN: 978-1-4654-6418-7

DK books are available at special discounts
when purchased in bulk for sales
promotions, premiums, fund-raising,
or educational use. For details, contact:
DK Publishing Special Markets, 345 Hudson
Street, New York, New York 10014
SpecialSales@dk.com

Printed in China

A WORLD OF IDEAS:
SEE ALL THERE IS TO KNOW

www.dk.com

CONTRIBUTORS

JACQUELINE MITTON, CONSULTANT EDITOR

Jacqueline Mitton is the author of more than 20 books on astronomy, including books for children. She has been a contributor, editor, and consultant for many other books. Becoming an astronomer was Jacqueline's childhood ambition. She studied physics at Oxford University and then earned her Ph.D. at Cambridge, where she still lives.

DAVID W. HUGHES

David W. Hughes is Emeritus Professor of Astronomy at the University of Sheffield, UK. He is an international authority on comets, asteroids, and the history of astronomy. He has spent more than 40 years explaining the joys of astronomy and physics to his students, and has published well over 200 research papers, as well as books on the moon, the solar system, the universe, and the Star of Bethlehem. He was a co-investigator on the European Space Agency's GIOTTO space mission to Halley's Comet and also on ESA's Smart 1 mission to the moon. David has served on a host of space and astronomy committees, and has been a vice president of both the Royal Astronomical Society and the British Astronomical Association.

ROBERT DINWIDDIE

Robert Dinwiddie is a science writer specializing in educational illustrated books on astronomy, cosmology, earth science, and the history of science. He has written or contributed to more than 50 books, including the DK titles *Universe*, *Space*, *The Stars*, *Science*, *Ocean*, *Earth*, and *Violent Earth*. He lives in southwest London and enjoys travel, sailing, and stargazing.

PENNY JOHNSON

Penny Johnson started out as an aeronautical engineer, working on military aircraft for 10 years, before becoming a science teacher, and then a publisher producing science courses for schools. Penny has been a full-time educational writer for the last 15 years.

TOM JACKSON

Tom Jackson is a science writer based in Bristol, UK. He has written about 150 books and contributed to many others, covering all kinds of subjects from fish to religion. Tom writes for adults and children, mostly about science and technology, with a focus on the histories of the sciences. He has worked on several astronomy books, including collaborations with Brian May and Patrick Moore.

CONTENTS

THE RISE OF ASTROPHYSICS
1850–1915

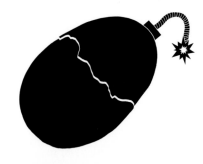

ATOMS, STARS, AND GALAXIES
1915–1950

NEW WINDOWS ON THE UNIVERSE
1950–1975

THE TRIUMPH OF TECHNOLOGY
1975–PRESENT

INTRODU

CTION

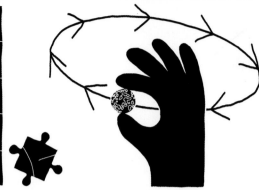

Throughout history, the aim of astronomy has been to make sense of the universe. In the ancient world, astronomers puzzled over how and why the planets moved against the backdrop of the starry sky, the meaning of the mysterious apparition of comets, and the seeming remoteness of the sun and stars. Today, the emphasis has changed to new questions concerning how the universe began, what it is made of, and how it has changed. The way in which its constituents, such as galaxies, stars, and planets, fit into the larger picture and whether there is life beyond Earth are some of the questions humans still endeavor to answer.

Understanding astronomy

The baffling cosmic questions of the day have always inspired big ideas to answer them. They have stimulated curious and creative minds for millennia, resulting in pioneering advances in philosophy, mathematics, technology, and observation techniques. Just when one fresh breakthrough seems to explain gravitational waves, another discovery throws up a new conundrum. For all we have learned about the universe's familiar constituents, as seen

through telescopes and detectors of various kinds, one of our biggest discoveries is what we do not understand at all: more than 95 percent of the substance of the universe is in the form of "dark matter" and "dark energy."

The origins of astronomy

In many of the world's most populated areas today, many of us are barely aware of the night sky. We cannot see it because the blaze of artificial lighting overwhelms the faint and delicate light of the stars. Light pollution on this scale has exploded since the mid-20th century. In past times, the starry patterns of the sky, the phases of the moon, and the meanderings of the planets were a familiar part of daily experience and a perpetual source of wonder.

Few people fail to be moved the first time they experience a clear sky on a truly dark night, in which the magnificent sweep of the Milky Way arches across the sky. Our ancestors were driven by a mixture of curiosity and awe in their search for order and meaning in the great vault of the sky above their heads. The mystery and grandeur of the heavens were explained by the spiritual and divine. At the same time, however, the orderliness and

predictability of repetitive cycles had vital practical applications in marking the passage of time.

Archaeology provides abundant evidence that, even in prehistoric times, astronomical phenomena were a cultural resource for societies around the world. Where there is no written record, we can only speculate as to the knowledge and beliefs early societies held. The oldest astronomical records to survive in written form come from Mesopotamia, the region that was between and around the valleys of the Tigris and Euphrates rivers, in present-day Iraq and neighboring countries. Clay tablets inscribed with astronomical information date back to about

Philosophy is written in this grand book, the universe, which stands continually open to our gaze.
Galileo Galilei

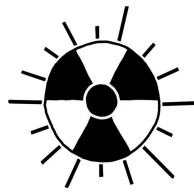

1600 BCE. Some of the constellations (groupings of stars) we know today have come from Mesopotamian mythology going back even earlier, to before 2000 BCE.

Astronomy and astrology

The Babylonians of Mesopotamia were greatly concerned with divination. To them, planets were manifestations of the gods. The mysterious comings and goings of the planets and unusual happenings in the sky were omens from the gods. The Babylonians interpreted them by relating them to past experience. To their way of thinking, detailed records over long periods were essential to establish connections between the celestial and the terrestrial, and the practice of interpreting horoscopes began in the 6th century BCE. Charts showed where the sun, moon, and planets appeared against the backdrop of the zodiac at some critical time, such as a person's birth.

For some 2,000 years, there was little distinction between astrology, which used the relative positions of celestial bodies to track the course of human lives and history, and the astronomy on which it relied. The needs of astrology, rather than pure curiosity, justified observation of the heavens. From the mid-17th

century onward, however, astronomy as a scientific activity diverged from traditional astrology. Today, astronomers reject astrology, because it is unfounded in scientific evidence, but they have good reason to be grateful to the astrologers of the past for leaving an invaluable historical record.

Time and tide

The systematic astronomical observations once used for astrology started to become increasingly important as a means of both timekeeping and navigation. Countries had highly practical reasons—civil, as well as military — to establish national observatories, as the world industrialized and international trade grew. For many centuries, only astronomers had the skills and equipment to preside over the world's timekeeping. This remained the case until the development of atomic clocks in the mid-20th century.

Human society regulates itself around three natural astronomical clocks: Earth's rotation, detectable by the apparent daily march of the stars around the celestial sphere to give us the day; the time our planet takes to make a circuit around the sun, otherwise known as a year; and the monthly cycle of the

moon's phases. The combined motion in space of Earth, the sun, and the moon also determines the timing and magnitudes of the oceanic tides, which are of crucial importance to coastal communities and seafarers.

Astronomy played an equally important role in navigation, the stars acting as a framework of reference points visible from anywhere at sea (cloud permitting). In 1675, British King Charles II commissioned an observatory, the Royal Observatory at Greenwich, near London. The instruction to its director, the first Astronomer Royal, John Flamsteed, was to apply himself diligently to making the observations needed "for the perfecting of the art of navigation." »

You have to have the imagination to recognize a discovery when you make one.
Clyde Tombaugh

Astronomy was largely discarded as the foundation of navigation in the 1970s, and replaced by artificial satellites, which created a global positioning system.

The purpose of astronomy

The practical reasons for pursuing astronomy and space science may have changed, but they still exist. For example, astronomy is needed to assess the risks our planet faces from space. Nothing illustrated Earth's apparent fragility more powerfully than the iconic images, such as "Earthrise" and "Blue Marble," taken from space by Apollo astronauts in the 1960s. These images reminded us that Earth is a small planet adrift in space. As

What a wonderful and amazing scheme have we here of the magnificent vastness of the universe.
Christiaan Huygens

surface inhabitants, the protection afforded by the atmosphere and Earth's magnetic field may make us feel secure, but in reality we are at the mercy of a harsh space environment, blasted by energetic particles and radiation, and at risk of colliding with rocks. The more we know about that environment, the better equipped we are to deal with the potential threats it presents.

A universal laboratory

There is another very important reason for doing astronomy. The universe is a vast laboratory in which to explore the fundamental nature of matter, and of time and space. The unimaginably grand scales of time, size, and distance, and the extremes of density, pressure, and temperature go far beyond the conditions we can readily simulate on Earth. It would be impossible to test the predicted properties of a black hole or watch what happens when a star explodes in an Earth-bound experiment.

Astronomical observations have spectacularly confirmed the predictions of Albert Einstein's general theory of relativity. As Einstein himself pointed out, his theory explained apparent anomalies in Mercury's orbit, where Newton's theory of gravity failed. In 1919,

Arthur Eddington took advantage of a total solar eclipse to observe how the paths of starlight deviated from a straight line when the light passed through the gravitational field of the sun, just as relativity predicted. Then, in 1979, the first example of a gravitational lens was identified, when the image of a quasar was seen to be double due to the presence of a galaxy along the line of sight, again as relativity had predicted. The most recent triumphant justification of Einstein's theory came in 2015 with the first detection of gravitational waves, which are ripples in the fabric of spacetime, generated by the merging of two black holes.

When to observe

One of the main methods scientists use to test ideas and search for new phenomena is to design experiments and carry them out in controlled laboratory conditions. For the most part, however, with the exception of the solar system—which is close enough for experiments to be carried out by robots—astronomers have to settle for a role as passive collectors of the radiation and elementary particles that happen to arrive on Earth. The key skill astronomers have mastered is that of making informed choices about

what, how, and when to observe. For instance, it was through the gathering and analysis of telescopic data that the rotation of galaxies could be measured. This, in turn, quite unexpectedly led to the discovery that invisible "dark matter" must exist. In this way, astronomy's contribution to fundamental physics has been immense.

Astronomy's scope
Up to the 19th century, astronomers could only chart the positions and movements of heavenly bodies. This led the French philosopher Auguste Comte to state in 1842 that it would never be possible to determine the compositions of planets or stars. Then, some two decades later, new techniques for the spectrum analysis of light began to open up the possibility of investigating the physical nature of stars and planets. A new word was invented to distinguish this new field from traditional astronomy: astrophysics.

Astrophysics became just one of many specialisms in the study of the universe in the 20th century. Astrochemistry and astrobiology are more recent branches. They join cosmology—the study of the origin and evolution of the universe as a whole—and celestial mechanics, which is the branch of astronomy concerned with the movement of bodies, especially in the solar system. The term "planetary science" encompasses every aspect of the study of planets, including Earth. Solar physics is another important discipline.

Technology and innovation
With the spawning of so many branches of enquiry connected with everything in space, including Earth as a planet, the meaning of the word "astronomy" has evolved once again to become the collective name encompassing the whole of the study of the universe. However, one closely related subject does not come under astronomy: "space

If astronomy teaches anything, it teaches that man is but a detail in the evolution of the universe.
Percival Lowell

science." This is the combination of technology and practical applications that blossomed with the establishment of the "space age" in the mid-20th century.

Collaboration of science
Every space telescope and mission to explore the worlds of the solar system makes use of space science, so sometimes it is hard to separate it from astronomy. This is just one example of how developments in other fields, especially technology and mathematics, have been crucial in propelling astronomy forward. Astronomers were quick to take advantage of the invention of telescopes, photography, novel ways of detecting radiation, and digital computing and data handling, to mention but a few technological advances. Astronomy is the epitome of "big science"—a large-scale scientific collaboration.

Understanding our place in the universe goes to the heart of our understanding of ourselves: the formation of Earth as a life-supporting planet; the creation of the chemical building blocks from which the solar system formed; and the origin of the universe as a whole. Astronomy is the means by which we tackle these big ideas. ∎

FROM M

TO SCIE

600 BCE—1550 CE

YTH
NCE

Anaximander of Miletus produces one of the earliest attempts at a **scientific explanation** of the universe.

In his *On the Heavens*, **Aristotle** outlines an **Earth-centered model** of the universe. Many of his ideas will dominate thinking for 2,000 years.

In Alexandria, **Eratosthenes** measures the **circumference of Earth** and estimates the distance to the sun.

c.550 BCE   **350 BCE**   **c.200 BCE**

c.530 BCE   **c.220 BCE**   **c.150 CE**

Pythagoras establishes a school in Croton, where he promotes the idea of a cosmos in which bodies move in **perfect circles**.

Aristarchus of Samos proposes a **sun-centered model** of the universe, but his idea does not gain wide acceptance.

Ptolemy writes the *Almagest*, which sets out an **Earth-centered model** of the universe that becomes widely accepted.

The traditions on which modern astronomy is built began in ancient Greece and its colonies. In nearby Mesopotamia, although the Babylonians had become highly proficient at celestial forecasting using complicated arithmetic, their astronomy was rooted in mythology, and their preoccupation was with divining the future. To them, the heavens were the realm of the gods, outside the scope of rational investigation by humans.

By contrast, the Greeks tried to explain what they observed happening in the sky. Thales of Miletus (c.624–c.546 BCE) is regarded as the first in a line of philosophers who thought that immutable principles in nature could be revealed by logical reasoning. The theoretical ideas put forward two centuries later by Aristotle (384–322 BCE) were to underpin the whole of astronomy until the 16th century.

Aristotle's beliefs

Aristotle was a pupil of Plato, and both were influenced by the thinking of Pythagoras and his followers, who believed that the natural world was a "cosmos" as opposed to "chaos." This meant that it is ordered in a rational way rather than incomprehensible.

Aristotle stated that the heavenly realms are unchanging and perfect, unlike the world of human experience, but he promoted ideas that were consistent with "common sense." Among other things, this meant Earth was stationary and at the center of the universe. Although it contained inconsistencies, his philosophy was adopted as the most acceptable overall framework of ideas for science and was later incorporated into Christian theology.

Geometrical order

Mathematically, much of Greek astronomy was based on geometry, particularly motion in circles, which were considered to be the most perfect shapes. Elaborate geometrical schemes were created for predicting the positions of the planets, in which circular motions were combined. In 150 CE, the Graeco–Egyptian astronomer Ptolemy, working in Alexandria, put together the ultimate compendium of Greek astronomy. However, by 500 CE, the Greek approach to astronomy had lost momentum. In effect, after Ptolemy, there were

In the *Aryabhatiya*, Indian astronomer **Aryabhata** suggests that the stars move across the sky because **Earth is rotating**.

Italian scholar **Gerard of Cremona** makes Arabic texts, including Ptolemy's *Almagest*, accessible in Europe by **translating them into Latin**.

Mongol ruler **Ulugh Beg** corrects many of the **postions of stars** found in the *Almagest*.

499 CE

c.1180

1437

1025

1279

1543

Arab scholar **Ibn al-Haytham** produces a work that **criticizes the Ptolomaic model** of the universe for its complexity.

Chinese astronomer **Guo Shoujing** produces an accurate measurement of the length of the **solar year**.

Nicolaus Copernicus's book *De revolutionibus orbium coelestium* is published, outlining a **sun-centered cosmos**.

---

no significant new ideas in astronomy in this tradition for nearly 1,400 years. Independently, great cultures in China, India, and the Islamic world developed their own traditions through the centuries when astronomy in Europe made little

It is the duty of an astronomer to compose the history of the celestial motions through careful and expert study.
Nicolaus Copernicus

progress. Chinese, Arab, and Japanese astronomers recorded the 1054 supernova in the constellation Taurus, which made the famous Crab nebula. Although it was much brighter than Venus, there is no record of its appearance being noted in Europe.

The spread of learning

Ultimately, Greek science returned to Europe via a roundabout route. From 740 CE, Baghdad became a great center of learning for the Islamic world. Ptolemy's great compendium was translated into Arabic, and became known as the *Almagest*, from its Arabic title. In the 12th century, many texts in Arabic were translated into Latin, so the legacy of the Greek philosophers, as well as the writings of the Islamic scholars, reached Western Europe.

The invention of the printing press in the mid-15th century widened access to books. Nicolaus Copernicus, who was born in 1473, collected books throughout his life, including the works of Ptolemy. To Copernicus, Ptolemy's geometrical constructions failed to do what the original Greek philosophers saw as their objective: describe nature by finding simple underlying principles. Copernicus intuitively understood that a sun-centered method could produce a much simpler system, but in the end his reluctance to abandon circular motion meant that real success eluded him. Nevertheless, his message that physical reality should underpin astronomical thinking arrived at a pivotal moment to set the scene for the telescopic revolution. ∎

IT IS CLEAR THAT EARTH DOES NOT MOVE

THE GEOCENTRIC MODEL

IN CONTEXT

KEY ASTRONOMER
Aristotle (384–322 BCE)

BEFORE
465 BCE Greek philosopher
Empedocles thinks that there
are four elements: earth, water,
air, and fire. Aristotle contends
that the stars and planets are
made of a fifth element, aether.

387 BCE Plato's student
Eudoxus suggests that the
planets are set in transparent
rotating spheres.

AFTER
355 BCE Greek thinker
Heraclides claims that the sky
is stationary and Earth spins.

12th century Italian Catholic
priest Thomas Aquinas begins
teaching Aristotle's theories.

1577 Tycho Brahe shows that
the Great Comet is farther
from Earth than the moon.

1687 Isaac Newton explains
force in his *Philosophiae
Naturalis Principia Mathematica*.

One of the most influential of all Western philosophers, Aristotle, from Macedonia in northern Greece, believed that the universe was governed by physical laws. He attempted to explain these through deduction, philosophy, and logic.

Aristotle observed that the positions of the stars appeared to be fixed in relation to each other, and that their brightness never changed. The constellations always stayed the same, and spun daily around Earth. The moon, sun, and planets, too, appeared to move in unchanging orbits around Earth. Their motion, he believed, was circular and their speed constant.

His observations of the shadow cast by Earth on the moon's surface during a lunar eclipse convinced him that Earth was a sphere. His conclusion was that a spherical Earth remained stationary in space, never spinning or changing its position, while the cosmos spun eternally around it. Earth was an unmoving object at the center of the universe.

Aristotle believed that Earth's atmosphere, too, was stationary. At the top of the atmosphere, friction occurred between the atmospheric gases and the rotating sky above. Episodic emanations of gases from volcanoes rose to the top of the atmosphere. Ignited by friction, these gases produced comets, and, if ignited quickly, they produced shooting stars. His reasoning remained widely accepted until the 16th century. ∎

Earth casts a circular shadow
on the moon during a lunar eclipse.
This convinced Aristotle that
Earth was a sphere.

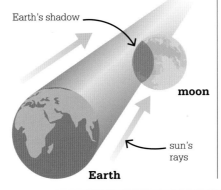

Earth's shadow

moon

sun's
rays

Earth

See also: Consolidating knowledge 24–25 ▪ The Copernican model 32–39 ▪ The Tychonic model 44–47 ▪ Gravitational theory 66–73

EARTH REVOLVES AROUND THE SUN ON THE CIRCUMFERENCE OF A CIRCLE
EARLY HELIOCENTRIC MODEL

IN CONTEXT

KEY ASTRONOMER
Aristarchus (310–230 BCE)

BEFORE
430 BCE Philolalus of Craton proposes that there is a huge fire at the center of the universe, around which the sun, moon, Earth, five planets, and stars revolve.

350 BCE Aristotle states that Earth is at the center of the universe and everything else moves around it.

AFTER
150 CE Ptolemy publishes his *Almagest*, describing an Earth-centered (geocentric) model of the universe.

1453 Nicolaus Copernicus proposes a heliocentric (sun-centered) universe.

1838 German astronomer Friedrich Bessel is the first to obtain an accurate measurement of the distance to a star, using a method known as parallax.

An astronomer and mathematician from the Greek island of Samos, Aristarchus is the first person known to have proposed that the sun, not Earth, is at the center of the universe, and that Earth revolves around the sun.

Aristarchus's thoughts on this matter are mentioned in a book by another Greek mathematician, Archimedes, who states in *The Sand Reckoner* that Aristarchus had formulated a hypothesis that "the fixed stars and sun remain unmoved" and "Earth revolves about the sun."

Unfashionable idea
Aristarchus persuaded at least one later astronomer—Seleucus of Seleucia, who lived in the second century BCE—of the truth of his heliocentric (sun-centered) view of the universe, but otherwise it seems his ideas did not gain wide acceptance. By the time of Ptolemy, in about 150 CE, the prevailing view was still a geocentric (Earth-centered) one, and this remained

Aristarchus was the real originator of the Copernican hypothesis.
Sir Thomas Heath
Mathematician and classical scholar

the case until the 15th century, when the heliocentric viewpoint was revived by Nicolaus Copernicus.

Aristarchus also believed that the stars were much farther away than had previously been imagined. He made estimates of the distances to the sun and moon, and their sizes relative to Earth. His estimates regarding the moon were reasonably accurate, but he underestimated the distance to the sun, mainly because of an inaccuracy in one of his measurements. ■

See also: The geocentric model 20 ▪ Consolidating knowledge 24–25 ▪ The Copernican model 32–39 ▪ Stellar parallax 102

THE EQUINOXES MOVE OVER TIME

SHIFTING STARS

I n about 130 BCE, the Greek
astronomer and mathematician
Hipparchus of Nicaea noticed
that a star named Spica had moved
2° east of a point on the celestial
sphere, called the fall equinox
point, compared to its position
recorded 150 years earlier. Further
research showed him that the
positions of all stars had shifted.
This shift became known as
"precession of the equinoxes."

The celestial sphere is an
imaginary sphere surrounding
Earth, in which stars are found at
specific points. Astronomers use

Industrious, and a
great lover of the truth.
Ptolemy
describing Hipparchus

exactly defined points and curves
on the surface of this sphere as
references for describing the
positions of stars and other celestial
objects. The sphere has north and
south poles, and a celestial equator,
which is a circle lying above Earth's
equator. The ecliptic is another
important circle on the sphere,
which traces the apparent path
of the sun against the background
of stars over the course of the year.
The ecliptic intersects the celestial
equator at two points: the spring
and fall equinox points. These mark
the positions on the celestial sphere
that the sun reaches on the
equinoxes in March and September.
The precession of the equinoxes
refers to the gradual drift of these
two points relative to star positions.

Hipparchus put this precession
down to a "wobble" in the movement
of the celestial sphere, which he
believed to be real and to rotate
around Earth. It is now known
that the wobble is actually in
the orientation of Earth's spin
axis, caused by the gravitational
influence of the sun and the moon. ■

See also: Gravitational theory 66–73 ▪ Halley's comet 74–77

THE MOON'S BRIGHTNESS IS PRODUCED BY THE RADIANCE OF THE SUN
THEORIES ABOUT THE MOON

IN CONTEXT

KEY ASTRONOMER
Zhang Heng (78–139 CE)

BEFORE
140 BCE Hipparchus discovers how to predict eclipses.

1st century BCE Jing Fang advances the "radiating influence" theory, stating that the light of the moon is the reflected light of the sun.

AFTER
150 CE Ptolemy produces tables for calculating the positions of celestial bodies.

11th century Shen Kuo's *Dream Pool Essays* explains that heavenly bodies are round like balls rather than flat.

1543 Nicolaus Copernicus's *On the Revolutions of the Celestial Spheres* describes a heliocentric system.

1609 Johannes Kepler explains the movements of the planets as free-floating bodies, describing ellipses.

The Chief Astrologer at the court of Chinese emperor An-ti, Zhang Heng was a skilled mathematician and a careful observer. He cataloged 2,500 "brightly shining" stars and estimated that there were a further 11,520 "very small" ones.

Also a distinguished poet, Zhang expressed his astronomical ideas through simile and metaphor. In his treatise *Ling Xian*, or *The Spiritual Constitution of the Universe*, he placed Earth at the center of the cosmos, stating that "the sky is like a hen's egg, and is as round as a crossbow pellet, and Earth is the yolk of the egg, lying alone at the center."

Shape but no light

Zhang concluded that the moon had no light of its own, but rather reflected the sun "like water." In this, he embraced the theories of his compatriot Jing Fang who, a century earlier, had declared that "the moon and the planets are Yin; they have shape but no light." Zhang saw that "the side that faces the

The sun is like fire and the moon like water. The fire gives out light and the water reflects it.
Zhang Heng

sun is fully lit, and the side that is away from it is dark." He also described a lunar eclipse, during which the sun's light cannot reach the moon because Earth is in the way. He recognized that the planets were similarly subject to eclipses.

Zhang's work was developed further in the 11th century by another Chinese astronomer, Shen Kuo. Shen demonstrated that the waxing and waning of the moon proved that the moon and sun were spherical. ■

See also: The Copernican model 32–39 ▪ Elliptical orbits 50–55

ALL MATTERS USEFUL TO THE THEORY OF HEAVENLY THINGS
CONSOLIDATING KNOWLEDGE

IN CONTEXT

KEY ASTRONOMER
Ptolemy (85–165 CE)

BEFORE
12th century BCE The Babylonians organize the stars into constellations.

350 BCE Aristotle asserts that the stars are fixed in place and Earth is stationary.

135 BCE Hipparchus produces a catalog of over 850 star positions and brightnesses.

AFTER
964 CE Persian astronomer al-Sufi updates Ptolemy's star catalog.

1252 The Alfonsine Tables are published in Toledo, Spain. These list the positions of the sun, moon, and planets based on Ptolemy's theories.

1543 Copernicus shows that it is far easier to predict the movement of the planets if the sun is placed at the center of the cosmos rather than Earth.

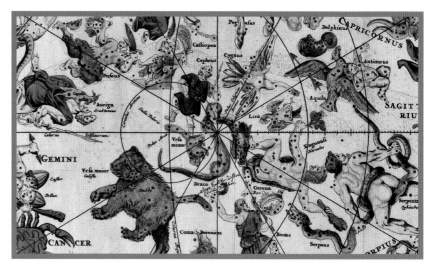

In his greatest known work, the *Almagest*, the Graeco-Egyptian astronomer Ptolemy produced a summary of all the astronomical knowledge of his time. Rather than producing radical new ideas of his own, Ptolemy mostly consolidated and built upon previous knowledge, particularly the works of the Greek astronomer Hipparchus, whose star catalog formed the basis of most of the calculations in the *Almagest*. Ptolemy also detailed the mathematics required to calculate the future positions of the planets. His system would be used by generations of astrologers.

The constellations devised by Ptolemy are used in this 17th-century star map. The number of stars per constellation ranges from two (Canis Minor) to 42 (Aquarius).

Ptolemy's model of the solar system had a stationary Earth at its center, with the heavens spinning daily around it. His model required complicated additions to make it match the data and allow it to be used to calculate the positions of the planets; nonetheless, it was largely unchallenged until Copernicus placed the sun at the center of the cosmos in the 16th century.

See also: The geocentric model 20 ▪ Shifting stars 22 ▪ The Copernican model 32–39 ▪ The Tychonic model 44–47 ▪ Elliptical orbits 50–55

Ptolemy produced a catalog of 1,022 star positions and listed 48 constellations in the part of the celestial sphere known to the Greeks—everything that could be seen from a northern latitude of about 32°. Ptolemy's constellations are still used today. Many of their names can be traced even further back to the ancient Babylonians, including Gemini (twins), Cancer (crab), Leo (lion), Scorpio (scorpion), and Taurus (bull). The Babylonian constellations are named on a cuneiform tablet called the *Mul Apin*, which dates back to the 7th century BCE, however, they are thought to have been compiled about 300 years earlier.

Early quadrant

To improve his measurements, Ptolemy built a plinth. One of the earliest examples of a quadrant, his plinth was a huge rectangular block of stone, one of whose vertical sides accurately aligned in the north–south plane. A horizontal bar protruded from the top of the stone, and its shadow gave a precise indication of the height of the sun at noon. Ptolemy took daily measurements to obtain accurate estimates of the time of the solstices and equinoxes, which confirmed previous measurements showing that the seasons were different lengths. He believed that the orbit of the sun around Earth was circular, but his calculations led him to the conclusion that Earth could not be at the exact center of that orbit.

Ptolemy the astrologist

Like most thinkers of his day, Ptolemy believed that the movements of the heavenly bodies profoundly affected events on Earth. His book on astrology, *Tetrabiblos*, rivaled the *Almagest* in popularity over the following 1,000 years. Ptolemy had not only provided a means to calculate planetary positions, but he had also produced a comprehensive interpretation of the ways those movements affected humans. ▪

Claudius Ptolemy

Ptolemy was a polymath and produced works on a wide range of topics, including astronomy, astrology, geography, music, optics, and mathematics.

Very little is known about him, but he probably spent all his life in Alexandria, the Egyptian seaport with a reputation for scholarship and a great library, where he was taught by the renowned mathematician Theon of Smyrna. Many of his prolific writings have survived. They were translated into Arabic and Latin, disseminating his ideas across the medieval world. *Geography* listed the locations of most of the places in the known world, and was carried by Christopher Columbus on his voyages of discovery in the 15th century. The *Almagest* remained in continual use in academia until about 1643, a century after Ptolemy's model of the universe had been challenged by Copernicus.

Key works

c.150 CE *Geography*
c.150 CE *Almagest*
c.150 CE *Tetrabiblos*

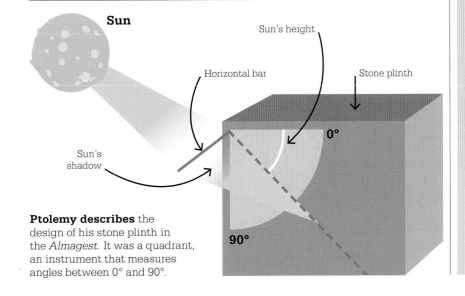

Ptolemy describes the design of his stone plinth in the *Almagest*. It was a quadrant, an instrument that measures angles between 0° and 90°.

THE UNMOVING STARS GO UNIFORMLY WESTWARD
EARTH'S ROTATION

From the 4th century BCE to the 16th century CE, the prevailing view throughout the Western world was that Earth is stationary and located at the center of the universe. Suggestions that Earth might be rotating were dismissed on the grounds that this would cause objects on Earth's surface to fly off into space. In India, however, an astronomer named Aryabhata was convinced that the movement of stars across the night sky was due not to the stars revolving in a distant sphere around Earth, but to Earth itself rotating.

An illusory movement

According to Aryabhata, the stars were stationary and their apparent movement toward the west was an illusion. His notion of a spinning Earth was not widely accepted until the mid-17th century—a century after Nicolaus Copernicus had endorsed the idea.

Aryabhata's achievements were considerable. His book *Aryabhatiya* was the most important work of astronomy in the 6th century.

He was the father of the Indian cyclic astronomy … that determines more accurately the true positions and distances of the planets.
Helaine Selin
Historian of astronomy

Essentially a compendium of the fundamentals of astronomy and relevant mathematics, it greatly influenced Arabic astronomy.

Among other achievements, Aryabhata calculated the length of the sidereal day (the time it takes Earth to rotate once in relation to the stars) to a high degree of accuracy, and devised original and accurate ways of compiling astronomical tables. ∎

See also: The geocentric model 20 ▪ The Copernican model 32–39 ▪ The Tychonic model 44–47 ▪ Elliptical orbits 50–55

A LITTLE CLOUD IN THE NIGHT SKY
MAPPING THE GALAXIES

IN CONTEXT

KEY ASTRONOMER
Abd al-Rahman al-Sufi
(903–986 CE)

BEFORE
400 BCE Democritus suggests that the Milky Way is made of a dense mass of stars.

150 CE Ptolemy records several nebulae (or cloudy objects) in the *Almagest*.

AFTER
1610 Galileo sees stars in the Milky Way using a telescope, confirming Democritus's theory.

1845 Lord Rosse makes the first clear observation of a spiral nebula, now known as the Whirlpool Galaxy.

1917 Vesto Slipher discovers that spiral nebulae are rotating independently of the Milky Way.

1929 Edwin Hubble shows that many spiral nebulae are far beyond the Milky Way and are galaxies themselves.

A bd al-Rahman al-Sufi, once better known in the West as Azophi, was a Persian astronomer who made the first record of what are now understood to be galaxies. To al-Sufi, these fuzzy, nebulous objects looked like clouds in the night's sky.

Al-Sufi made most of his observations in Isfahan and Shiraz, in what is now central Iran, but he

The Large Magellanic Cloud, seen here above the ESO's Paranal observatory in Chile, can be easily observed with the naked eye from the southern hemisphere.

also consulted Arab merchants who traveled to the south and east, and who saw more of the sky. His work centered on translating Ptolemy's *Almagest* into Arabic. In the process, al-Sufi tried to merge the Hellenistic constellations (which dominate star maps today) with their Arab counterparts, most of which were totally different.

The fruit of this labor was *Kitab suwar al-kawakib*, or the *Book of Fixed Stars*, published in 964 CE. The work contained an illustration of "a little cloud," which is now know to be the Andromeda Galaxy. This object was probably known to earlier Persian astronomers, but al-Sufi's mention is the earliest record. Similarly, *The Book of Fixed Stars* includes the White Ox, another cloudy object. This is now named the Large Magellanic Cloud and is a small galaxy that orbits the Milky Way. Al-Sufi would not have been able to observe this object himself, but would have received reports of it from astronomers in Yemen and sailors who crossed the Arabian Sea. ∎

See also: Consolidating knowledge 24–25 ∎ Examining nebulae 104–05 ∎ Spiral galaxies 156–61 ∎ Beyond the Milky Way 172–77

A NEW CALENDAR FOR CHINA

THE SOLAR YEAR

IN CONTEXT

KEY ASTRONOMER
Guo Shoujing (1231–1314)

BEFORE
100 BCE Emperor Wu of
the Han Dynasty establishes
the Chinese calendar based
on a solar year.

46 BCE Julius Caesar reforms
the Roman calendar using a
year-length of 365 days and
6 hours, and adds a leap day
every four years.

AFTER
1437 The Timurid astronomer
Ulugh Beg measures the
solar year as 365 days,
5 hours, 49 minutes, and
15 seconds using a 164-ft
(50-m) gnomon (the central
column of a sundial).

1582 Pope Gregory adopts
the Gregorian calendar as a
reform of the ancient Julian
calendar by using a 365.25-day
year, the same year as Guo's
Shoushi calendar.

The traditional Chinese calendar is a complex blend of lunar and solar cycles, with 12 or 13 lunar months matched up to the solar-derived seasons. It had first been formalized in the 1st century BCE during the Han Dynasty, and used a solar year of 365.25 days (365 days and 6 hours).

A trained engineer, Guo Shoujing invented a water-powered version of an armillary sphere, which is an instrument used to model the positions of celestial bodies.

China's calculations were ahead of the West's: 50 years later, this same period was used by Julius Caesar to create the Roman Empire's Julian system.

By the time the Mongol leader Kublai Khan conquered most of China in 1276, a variant of the original calendar, the Daming calendar, was in use, but was centuries old and in need of correction. The khan decided to impose his authority with a new, more accurate calendar, which became known as the Shoushi ("well-ordered") calendar. The task of creating it was entrusted to Guo Shoujing, the khan's brilliant Chinese chief astronomer.

Measuring the year

Guo's job was to measure the length of the solar year, and to this end he set up an observatory in Khanbaliq (the "City of the Khan"), a new imperial capital that would one day become known as Beijing. The observatory may have been the largest anywhere in the world at the time.

Working with mathematician Wang Chun, Guo began a series of observations tracking the motion of the sun throughout the year.

See also: Shifting stars 22 ▪ Improved instruments 30–31 ▪ Zu Chongzhi (Directory) 334

The two men traveled widely, setting up another 26 observatories across China. In 1279, the pair announced that there were 29.530593 days to a month, and that the true solar year was 365.2524 days long (365 days, 5 hours, 49 minutes, and 12 seconds). This is just 26 seconds longer than the current accepted measurement. Again, China was ahead of the West. The same figure was not independently measured and adopted for the universal Gregorian calendar in Europe until 300 years later.

Enduring calendar

A great technological innovator, Guo invented several new observational devices and made enhancements to the Persian equipment that had begun to arrive in China under Kublai Khan's rule. Most importantly, he built a giant gnomon to a height of 44 ft (13.3 m), which was five times taller than the previous Persian design and featured a horizontal crossbar marked with

The calendar has **365 days and 6 hours** in the year, but **does not match** the motion of the sun through the year.

There is a need to create a new calendar that **matches the solar year**.

To measure the length of the year, **better instruments** must be created.

The solar year is found to be **365 days, 5 hours, 49 minutes, and 12 seconds**. There is a **new calendar for China**.

measurements. This allowed Guo to measure the angle of the sun with far greater accuracy.

The Shoushi calendar was widely regarded as the most accurate calendar in the world at the time. As a testament to its success, it continued to be used for 363 years, making it the longest-serving official calendar in Chinese history. China officially adopted the Gregorian calendar in 1912, but the traditional calendar, today known as the rural or former calendar, still plays a role in Chinese culture, determining the most propitious dates to hold weddings, family celebrations, and public holidays. ▪

Guo Shoujing

Guo Shoujing was born into a poor family in the north of China, in the years when the Mongols were consolidating their control over the region. A child prodigy who had built a highly advanced water clock by the age of 14, Guo was taught mathematics, astronomy, and hydraulics by his grandfather. He became an engineer, working for the emperor's chief architect Liu Bingzhong. In the late 1250s, Kublai Khan took the throne and chose the region around the town of Dadu near the Yellow River to build the new capital of Khanbaliq, now known as Beijing.

Guo was tasked with building a canal to bring spring water from the mountains to the new city. In the 1290s, Guo—by now the khan's chief science and engineering adviser—connected Khanbaliq to the ancient Grand Canal system that linked to the Yangtze and other major rivers. In addition to continuing his astronomical work, Guo oversaw similar irrigation and canal projects across China, and his theoretical and technological innovations continued to influence Chinese society for centuries after his death.

WE HAVE RE-OBSERVED ALL OF THE STARS IN PTOLEMY'S CATALOG
IMPROVED INSTRUMENTS

IN CONTEXT

KEY ASTRONOMER
Ulugh Beg (1384–1449)

BEFORE
c.130 BCE Hipparchus publishes a star catalog giving the positions of more than 850 stars.

150 CE Ptolemy publishes a star catalog in the *Almagest*, which builds on the work of Hipparchus and is seen as the definitive guide to astronomy for more than a millennium.

964 CE Abd al-Rahman al-Sufi adds the first references to galaxies in his star catalog.

AFTER
1543 Nicolaus Copernicus places the sun as the center of the universe, not Earth.

1577 Tycho Brahe's star catalog records a nova, showing that the "fixed stars" are not eternal and do change.

For more than 1,000 years, Ptolemy's *Almagest* was the world's standard authority on star positions. Translated into Arabic, Ptolemy's work was also influential in the Islamic world up until the 15th century, when the Mongol ruler Ulugh Beg showed that a lot of the *Almagest*'s data were wrong.

A grandson of the Mongol conqueror Timur, Ulugh Beg was just 16 years old when he became ruler of the family's ancestral seat at Samarkand (in present-day Uzbekistan) in 1409. Determined to turn the city into a respected place of learning, Ulugh Beg invited scholars of many disciplines from far and wide to study at his new madrasa, an educational institution.

Ulugh Beg's own interest was in astronomy, and it may have been his discovery of serious errors in the star positions of the *Almagest* that inspired him to order the building of a gigantic observatory, the largest in the world at the time. Located on a hill to the north of the city, it took five years to construct and was

Ulugh Beg

The name Ulugh Beg means "Great Leader." The sultan–astronomer's birth name was Mirza Muhammad Taraghay bin Shahrukh. He was born on the move, as Timur's army traveled through Persia.

His grandfather's death in 1405 brought the army to a halt in western China. The ensuing fight for control of his lands was eventually won by Ulugh Beg's father, Shah Rukh. In 1409, Ulugh Beg was sent to Samarkand as his father's regent, and by 1411, as he turned 18, his rule over the city was extended to include the surrounding province.

Ulugh Beg's flair for mathematics and astronomy was not matched by his leadership skills. When Shah Rukh died in 1447, Ulugh Beg assumed the imperial throne, but he did not command enough authority to keep it. In 1449, he was beheaded by his own son.

Key work

1437 *Zij-i Sultani*

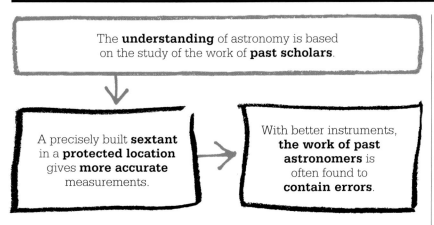

The **understanding** of astronomy is based on the study of the work of **past scholars**.

A precisely built **sextant** in a **protected location** gives **more accurate** measurements.

With better instruments, **the work of past astronomers is** often found to **contain errors**.

completed in 1429. It was there, with his team of astronomers and mathematicians, that he set about compiling a new star catalog.

Giant instruments

Ptolemy's catalog had largely been derived from the work of Hipparchus, and many of its star positions were not based on fresh observations.

To measure accurately, Ulugh Beg built the observatory on an immense scale. Its most impressive instrument was the so-called Fakhri sextant. In fact, more like a quadrant (a quarter-circle rather

than a sixth), it is estimated to have had a radius of more than 130 ft (40 m) and would have been three stories high. The instrument was kept underground to protect it from earthquakes and rested in a curved trench along the north–south meridian. As the sun and the moon passed overhead, their light focused into the dark trench, and their positions could be

measured to within a few hundredths of a degree, as could the positions of the stars.

In 1437, *Zij-i Sultani* ("The Sultan's Catalog of Stars") was published. Of the 1,022 stars included in the *Almagest*, Ulugh Beg corrected the positions of 922. *Zij-i Sultani* also contained new measurements for the solar year, planetary motion, and the axial tilt of Earth. These data became very important, enabling the prediction of eclipses, the time of sunrise and sunset, and the altitude of celestial bodies, which were needed to navigate. Ulugh Beg's work remained the definitive star catalog until Tycho Brahe's, nearly 200 years later. ▪

All that remains of the Fakhri sextant is a 6½-ft (2-m) wide trench gouged in a hillside. The observatory was destroyed after Ulugh Beg's death in 1449 and not discovered until 1908.

The religions disperse, kingdoms fall apart, but works of science remain for all ages.
Ulugh Beg

FINALLY WE SHALL PLACE THE SUN HIMSELF AT THE CENTER OF THE UNIVERSE

THE COPERNICAN MODEL

KEY ASTRONOMER
Nicolaus Copernicus
(1473–1543)

BEFORE

c.350 BCE Aristotle places Earth at the center of the universe.

c.270 BCE Aristarchus proposes a sun-centered (heliocentric) universe, with the stars a vast distance away.

c.150 CE Ptolemy publishes the *Almagest*.

AFTER

1576 English astronomer Thomas Digges suggests modifying the Copernican system, removing its outer edge and replacing it with a star-filled unbound space.

1605 Johannes Kepler discovers that orbits are elliptical.

1610 Galileo Galilei discovers the phases of Venus, and Jupiter's moons, strengthening the heliocentric viewpoint.

To most people in mid-15th century Europe, questions about Earth's place in the cosmos had been answered in the 2nd century by the Greco-Egyptian mathematician Ptolemy, who had modified ideas first put forward by Aristotle. These ideas placed Earth at the center of the cosmos, and they carried an official stamp of approval from the Church. Yet the first convincing challenge to this orthodoxy was to come from a figure within the Church, the Polish canon Nicolaus Copernicus.

A stationary Earth
According to the version of the universe described by Aristotle and Ptolemy, Earth was a stationary point at the center of the universe, with everything else circling around it, and stars were fixed in a large, invisible, distant sphere, which rotated rapidly around Earth. The sun, moon, and planets also revolved at different speeds around Earth.

This idea of the universe seemed like common sense. After all, one only had to stand outside and look up at the sky, and it appeared obvious that

Of all discoveries and opinions, none may have exerted a greater effect on the human spirit than the doctrine of Copernicus.
Johann von Goethe

Earth stayed in one place, while everything else rose in the east, swung across the sky, and set in the west. Furthermore, the Bible seemed to state that the sun moves, whereas Earth does not, so anyone who contradicted this view risked being accused of heresy.

Nagging doubts
The Earth-centered, or geocentric, model of the universe had never convinced everyone—in fact, doubts about it had surfaced from time to time for more than 1,800 years. The most serious

Nicolaus Copernicus

Nicolaus Copernicus was born in Torun, Poland, in 1473. From 1491 to 1495, he studied mathematics, astronomy, and philosophy at the University of Kraków, then from 1496, canon (religious) law and astronomy at the University of Bologna, Italy. In 1497, he was appointed canon of the cathedral of Frombork, Poland, a post he retained for life. From 1501 to 1505, he studied law, Greek, and medicine at the University of Padua, Italy. Subsequently, he returned to Frombork, where he spent much of the rest of his life. By 1508, he had begun

developing his sun-centered model of the universe. He did not complete this work until 1530, although he did publish a summary of his ideas in 1514. Realizing that he risked being ridiculed or persecuted, Copernicus delayed publishing the full version of his theory until the last weeks of his life.

Key works

1514 *Commentariolus*
1543 *De revolutionibus orbium coelestium* (*On the Revolutions of the Celestial Spheres*)

concern related to predicting the movements and appearances of the planets. According to the Aristotelian version of geocentrism, the planets—like all other celestial bodies—were embedded in invisible concentric spheres that revolved around Earth, each rotating at its own steady speed. But if this were true, each planet should move across the sky at a constant pace and with an unvarying brightness—and this wasn't what was observed.

Ptolemy's fixes

The most glaring anomaly was Mars, which had been carefully observed in ancient times by both the Babylonians and the Chinese. It appeared to speed up and slow down from time to time. If its movements were compared to those of the rapidly rotating outer sphere of fixed stars, Mars usually moved in a particular direction, but occasionally it reversed direction— a strange behavior described as "retrograde motion." In addition, its brightness varied greatly over the course of a year. Similar, but less dramatic, irregularities were also observed in the other planets. To

In so many and such important ways, then, do the planets bear witness to the Earth's mobility.
Nicolaus Copernicus

Ptolemy tried to fix some of the anomalies in Aristotle's geocentric model by proposing that each planet moved in a small circle called an epicycle. Each epicycle was embedded in a sphere called a deferent. Each planet's deferent rotated around a point slightly displaced from Earth's position in space. This point, in turn, continuously rotated around another point called an equant. Each planet had its own equant.

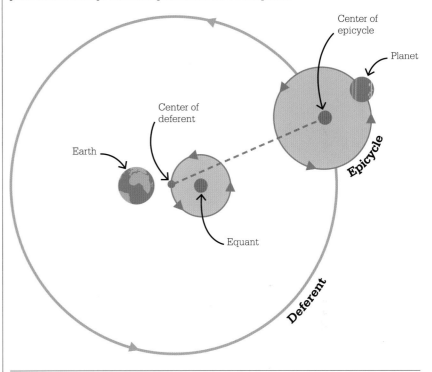

Center of epicycle

Planet

Center of deferent

Earth

Equant

Epicycle

Deferent

address these problems, Ptolemy modified the original Aristotelian geocentric model. In his revised model, the planets were attached not to the concentric spheres themselves, but to circles attached to the concentric spheres. He called these circles "epicycles." These were suborbits around which the planets circled while the central pivot points of these suborbits were carried around the sun. These modifications, Ptolemy thought, sufficed to explain the anomalies observed and matched observational data. However, his model became hugely complicated, as further epicycles needed to be added to keep prediction in line with observation.

Alternative views

From about the 4th century BCE, a number of astronomers had suggested theories refuting the geocentric model. One of these ideas was that Earth spins on its own axis, which would account for a large proportion of the daily movements of celestial objects. The concept of a rotating Earth had initially been put forward by a Greek, Heraclides Ponticus, in about 350 BCE and later by various »

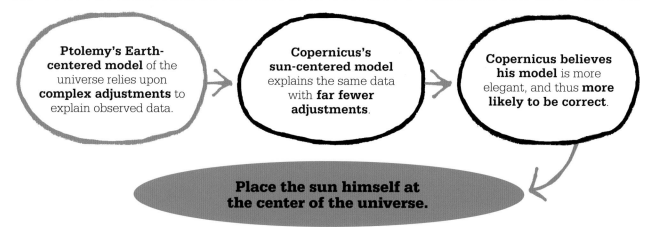

Ptolemy's Earth-centered model of the universe relies upon **complex adjustments** to explain observed data.

Copernicus's **sun-centered model** explains the same data with **far fewer adjustments**.

Copernicus believes **his model** is more elegant, and thus **more likely to be correct**.

Place the sun himself at the center of the universe.

Arabic and Indian astronomers. Supporters of geocentrism rejected his idea as absurd, believing a spinning Earth would create huge winds, such that objects on Earth's surface would simply fly off. Another idea, first proposed by Aristarchus of Samos in about 250 BCE, was that Earth might move around the sun. Not only did this go against deeply ingrained Aristotelian ideas, but supporters of geocentrism had also for centuries cited what seemed a scientifically valid reason for ruling it out—the "lack of stellar parallax." They argued that if Earth moved around the sun, it would be possible to observe some variation in the relative positions of stars. No such variation could ever be detected so, they said, Earth could not move.

In the face of such an established philosophical tradition with little observational evidence to contradict it, and the theological arguments in favor of it, the geocentric view of the universe went unchallenged for centuries. However, in about 1545, rumors began circulating in Europe of a highly convincing challenge that had appeared in the form of a book entitled *De revolutionibus orbium coelestium (On the Revolutions of the Celestial Spheres)*, by a Polish scholar, Nicolaus Copernicus.

Copernican revolution

The work was extremely comprehensive, and proposed a new, detailed, mathematical, and geometrical model of how the universe works, based on years of astronomical observations.

Copernicus's theory was based on a number of basic propositions. First, Earth rotates on its axis daily, and this rotation accounts for most of the daily movements of the stars, sun, and planets across the sky.

In his 1660 star atlas, German mapmaker Andreas Cellarius illustrated the cosmic systems of Ptolemy, Tycho Brahe, and Copernicus (shown here). All three still had their champions.

Copernicus thought it was just too unlikely that thousands of stars were spinning rapidly around Earth every 24 hours. Instead, he considered them to be fixed and immovable in their distant, outer sphere, and that their apparent movement was actually an illusion caused by Earth's spin. To refute the idea that a spinning Earth would create huge winds, and that objects on its surface would fly off, Copernicus pointed out that Earth's oceans and atmosphere were part of the planet and were naturally part of this spinning motion. In his own words: "We would only say that not merely the Earth and the watery element joined with it have this motion, but also no small part of the air and whatever is linked in the same way to the Earth."

Second, Copernicus proposed that it is the sun that is at the center of the universe, not Earth, which is simply one of the planets, all of which circle the sun at differing speeds.

Elegant solution

These two central tenets of Copernicus's theory were of utmost importance because they explained the movements and variation in brightness of the planets without recourse to Ptolemy's complicated adjustments. If Earth and another planet, such as Mars, both circle the sun and do so at different speeds, taking a different amount of time to complete each revolution, they will sometimes be close to each other on the same side as the sun and sometimes far from each other, on opposite sides to the sun. This, at a stroke, explained the observed variations in brightness of Mars and the other planets. The heliocentric system also elegantly explained apparent retrograde motion. In place of Ptolemy's »

In the Ptolemaic model (top), Earth is at the center and other celestial bodies go around Earth. In the Copernican system (bottom), Earth together with the moon have swapped position with the sun; the sphere of the fixed stars is much farther out.

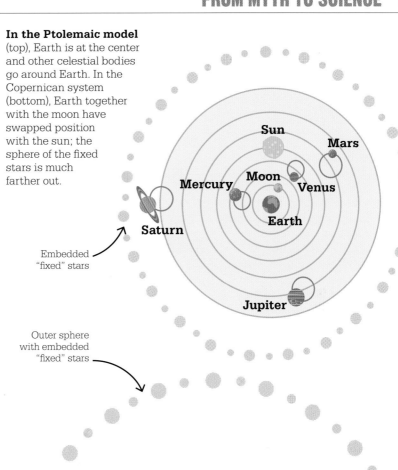

Embedded "fixed" stars

Outer sphere with embedded "fixed" stars

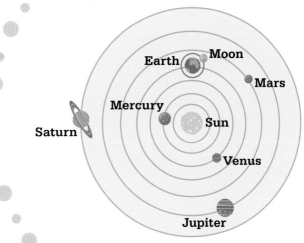

complicated epicycles, Copernicus explained that such motion could be attributed to changes in perspective caused by Earth and the other planets moving at different speeds.

Distant stars

Another of Copernicus's tenets was that the stars are much farther away from Earth and the sun than had previously been believed. He said: "The distance between Earth and the sun is an insignificant fraction of the distance from Earth and sun to the stars." Earlier astronomers knew that the stars were distant, but few suspected just how far away they were, and those who did, such as Aristarchus, had not managed to convince anyone. Even Copernicus probably

> Those things which I am saying now may be obscure, yet they will be made clearer in their proper place.
> **Nicolaus Copernicus**

never realized quite how far away the stars are—it is now known that the very closest are about 260,000 times more distant than the sun. But his assertion was extremely important because

of its implications for stellar parallax. For centuries, supporters of geocentrism had argued that the absence of parallax could only be due to Earth not moving. Now, there was an alternative explanation: the parallax was not absent, but because of the great distance to the stars, it was simply too tiny to be measured with the instruments of the time.

Copernicus additionally proposed that Earth is at the center of the lunar sphere. Copernicus maintained that the moon circled Earth, as it did in the geocentric model. In his heliocentric model, the moon moved with Earth as it circled the sun. In this system, the moon was the only celestial object that did not primarily move around the sun.

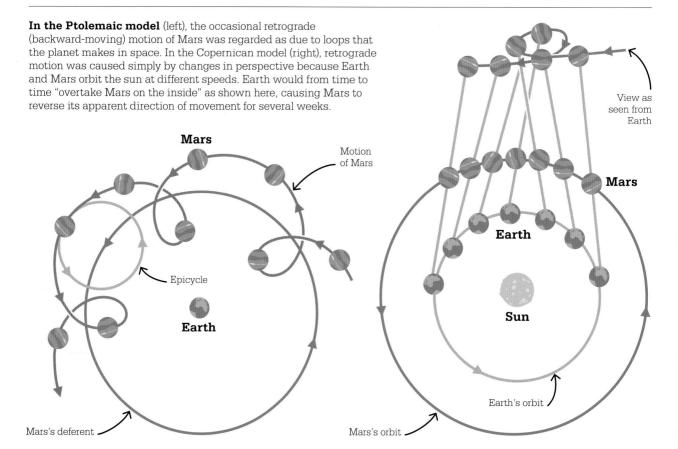

In the Ptolemaic model (left), the occasional retrograde (backward-moving) motion of Mars was regarded as due to loops that the planet makes in space. In the Copernican model (right), retrograde motion was caused simply by changes in perspective because Earth and Mars orbit the sun at different speeds. Earth would from time to time "overtake Mars on the inside" as shown here, causing Mars to reverse its apparent direction of movement for several weeks.

Mars

Motion of Mars

Epicycle

Earth

Mars's deferent

View as seen from Earth

Mars

Earth

Sun

Earth's orbit

Mars's orbit

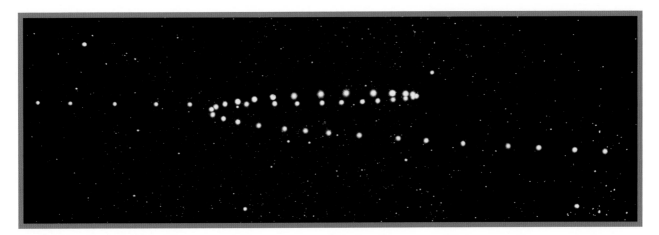

Though Copernicus's work was widely circulated, it took a century or more before its basic ideas were accepted by most other astronomers, let alone the public at large. One difficulty was that, although it resolved many of the problems of the Ptolemaic system, his model also contained faults that had to be amended by later astronomers. Many of these faults were due to the fact that, for philosophical reasons, Copernicus clung to the belief that all the movements of celestial bodies occurred with the objects embedded in invisible spheres

I am deterred by the fate of our teacher Copernicus who, although he had won immortal fame with a few, was ridiculed and condemned by countless people (for very great is the number of the stupid).
Galileo Galilei

and that these movements must be perfect circles. This therefore forced Copernicus to retain some of Ptolemy's epicycles in his model. The work of Johannes Kepler later replaced the idea of circular orbits with that of elliptical orbits, eliminating most of the remaining faults in Copernicus's model. It wasn't until the 1580s and the work of Danish astronomer Tycho Brahe that the idea of celestial spheres was abandoned in favor of free orbits.

Banned by the Church

De revolutionibus initially met with little or no resistance from the Roman Catholic Church, although some Protestants denounced it as heretical. In 1616, however, the Catholic Church condemned Copernicus's book and it remained proscribed reading for more than 200 years. The Church's decision coincided with a dispute it was having at the time with the astronomer Galileo Galilei. Galileo was an avid champion of the Copernican theory and had made discoveries in 1610 that strongly supported the heliocentric view. The dispute with Galileo caused the Church authorities to examine *De revolutionibus* with intense scrutiny, and the fact that

Mars's apparent retrograde motion occurs about every 26 months and lasts for 72 days. Its orbit is on a slightly different plane from Earth's, contributing to the apparent loop.

some of its propositions went against Biblical texts probably led to the ban.

Viewed somewhat ambivalently at first by astronomers, and prohibited by the Catholic Church, Copernicus's heliocentric model therefore took considerable time to catch on. Several centuries passed before some of its basic propositions were demonstrated to be true beyond dispute: that Earth moves in relation to the stars was eventually proved conclusively by English astronomer James Bradley in 1729. Proof that Earth spins came with the first demonstration of Foucault's pendulum in 1851.

Copernicus's theory was a serious blow to old ideas about how the world and wider universe work—many of them dating from the time of Aristotle. As such, it is often cited as ushering in the "Scientific Revolution"—a series of sweeping advances in many areas of science that occurred between the 16th and 18th centuries. ■

THE TE
REVOLU
1550–1750

ESCOPE
TION

Tycho Brahe builds
a large **observatory**
on the island of
Hveen, from where he
makes observations
for 20 years.

Dutch eyeglass-maker
Hans Lippershey
applies for a patent for
a **telescope** with
three-times
magnification.

Johannes Kepler describes
the **elliptical orbits** of
planets with his three laws
of planetary motion.

1576

1608

1619

1600

1610

1639

Italian friar **Giordano Bruno** is
burned at the stake as a heretic
after expressing a view that the
sun and Earth are not central
or special in the universe.

Using a telescope with
33-times magnification,
Galileo Galilei
discovers **four moons**
orbiting Jupiter.

English astronomer
Jeremiah Horrocks
observes the **transit
of Venus** across the
face of the sun.

The Dane Tycho Brahe was the last great astronomer of the pre-telescope era. Realizing the importance of trying to record more accurate positions, Tycho built some high-precision instruments for measuring angles. He accumulated an abundance of observations, far superior to those available to Copernicus.

Magnifying the image
The realm of heavenly bodies still seemed remote and inaccessible to astronomers at the time of Tycho's death in 1601. However, the invention of the telescope around 1608 suddenly brought the distant universe into much closer proximity.

Telescopes have two important advantages over eyes on their own: they have greater light-gathering

power, and they can resolve finer detail. The bigger the main lens or mirror, the better the telescope on both counts. Starting in 1610, when Galileo made his first telescopic observations of the planets, the moon's rugged surface, and the star clouds of the Milky Way, the telescope became the primary tool of astronomy, opening up unimagined vistas.

Planetary dynamics
After Tycho Brahe died, the records of his observations passed to his assistant Johannes Kepler, who was convinced by Copernicus's arguments that the planets orbit the sun. Armed with Tycho's data, Kepler applied his mathematical ability and intuition to discover that planetary orbits are elliptical, not circular. By 1619, he had

formulated his three laws of planetary motion describing the geometry of how planets move.

Kepler had solved the problem of how planets move, but there remained the problem of why they move as they do. The ancient Greeks had imagined

If I have seen further it
is by standing on the
shoulders of giants.
Isaac Newton

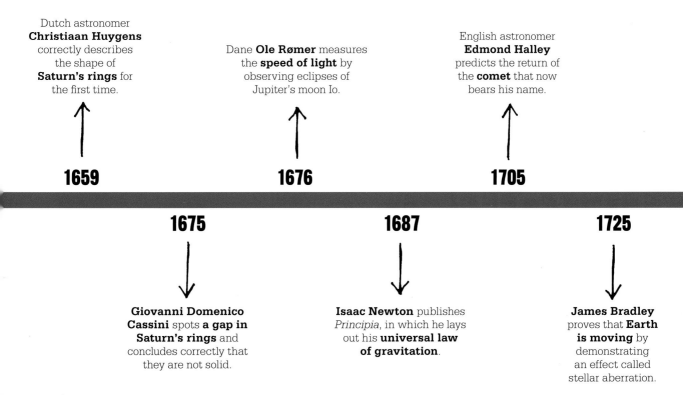

Dutch astronomer **Christiaan Huygens** correctly describes the shape of **Saturn's rings** for the first time.

Dane **Ole Rømer** measures the **speed of light** by observing eclipses of Jupiter's moon Io.

English astronomer **Edmond Halley** predicts the return of the **comet** that now bears his name.

1659

1676

1705

1675

1687

1725

Giovanni Domenico Cassini spots **a gap in Saturn's rings** and concludes correctly that they are not solid.

Isaac Newton publishes *Principia*, in which he lays out his **universal law of gravitation**.

James Bradley proves that **Earth is moving** by demonstrating an effect called stellar aberration.

that the planets were carried on invisible spheres, but Tycho had demonstrated that comets travel unhindered through interplanetary space, seeming to contradict this idea. Kepler thought that some influence from the sun impelled the planets, but he had no scientific means to describe it.

Universal gravitation

It fell to Isaac Newton to describe the force responsible for the movement of the planets, with a theory that remained unchallenged until Einstein. Newton concluded that celestial bodies pull on each other and he showed mathematically that Kepler's laws follow as a natural consequence if the pulling force between two bodies decreases in proportion to the square of the distance between them. Writing

about this force, Newton used the word *gravitas*, Latin for weight, from which we get the word gravity.

Improving telescopes

Newton not only created a new theoretical framework for astronomers with his mathematical way of describing how objects move, but he also applied his genius to practical matters. Early telescope makers found it impossible to obtain images free from colored distortion with their simple lenses, although it helped to make the telescope enormously long. Giovanni Domenico Cassini, for example, used long "aerial" telescopes without a tube to observe Saturn in the 1670s.

In 1668, Newton designed and made the first working version of a reflecting telescope, which did

not suffer from the color problem. Reflecting telescopes of Newton's design were widely used in the 18th century, after English inventor John Hadley developed methods for making large curved mirrors of precisely the right shape from shiny speculum metal. James Bradley, Oxford professor and later Astronomer Royal, was one astronomer who was impressed and acquired a reflector.

There were also developments in lens-making. In the early-18th century, English inventor Chester Moore Hall designed a two-part lens that greatly reduced color distortion. The optician John Dollond used this invention to build much-improved refracting telescopes. With high-quality telescopes now widely available, practical astronomy was transformed. ∎

I NOTICED A NEW AND UNUSUAL STAR

THE TYCHONIC MODEL

IN CONTEXT

KEY ASTRONOMER
Tycho Brahe (1546–1601)

BEFORE
1503 The most accurate star positions to date are recorded by Bernhard Walther at Nuremberg.

1543 Copernicus introduces the idea of a sun-centered cosmos, improving the prediction of planetary positions. These, however, are still inaccurate.

AFTER
1610 Galileo's use of the telescope starts a revolution that eventually supersedes naked-eye astronomy.

1620 Johannes Kepler completes his laws of planetary motion.

1670s Major observatories are established in all the capitals of Europe.

In the 16th century, the exact orbits of the planets were a mystery. Danish nobleman Tycho Brahe realized that accurate observations would need to be taken over an extended period of time if this problem were to be solved. The need for better data was underlined by the fact that a conjunction of Jupiter and Saturn in 1562, when Tycho was 17, occurred days away from the time predicted by the best available astronomical tables. Tycho undertook to take measurements along the entirety of the planets' visible paths.

The astronomy of Tycho's time still followed the teachings that Aristotle had laid down nearly

See also: The geocentric model 20 ▪ Consolidating knowledge 24–25 ▪ The Copernican model 32–39 ▪ Elliptical orbits 50–55 ▪ Hevelius (Directory) 335

The appearance of a new star challenges Aristotle's insistence that the stars never change.

↓

Careful measurement shows that the new star is **not an atmospheric phenomenon**.

↓

Further careful measurements of **the Great Comet** show that it is much **farther away than the moon**.

↓

Careful measurements are the **key to accurate models** of the solar system.

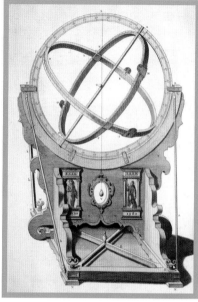

Tycho used his immense wealth to design and build fine instruments, such as this armillary sphere, which was used to model the night sky as seen from Earth.

1,900 years earlier. Aristotle had stated that the stars in the heavenly firmament were fixed, permanent, and unchanging. In 1572, when Tycho was 26, a bright new star was seen in the sky. It was in the constellation of Cassiopeia and stayed visible for 18 months before fading from view. Influenced by the prevailing Aristotelian dogma, most observers assumed that this was an object high in the atmosphere, but below the moon. Tycho's careful measurements of the new object convinced him that it did not move in relation to nearby stars, so he concluded that it was not an atmospheric phenomenon but a real star. The star was later discovered to be a supernova, and the remnant of this stellar explosion is still visible in the sky as Cassiopeia B. The observation of a new star was an extremely rare event. Only eight naked-eye observations of supernovae have ever been recorded. This sighting showed that the star catalogs in use at the time did not tell the whole story. Greater precision was needed, and Tycho led the way.

Precision instruments

To accomplish his task, Tycho set about constructing a collection of reliable instruments (quadrants and sextants (p.31), and armillary spheres) that could measure the position of a planet in the sky to an accuracy of about 0.5 arcminute ($\pm \frac{1}{120}°$). He personally measured planetary positions over a period of around 20 years, and for this purpose in 1576 he oversaw the building of a large complex on the small island of Hven in the Øresund Strait, between what is now Denmark and Sweden. This was one of the first research institutes of its kind.

Tycho carefully measured the positions of the stars and recorded them on brass plates on a spherical wooden globe about 5 ft 3 in (1.6 m) in diameter at his observatory on Hven. By 1595, his globe had around 1,000 stars recorded on it. It could spin around a polar axis, and a horizontal ring was used so that stars positioned above the horizon at any given time could be distinguished from those below the horizon. Tycho carried the globe with him on his travels, but it was destroyed in a fire in Copenhagen in 1728. »

Further evidence of a changing cosmos came from Tycho's observation of the Great Comet in 1577. Aristotle had claimed that comets were atmospheric phenomena, and this was still generally believed to be the case in the 16th century. Tycho compared measurements of the comet's position that he had taken on Hven with those that had been taken at the same time by Bohemian astronomer Thaddaeus Hagecius in Prague. In both instances, the comet was observed in roughly the same place, but the moon was not, suggesting that the comet was much farther away.

Tycho's observations of the way the comet moved across the sky over the months also convinced him that it was traveling through the solar system. This overturned another theory that had been believed for the previous 1,500 years. The great Graeco-Egyptian astronomer Ptolemy had been convinced that the planets were embedded in real, solid, ethereal, transparent crystalline spheres, and that the spinning of these spheres moved the planets across the sky. However, Tycho observed that the comet seemed to move unhindered, and he concluded that the spheres could not exist. He therefore proposed that the planets moved unsupported through space, a daring concept at the time.

No parallax

Tycho was also very interested in Copernicus's proposition that the sun, rather than Earth, was at the center of the cosmos. If Copernicus was right, the nearby stars should appear to swing from side to side as Earth traveled annually on its orbit around the sun—a phenomenon known as parallax. Tycho searched hard, but could not find any stellar parallax. There were two possible conclusions. The first was that the stars were too far away, meaning that the change in their position was too small for Tycho to measure with the instruments of the day. (It is now known that the parallax of even the closest star is about 100 times smaller than the typical accuracy of Tycho's observations.) The second possibility was that

Tycho Brahe's observatory complex on the island of Hven attracted scholars and students from all over Europe between its founding in 1576 and its closure in 1597.

Copernicus was wrong and that Earth did not move. This was Tycho's conclusion.

The Tychonic model

In reaching this conclusion, Tycho trusted his own direct experience. He did not feel Earth moving. In fact, nothing that he observed convinced him that the planet was moving. Earth appeared to be stationary and the external universe was the only thing that appeared to be in motion. This led Tycho to discard the Copernican cosmos and introduce his own. In his model of the cosmos, all the planets except Earth orbited the sun, but the sun and the moon orbited a stationary Earth.

For many decades after his death in 1601, Tycho's model was popular among astronomers who were dissatisfied with Ptolemy's Earth-centric system but who did not wish to anger the Catholic Church by adopting the proscribed Copernican model. However, Tycho's own insistence on observational accuracy provided the data that would lead to his idea being discredited shortly after his death. His accurate observations helped Johannes Kepler

The Tychonic model kept Earth at the center of the cosmos as in the Ptolemaic model, but the five known planets were now orbiting the sun. Although he was impressed by the Copernican model, Tycho believed that Earth did not move.

Mars

Jupiter

Venus

Sun — Mercury — Saturn

Earth

Moon

Outer ring of stars

to demonstrate that the planets' orbits are ellipses and to create a model that would displace both the Tychonic and Copernican models.

Tycho's improved measurements would also allow English astronomer Edmond Halley to discover the proper motion of stars (the change in position due to the stars' motion through space) in 1718. Halley realized that the bright stars Sirius, Arcturus, and Aldebaran had, by Tycho's time, moved over half a degree away from the positions recorded by Hipparchus 1,850 years earlier. Not only were the stars not fixed in the sky, but the changing positions of the closer stars could also be measured. Stellar parallax was not detected until 1838. ∎

Tycho Brahe

Born a nobleman in 1546 in Scania (then Denmark, but now Sweden), Tyge Ottesen Brahe (Tycho is the Latinized version of his first name) became an astronomer after sighting a predicted solar eclipse in 1560.

In 1575, King Frederick II gave Tycho the island of Hven in the Øresund Strait, where he built an observatory. Tycho later fell out with Frederick's successor, Christian IV, over the potential transfer of the island to his children and closed the observatory. In 1599, he was appointed Imperial Mathematician to Emperor Rudolph II in Prague. There, Tycho appointed Johannes Kepler as his assistant.

Tycho was famed for his distinctive metal nose, the legacy of a duel he fought as a student. He died in 1601, allegedly of a burst bladder, having refused out of politeness to take a toilet break during a long royal banquet.

Key work

1588 *Astronomiæ Instauratæ Progymnasmata* (*Introduction to the New Astronomy*)

MIRA CETI IS A VARIABLE STAR

A NEW KIND OF STAR

IN CONTEXT

KEY ASTRONOMER
David Fabricius (1564–1617)

BEFORE
350 BCE Greek philosopher
Aristotle asserts that the stars
are fixed and unchanging.

AFTER
1667 Italian astronomer
Geminiano Montanari notes
that the star Algol varies
in brightness.

1784 John Goodricke discovers
Delta Cephei, a star that varies
in brightness over five days;
English astronomer Edward
Pigott discovers the variable
Eta Aquilae.

19th century Different kinds
of variable star are discovered,
including long-period variables,
cataclysmic variables, novae,
and supernovae stars.

1912 Henrietta Swan Leavitt
discovers a relationship
between the periods and the
brightness of variable stars
such as Delta Cephei.

The star **Mira Ceti** is
observed to **change in
brightness** periodically.

⬇

**Mira Ceti is a
variable star.**

⬇

Some stars are
variable.

⬇

**Aristotle was
wrong** when he asserted
that the stars are fixed
and eternal.

Before the work of German
astronomer David Fabricius,
it was thought that there
were only two types of star.
The first were those of constant
brightness, such as the 2,500 or
so that can be seen with the naked
eye above the horizon on a clear
dark night. The second type were
the "new stars," such as those
seen by Tycho Brahe in 1572
and Johannes Kepler in 1604.

The constant stars were
synonymous with the fixed,
permanent stars in the ancient
Greek cosmos—those that
mapped out the patterns in the
constellations and never changed.
The new stars, by contrast, would
appear unexpectedly, apparently
from nowhere, then fade away,
never to be seen again.

A third kind of star

While observing the star Mira Ceti
(also called Omicron Ceti), in the
constellation of Cetus the whale,
Fabricius realized that there was
a third type of star in the sky—one
that regularly varied in brightness.
He made his discovery in August
1596 as he was plotting the
movement of Jupiter across the
sky in relation to a nearby star.

See also: The geocentric model 20 ▪ The Tychonic model 44–47 ▪ Elliptical orbits 50–55 ▪ Variable stars 86 ▪ Measuring the universe 130–37

An artist's impression shows material flowing from Mira A (right) onto the hot disk around its companion white dwarf Mira B (left). The hot gas in the system emits X-rays.

To Fabricius's amazement, a few days later, the brightness of this star had increased by a factor of about three. After a few weeks, it disappeared from view altogether, only to reappear some years later. In 1609, Fabricius confirmed that Mira Ceti was a periodic variable star, showing that, contrary to the prevailing Greek philosophy that the cosmos was unchanging, stars were not constant.

Working with his son Johannes, Fabricius also used a camera obscura to look at the sun. They studied sunspots, observing that the spots moved across the sun's disk from east to west at a constant speed. They then disappeared, only to reappear on the other side, having been out of sight for the same time that it had taken them to move across the sun's disk. This was the first concrete evidence that the sun rotated, providing further proof of the variable nature of heavenly bodies. However, the book they published on the subject in 1611 was mostly overlooked, and the credit for describing the movement of sunspots went to Galileo, who published his results in 1613.

Double-star system

It is now known that Mira Ceti is a double-star system 420 light-years away. Mira A is an unstable red giant star, about 6 billion years old and in a late phase of its evolution. It pulses in and out, changing not only its size but also its temperature. During the cooler part of its cycle, it emits much of its energy as infrared radiation rather than light, so its brightness diminishes dramatically. Mira B is a white dwarf star surrounded by a disk of hot gas that is flowing from Mira A. ▪

In short, this new star signifies peace ... as well as change in the [Holy Roman] Empire for the better.
David Fabricius
in a letter to
Johannes Kepler

David Fabricius

David Fabricius was born in 1564 in Esens, Germany, and studied at the University of Helmsted. He later became a Lutheran pastor for a group of churches in Frisia.

Together with his son Johannes (1587–1615), he was fascinated by astronomy and an avid user of early telescopes, which his son had brought back with him from a trip to the Netherlands. Fabricius corresponded extensively with Johannes Kepler, with whom Fabricius pioneered the use of a camera obscura to observe the sun.

Little is known of Fabricius's life beyond his letters and publications. He died in 1617 after he was struck on the head with a shovel by a local goose thief, whom he had denounced from the pulpit.

Key work

1611 *Narration on Spots Observed on the Sun and their Apparent Rotation with the Sun* (with his son Johannes)

THE MOST TRUE PATH OF THE PLANET IS AN ELLIPSE

ELLIPTICAL ORBITS

IN CONTEXT

KEY ASTRONOMER
Johannes Kepler (1571–1630)

BEFORE
530–400 BCE The works of
Plato and Pythagoras convince
Kepler that the cosmos can be
explained using mathematics.

1543 Copernicus's sun-
centered cosmos helps
astronomers to visualize a
physical solar system but still
gives no indication as to the
true shape of a planetary orbit.

1600 Tycho Brahe convinces
Kepler of the reliability of his
planetary observations.

AFTER
1687 Isaac Newton realizes
that an inverse square law of
gravitational force explains why
the planets obey Kepler's laws.

1716 Edmond Halley uses
observations of the transit of
Venus to convert Kepler's ratios
of planetary distance from
the sun into absolute values.

Kepler was never satisfied by a
moderate agreement between
theory and observation.
The theory had to fit exactly
otherwise some new
possibility had to be tried.
Fred Hoyle

Before the 17th century,
all astronomers were also
astrologers. For many,
including German astronomer
Johannes Kepler, casting horoscopes
was the main source of their income
and influence. Knowing where
the planets had been in the sky
was important, but of greater
significance for constructing
astrological charts was the ability
to predict where the planets would
be over the next few decades.

To make predictions, astrologers
assumed that the planets moved
on specific paths around a central
object. Before Copernicus, in the
16th century, this central body
was thought by most to be Earth.
Copernicus showed how the
mathematics of planetary
prediction could be simplified
by assuming that the central body
was the sun. However, Copernicus
assumed that orbits were circular,
and to provide any reasonable
predictive accuracy, his system
still required the planets to

Kepler's most productive years
came in Prague under the patronage
of Holy Roman Emperor Rudolf II
(r.1576–1612). Rudolf was particularly
interested in astrology and alchemy.

move around a small circle, the
center of which moved around
a larger circle. These circular
velocities were always assumed
to be constant.

Kepler supported the Copernican
system, but the planetary tables it
produced could still easily be out by
a day or two. The planets, the sun,
and the moon always appeared in
a certain band of the sky, known
as the ecliptic, but actual paths of
individual planets around the sun
were still a mystery, as was the
mechanism that made them move.

Finding the paths
To improve the predictive tables,
Danish astronomer Tycho Brahe
spent more than 20 years observing
the planets. He next tried to
ascertain a path of each planet

See also: The Copernican model 32–39 ▪ The Tychonic model 44–47 ▪ Galileo's telescope 56–63 ▪ Gravitational theory 66–73 ▪ Halley's comet 74–77

through space that would fit the observational data. This is where the mathematical abilities of Kepler, Brahe's assistant, came into play. He considered specific models for the solar system and the paths of the individual planets in turn, including circular and ovoid (egg-shaped) orbits. After many calculations, Kepler determined whether or not the model led to predictions of planetary positions that fit into Tycho's precise observations. If there was not exact agreement, he would discard the idea and start the process again.

Abandoning circles

In 1608, after 10 years of work, Kepler found the solution, which involved abandoning both circles and constant velocity. The planets made an ellipse—a kind of stretched-out circle for which the amount of stretching is measured by a quantity called an eccentricity (p.54). Ellipses have two foci. The distance of a point on an ellipse from one focus plus the

distance from the other focus is always constant. Kepler found that the sun was at one of these two foci. These two facts made up his first law of planetary motion: the motion of the planets is an ellipse with the sun as one of the two foci.

Kepler also noticed that the speed of a planet on its ellipse was always changing, and that this change followed a fixed law (his second): a line between the planet and the sun sweeps out equal areas in equal times (p.54). These two laws were published in his 1609 book *Astronomia Nova*.

Kepler had chosen to investigate Mars, which had strong astrological significance, thought to influence human desire and action. Mars took variable retrograde loops—periods during which it would reverse its direction of movement—and large variations in brightness. It also had an orbital period of only 1.88 Earth years, meaning that Mars went around the sun about 11 times in Tycho's data »

Johannes Kepler

Born prematurely in 1571, Kepler spent his childhood in Leonberg, Swabia, in his grandfather's inn. Smallpox affected his coordination and vision. A scholarship enabled him to attend the Lutheran University of Tübingen in 1589, where he was taught by Michael Maestlin, Germany's top astronomer at the time. In 1600, Tycho Brahe invited Kepler to work with him at Castle Benátky near Prague. On Tycho's death in 1601, Kepler succeeded him as Imperial Mathematician.

In 1611, Kepler's wife died, and he became a teacher in Linz. He remarried and had seven more children, five of whom died young. His work was then disrupted between 1615 and 1621 while he defended his mother from charges of witchcraft. The Catholic Counter-Reformation in 1625 caused him further problems, and prevented his return to Tübingen. Kepler died of a fever in 1630.

Key works

1609 *Astronomia Nova*
1619 *Harmonices Mundi*
1627 *Rudolphine Tables*

Neither circular nor ovoid orbits fit **Tycho Brahe's data** on Mars.

An ellipse fits the data, so the path of Mars is an ellipse.

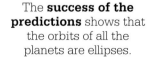

The **success of the predictions** shows that the orbits of all the planets are ellipses.

The **Three Laws of Planetary Motion** allow for new, improved predictive tables.

When just one body

goes around a larger body undisturbed, the paths it can follow are known as Kepler orbits. These are a group of curves called conic sections, which include ellipses, parabolas, and hyperbolas. The shape of the orbit is defined by a property called eccentricity. An eccentricity of 0 is a circle (A). Eccentricity between 0 and 1 is an ellipse (B). Eccentricity equal to 1 produces a parabola (C), and greater than 1 a hyperbola (D).

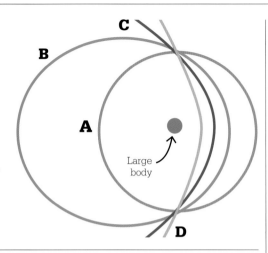

Large body

searched for a divine purpose within his scientific work. Since he saw six planets, he presumed that the number six must have a profound significance. He produced an ordered geometric model of the solar system in which the sun-centered spheres that contained each planetary orbit were inscribed and circumscribed by a specific regular "platonic" solid (the five possible solids whose faces and internal angles are all equal). The sphere containing the orbit of Mercury was placed inside an octahedron. The sphere that just touched the points of this regular solid contained the orbit of Venus. This in its turn was placed inside an icosahedron. Then followed the orbit of Earth, a dodecahedron, Mars, a tetrahedron, Jupiter, a cube, and finally Saturn. The system was beautifully ordered, but inaccurate.

set. Kepler was lucky to have chosen Mars, since its orbit has a high eccentricity, or stretch: 0.093 (where 0 is a circle and 1 is a parabola). This is 14 times the eccentricity of Venus. It took him another 12 years to show that the other planets also had elliptical orbits.

Studying Brahe's observations, Kepler was also able to work out the planets' orbital periods. Earth goes around the sun in one year, Mars in 1.88 Earth years, Jupiter in 11.86, and Saturn in 29.45. Kepler realized that the square of the orbital period was proportional to the cube of the planet's average distance from the sun. This became his third law and he published it in 1619 in his book *Harmonices Mundi*, alongside lengthy tracts on astrology, planetary music, and platonic figures. The book had taken him 20 years to produce.

Searching for meaning

Kepler was fascinated by patterns he found in the orbits of the planets. He noted that, once you accepted the Copernican system for the cosmos, the size of the orbits of the six planets—Mercury, Venus, Earth, Mars, Jupiter, and Saturn—appeared in the ratios 8 : 15 : 20 : 30 : 115 : 195.

Today, astronomers might look at a list of planetary orbital sizes and eccentricities and regard them as the result of the planetary formation process coupled with a few billion years of change. To Kepler, however, the numbers needed explanation. A deeply religious man, Kepler

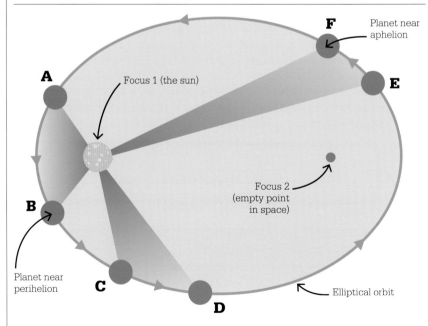

According to Kepler's second law, the line joining a planet to the sun sweeps out equal areas in equal times. This is also known as the law of equal areas. It is represented by the equal areas of the three shaded areas ABS, CDS, and EFS. It takes as long to travel from A to B as from C to D and from E to F. A planet moves most rapidly when it is nearest the sun, at perihelion; a planet's slowest motion occurs when it is farthest from the sun, at aphelion.

> Kepler was convinced that God created the world in accordance with the principle of perfect numbers, so that the underlying mathematical harmony … is the real and discoverable cause of the planetary motion.
> **William Dampier**
> *Science historian*

Kepler's great breakthrough was his calculation of the actual form of the planetary orbits, but the physics behind his three laws did not seem to concern him. Rather, he suggested that Mars was carried on its orbit by an angel in a chariot, or swept along by some magnetic influence emanating from the sun. The idea that the movements were due to a gravitational force only arrived with the ideas of Isaac Newton some 70 years later.

Wider contributions

Kepler also made important advances in the study of optics, and his 1604 book *Astronomiae Pars Optica* is regarded as the pioneer tome in the subject. Galileo's telescope interested him greatly and he even suggested an improved design using convex lenses for both the objective and the magnifying eyepiece. He wrote, too, about the supernova that was first seen in October 1604, today commonly called Kepler's supernova. Following Tycho, Kepler realized that the heavens could

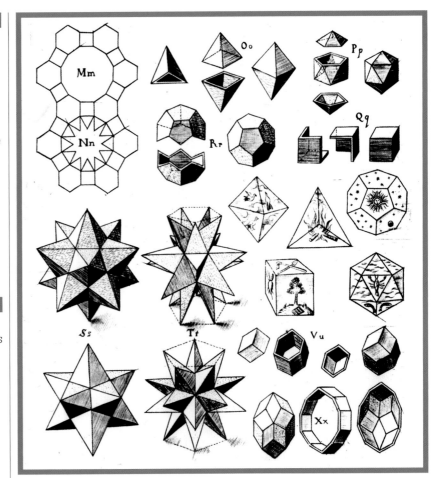

change, contradicting Aristotle's idea of a "fixed cosmos." A recent planetary conjunction coupled with this new star led him to speculate about the Biblical "Star of Bethlehem." Kepler's fervent imagination also produced the book *Somnium,* in which he discusses space travel to the moon and the lunar geography a visitor might expect on arrival. Many regard this as the first work of science fiction.

Kepler's most influential publication, however, was a textbook on astronomy called *Epitome Astronomiae Copernicanae,* and this became the most widely used astronomical work between 1630 and 1650. He ensured that

In *Harmonices Mundi*, Kepler experimented with regular shapes to unlock the secrets of the cosmos. He linked these shapes with harmonics to suggest a "music of the spheres."

the Rudolphine Tables (named after Emperor Rudolf, his patron in Prague) were eventually published, and these tables of predicted planetary positions helped him greatly with the well-paid calendars that he published between 1617 and 1624. The accuracy of his tables, proven over a few decades, also did much to encourage the acceptance of both the Copernican sun-centered solar system and Kepler's own three laws. ∎

OUR OWN EYES SHOW US FOUR STARS TRAVELING AROUND JUPITER

GALILEO'S TELESCOPE

IN CONTEXT

KEY ASTRONOMER
Galileo Galilei (1564–1642)

BEFORE
1543 Nicolaus Copernicus
proposes a theory of a sun-
centered cosmos, but proof
is needed because Earth does
not seem to move.

1608 Dutch eyeglass-makers
develop the first telescopes.

AFTER
1656 Dutch scientist
Christiaan Huygens builds
ever-bigger telescopes that
are capable of detecting more
detail and fainter objects.

1668 Isaac Newton produces
the first reflecting telescope,
an instrument that is far less
affected by the distortion of
chromatic aberration (p.60).

1733 The first flint glass/crown
glass achromatic lens is made.
This greatly improves the
potential image quality of
refracting telescopes.

Galileo Galilei's effective
use of a telescope marked
a watershed in the history
of astronomy. There have been
other turning points—such as the
introduction of photography, the
discovery of cosmic radio waves,
and the invention of the electronic
computer—but the invention of the
telescope was fundamental to the
advancement of the subject.

Limits of the naked eye
Before Galileo, the naked eye was
all that was available to observe
the sky. The naked eye is limited
in two main ways: it is unable to
record detail, and it can only detect
objects that are reasonably bright.

When looking at a full moon, the
lunar diameter subtends (spans) an
angle of $\frac{1}{2}°$ at Earth's surface. This
means that two lines extending
from opposite sides of the moon
meet at the eye to make an angle of
$\frac{1}{2}°$. However, the naked eye can only
detect separate objects that are more
than about $\frac{1}{60}°$ apart. This is the
eye's resolution, and determines
the level of detail it can detect.
Looking at the full moon with
the naked eye, the lunar diameter
is resolved into only 30 picture

The Milky Way is nothing
else but a mass of
innumerable stars planted
together in clusters.
Galileo Galilei

elements, analogous to individual
pixels in a digital photograph (see
below). Dark lunar seas and lighter
lunar highland are discernible, but
individual mountains and their
shadows are beyond detection.

Looking up at the night sky
on a cloud-free, moonless night in
Galileo's Italian countryside, 2,500
stars would be visible above the
horizon. The Milky Way—the disk
of the solar system seen side-on—
looks like a river of milk to the
naked eye. Only a telescope shows
that the Milky Way seems to be
made up of individual stars; the
bigger the telescope, the more stars

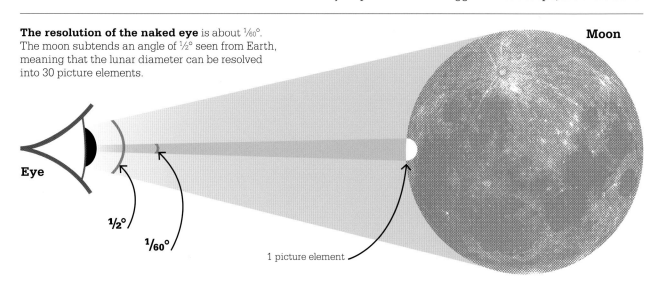

The resolution of the naked eye is about $\frac{1}{60}°$.
The moon subtends an angle of $\frac{1}{2}°$ seen from Earth,
meaning that the lunar diameter can be resolved
into 30 picture elements.

Moon

Eye

$\frac{1}{2}°$

$\frac{1}{60}°$

1 picture element

See also: The Copernican model 32–39 ▪ The Tychonic model 44–47 ▪ Elliptical orbits 50–55 ▪ Barnard (Directory) 337

Galileo demonstrates his telescope to Leonardo Donato, the Doge of Venice. Like other astronomers of his time, Galileo relied on patronage to fund and legitimize his work.

To the **naked eye**, Jupiter looks like a **bright star**.

The **telescope** shows a **finer resolution** than the naked eye.

This reveals that **Jupiter** is a **disk** with **four stars** around it.

The **four stars** can be seen to be **orbiting Jupiter**.

Jupiter has at least **four moons**.

are visible. By turning his new telescope to the night sky, Galileo would become one of the very first people to appreciate the true nature of this band of stars across the sky.

Building a telescope

Galileo did not invent the telescope himself. The idea of combining two lenses—a large one at the front of a tube to collect the light, and a small one at the back to magnify the image—had come from the Dutchmen Hans Lipperhey, Jacob Metius, and Sacharias Janssen in around September 1608. (It had taken over 300 years to progress from the invention of reading glasses to the invention of a

telescope.) After hearing about this new instrument, Galileo had resolved to make one for himself.

A telescope does two important things. Its resolution (the detail a telescope can detect) is proportional to the diameter of the objective lens—the large lens at the front that collects the light. The larger the objective lens, the better the resolution. An eye that has fully adapted to the dark has a pupil that is about ¼ in (0.5 cm) across, and a resolution of around $1/60°$. Put the eye at the back of a telescope with an objective lens of 1, 2, or 4 cm diameter, and the resolution improves to $1/120°$, $1/240°$, and $1/480°$ respectively. Details then spring into view. Jupiter, for example, looks like a disk and not just a point.

A telescope also acts as a "light bucket." Every time the diameter of the objective lens is doubled, the light gathered increases by a factor

of four, and objects of similar light output can be detected if they are twice as far away. Objective lenses of 1, 2, and 4 cm enable the eye to discern 20,000, 160,000, and 1,280,000 stars respectively.

Galileo was not satisfied with his first instrument, which only magnified three times. He realized that a telescope's magnification »

> My dear Kepler, what would you say of the learned who ... have steadfastly refused to cast a glance through the telescope?
> **Galileo Galilei**

was directly related to the ratio of the focal length of the objective lens to the focal length of the eyepiece. A longer-focus convex lens for the objective, or a shorter-focus concave lens for the eyepiece was required. Since these were not available, Galileo taught himself to grind and polish lenses and made them for himself. Living in northern Italy, the glassmaking center of the world at the time, helped him considerably. He eventually

developed a new telescope with 33 times magnification, and it was with this improved instrument that he discovered the Jovian ("of Jupiter") moons.

"Three little stars"

Galileo discovered the moons of the planet Jupiter on the night of January 7, 1610. At first, he thought he was looking at distant stars, but he quickly realized that the new bodies were moving around Jupiter, At the time, Galileo was a 45-year-old professor of mathematics at the University of Padua near Venice. When he published his pioneering telescopic observations, he wrote: "Through a spyglass, Jupiter presented himself. And since I had prepared for myself a superlative instrument, I saw (which earlier had not happened because of the weakness of other instruments) that three little stars were positioned near him—small but yet very bright. Although I believed them to be among the number of fixed stars, they nevertheless intrigued me because

> Galileo had the experience of beholding the heavens as they actually are for perhaps the first time.
> **I Bernard Cohen**

they appeared to be arranged exactly along a straight line and parallel to the ecliptic ...."

Repeated observations

Galileo's unexpected discovery fascinated him. As he observed Jupiter night after night, it soon became clear that the new stars were not beyond Jupiter, in the distant heavens. They not only accompanied the planet as it moved along its path across the sky, but also moved around the planet.

Just as the moon orbits Earth every month, Galileo realized that there were four moons in orbit around Jupiter, staying with it as it orbited the sun. The more distant moons took longer to complete their orbits than the closer ones. The time to complete one orbit from the inner to the outer moon is 1.77, 3.55, 7.15, and 16.69 days, respectively. The Jovian moon system looked like a small model of the sun's planetary system. It was proof that not everything in the cosmos orbited Earth, as had been thought in pre-Copernican days. The observation of these four moons was a boost to the theory of the sun-centered cosmos.

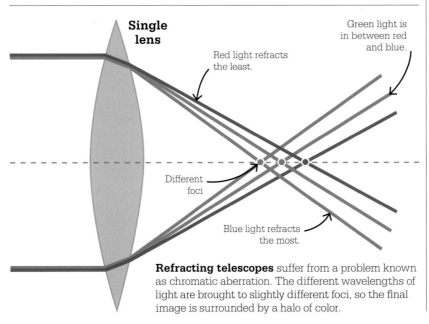

Single lens

Green light is in between red and blue.

Red light refracts the least.

Different foci

Blue light refracts the most.

Refracting telescopes suffer from a problem known as chromatic aberration. The different wavelengths of light are brought to slightly different foci, so the final image is surrounded by a halo of color.

Galileo's telescope had a concave lens as an eyepiece. When viewing a celestial object a great distance away, the distance between the two lenses would equal the focal length of the objective lens minus the focal length of the eyepiece.

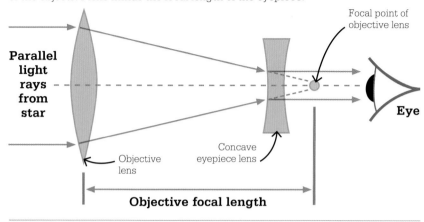

Focal point of objective lens

Parallel light rays from star

Eye

Objective lens

Concave eyepiece lens

Objective focal length

Refracting telescopes

Kepler's telescope, developed soon after, had a convex lens as an eyepiece. The length of the telescope was equal to the objective focal length plus the focal length of the eyepiece.

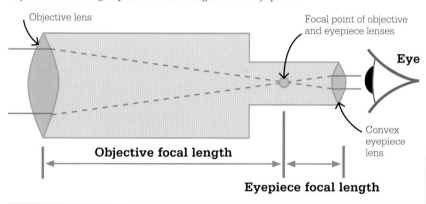

Objective lens

Focal point of objective and eyepiece lenses

Eye

Convex eyepiece lens

Objective focal length

Eyepiece focal length

There were two kinds of early refracting telescope: the Galilean, and the Keplerian, developed in 1611 by Johannes Kepler (see left). They both had a long-focus, large diameter lens at the front, called the objective. This collected the light and brought it to a focus. The image at the focus was magnified using the smaller, short-focus eyepiece lens.

The magnification of the instrument is equal to the focal length of the objective lens divided by the focal length of the eyepiece. A flatter convex objective lens reduced chromatic aberration (see opposite), gave a longer focal length, and, for a fixed eyepiece, greater magnification. For this reason, telescopes became longer in the 17th century. The minimum focal length of eyepieces at the time of Galileo and Kepler was about 1–1½ in (2–4 cm). This meant that, for a magnification of x30, an objective lens with a focal length of 24–48 in (60–120 cm) was needed. Built in 1888, the huge James Lick Telescope on Mount Hamilton, California (above), has a 36-in (90-cm) lens and a focal length of 57 ft (17.37 m).

Galileo quickly published his discovery in his book *Siderius Nuncius (The Starry Messenger)*, published on March 10, 1610. In the hope of advancement, Galileo dedicated the book to a former pupil of his who later became the Grand Duke of Tuscany, Cosimo II de' Medici. He named the moons the Medicean Stars in honor of the four royal Medici brothers. This political thoughtfulness won him the position of Chief Mathematician and Philosopher to the Medici at the University of Pisa. However, the name did not catch on.

At first, many were sceptical, suggesting that the moons were no more than defects in the telescope lens. However, other pioneering telescopic astronomers such as Thomas Harriot, Joseph Gaultier de la Vatelle, and Nicolas-Claude Fabri de Peiresc confirmed their existence when Jupiter returned to the night sky later in 1610, after passing behind the sun.

Disputed priority

In 1614, German astronomer Simon Marius published *Mundus Iovialis*, in which he described Jupiter's »

moons and claimed to have discovered them before Galileo. Galileo would later accuse Marius of plagiarism, but it is now generally accepted that he made his discovery independently at around the same time. Marius named the moons Io, Europa, Ganymede, and Callisto after the Roman god Jupiter's love conquests, and these names are still used. They are now known collectively as the Galilean moons.

A Jovian clock

Galileo carefully studied the changing positions of the Jovian moons from day to day. He concluded that, like the planets, their positions could be calculated in advance. Galileo saw that, if this could be done accurately, the system would act as a universal clock and could solve the problem of measuring longitude at sea. To establish longitude requires the ability to tell the time, but in Galileo's day, there were no timepieces that would work on a boat. Because Jupiter is at least

four times farther away from Earth than the sun, the Jovian system looks the same from anywhere on Earth, so a "Jovian clock" would work from anywhere. The longitude problem was finally solved with the introduction of accurate chronometers by the English clockmaker John Harrison around 1740. This was well before the orbits of Jupiter's moons had been worked out in detail.

Galileo's discovery of four satellites around Jupiter had another interesting consequence.

The Bible shows the way to go to heaven, not the way the heavens go.
Galileo Galilei

When Jonathan Swift published *Gulliver's Travels* in 1726, he predicted, in the chapter on Laputa, that Mars would have two moons simply because Earth had one and Jupiter four. In 1877, this prediction was fortuitously proved to be correct when Asaph Hall discovered Mars's two small moons, Phobos and Deimos, using a new 26-in (66-cm) refracting telescope at the US Naval Observatory in Washington.

Support for Copernicus

In Galileo's time, there was still a heated debate between believers of the old biblical theory that Earth was stationary at the center of the cosmos and Copernicus's new idea that the Earth was in orbit around the sun. The geocentric (Earth-centered) idea stressed the uniqueness of the planet, while the heliocentric (sun-centered) proposal made Earth just one of a family of planets. The assumption that Earth does not occupy a privileged place in the cosmos is now known as the Copernican principle.

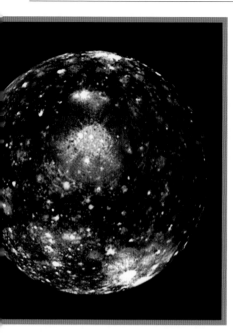

Starting with the closest to Jupiter, the Galilean moons are, from left to right, Io, Europa, Ganymede, and Callisto. Ganymede is larger than the planet Mercury.

The challenge now was to find observations to prove that one theory was correct and the other false. The discovery of moons around Jupiter was great support for a sun-centered system. It was now clear that everything did not orbit around a central Earth, but there were still unanswered questions. If the sun-centered system was correct, Earth must be moving. If Earth had to travel around the sun

every year, it had to have an orbital speed of 20 miles/sec (30 km/sec). In Galileo's time, the exact distance from Earth to the sun was not known, but it was clearly far enough that Earth would need to be moving quickly, and humans cannot apprehend this movement. Also, this orbital motion should make the stars appear to swing from side to side every year in a phenomenon called stellar parallax (p.102). This again was not observed at the time. Galileo and his contemporaries did not suspect that the typical distance between stars in the Milky Way was about 500,000 times larger than the distance between Earth and the sun, which makes stellar parallax so small that it is difficult to measure. It was not until the mid-19th century that vastly improved instruments made it possible to detect this swing.

Despite these questions, Galileo considered that his findings had proved Copernicus correct

beyond reasonable doubt. His discoveries also included the phases of Venus, which are best explained if the planet is in orbit around the sun, and the fact that the sun is spinning, shown by the movement of sunspots. By 1619, Galileo's pugnacious defense of Copernicus had drawn him into conflict with the Church, which had declared in 1616 that heliocentricism was heretical. In 1633, he appeared before the Inquisition. His books were banned, and he spent the last 10 years of his life under house arrest.

New moons

Jupiter only had four known moons for 283 years. A fifth satellite, Amalthea, was discovered by the American astronomer E. E. Barnard in 1892, using the 36-in (91-cm) refractor at the Lick Observatory in California. It was the last solar system satellite to be discovered by direct observation. Subsequently, satellites have been found by the meticulous examination of photographs. The number of known Jupiter satellites had crept up to 12 by the mid-1950s, and has now reached 67. Many smaller moons may be found in the future. ■

Galileo Galilei

Galileo Galilei was born in Pisa, Italy on 15 February 1564. He was appointed to the Chair of Mathematics at the University of Pisa in 1589, moving to the University of Padua in 1590. Galileo was an astronomer, physicist, mathematician, philosopher, and engineer, who played a pivotal role in the process of intellectual advances in Europe now known as the Scientific Revolution.

He was the first person to effectively turn the refractor telescope on the heavens. During 1609–10, he discovered

that the planet Jupiter had four moons, Venus underwent phase changes, the moon was mountainous, and the sun was spinning round once in about a month. He was a prolific writer and made his findings accessible to a wide audience.

Key works

1610 *The Starry Messenger*
1632 *Dialogue Concerning the Two Chief World Systems*
1638 *The Discourses and Mathematical Demonstrations Relating to Two New Sciences*

A PERFECTLY CIRCULAR SPOT CENTERED ON THE SUN
THE TRANSIT OF VENUS

IN CONTEXT

KEY ASTRONOMER
Jeremiah Horrocks
(1618–1641)

BEFORE
c.150 CE Ptolemy estimates the Earth–sun distance at 1,210 times Earth's radius—around 5 million miles (8 million km).

1619 Kepler's third law gives the ratio of the sizes of planetary orbits but the absolute values are not known.

1631 French astronomer Pierre Gassendi observes a transit of Mercury across the solar disk, the first planetary transit to be recorded in history.

AFTER
1716 Edmond Halley suggests that an accurate timing of the transit of Venus could lead to an accurate Earth–sun distance.

2012 The most recent transit of Venus takes place. The next two will be in 2117 and 2125.

In 1639, a 20-year-old English astronomer named Jeremiah Horrocks predicted a transit of Venus across the face of the sun after finding errors in tables made by Johannes Kepler. Because the transit was only four weeks away, Horrocks wrote to his collaborator, William Crabtree, urging him to observe it. On December 4, 1639, Horrocks and Crabtree independently set up helioscopes that focused an image of the sun from a telescope onto a plane. They became the first people to witness a transit of Venus.

As it crossed the sun's disk, Horrocks tried to calculate Venus's size and distance. He noted that it subtended, or spanned, an angle of 76 arcseconds ($^{76}/_{3600}°$) at Earth (p.58), which was smaller than the value guessed by Kepler. Using the ratios of planetary distances known from Kepler's third law, Horrocks calculated that the disk of Venus subtended an angle of about 28 arcseconds as seen from the sun.

Using data from a transit of Mercury that had taken place in 1631, Horrocks calculated that

The most recent transit of Venus in 2012 (the tiny dot in the top right of the sun's disk) was captured by NASA's Solar Dynamics Observatory.

Mercury subtended the same angle as Venus. He guessed that all the planets subtend the same angle at the sun, and calculated the distance from Earth to the sun to be 59 million miles (95 million km).

Horrocks's guess is now known to be wrong: Earth subtends 17.8 arcseconds at the sun, which is 93 million miles (150 million km) away. Nevertheless, he was the first to have a reasonably accurate idea of the size of the solar system. ∎

See also: Elliptical orbits 50–55 ▪ Halley's comet 74–77

NEW MOONS AROUND SATURN
OBSERVING SATURN'S RINGS

IN CONTEXT

KEY ASTRONOMER
Giovanni Domenico Cassini
(1625–1712)

BEFORE
1610 Galileo announces the discovery of four moons around Jupiter.

1655 Christiaan Huygens discovers Titan, a moon orbiting Saturn that is 50 percent larger than Earth's moon.

AFTER
1801 The first asteroid is discovered in an orbit between Mars and Jupiter.

1859 Scottish physicist James Clerk Maxwell proves that Saturn's rings cannot be solid, since they would break apart under the force of gravity.

1960s onward In recent decades, spacecraft have gone into orbit around Jupiter and Saturn, and Voyager 2 flew by Uranus and Neptune. Large numbers of moons have been discovered.

Working at the Panzano Observatory near Bologna, Italian astronomer Giovanni Cassini was provided in 1664 with a state-of-the-art refracting telescope made by Guiseppe Campini of Rome. With it, he discovered the bands and spots on Jupiter, measured the planet's spin period and polar flattening, and made observations of the orbits of Jupiter's four known moons.

Observing Saturn
Cassini's reputation as a brilliant observer led to an invitation to oversee the completion of the new Paris observatory. There, he turned his telescope on Saturn, the largest moon of which, Titan, had been discovered in 1655 by Christiaan Huygens. Cassini discovered two more satellites: Iapetus in 1671 and Rhea in 1672. In 1675, he noticed a large gap in the Saturnian rings and concluded, correctly, that the rings were not solid but made up of a multitude of small orbiting bodies. In 1684, he discovered two fainter satellites, Tethys and Dione.

With these observations, Cassini single-handedly nearly doubled the number of known satellites in the solar system. This number has since increased dramatically.

Jupiter and Saturn have more than 60 known satellites each. The gas giants in the outer solar system have two types of moon—large ones that were formed at the same time as the planet and smaller ones captured from the asteroid belt. In the inner solar system, Mars has two small captured asteroidal moons, while Mercury and Venus have no moons. Earth has one huge moon, $\frac{1}{81}$ its mass, and astronomers are still not certain how it formed. ■

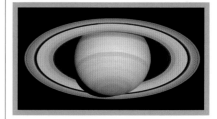

The largest gap in Saturn's rings, called the Cassini Division, separates the outer A-ring from the inner B-ring. It is 3,000 miles (4,800 km) wide.

See also: Galileo's telescope 56–63 ▪ The origin of the moon 186–87 ▪ Huygens (Directory) 335

GRAVITY EXPLAINS THE MOTIONS OF THE PLANETS

GRAVITATIONAL THEORY

IN CONTEXT

KEY ASTRONOMER
Isaac Newton (1642–1726)

BEFORE
1609 Johannes Kepler shows that Mars has an elliptical orbit.

AFTER
1798 Henry Cavendish measures the gravitational constant for the first time.

1846 French mathematician Urbain Le Verrier uses Newton's laws to calculate the planet Neptune's position.

1915 Albert Einstein introduces general relativity and explains the gravitational force as a function of the curvature of spacetime.

2014 The gravitational constant is measured by studying the behavior of atoms. The latest figure is given as 6.6719×10^{-11} $m^3 kg^{-1} s^{-2}$. This is about 1 percent less than the value Henry Cavendish calculated.

G ravity is the name given to the force of attraction between any two masses. It is the force that attracts all objects to Earth, giving them weight. It draws objects downward, toward the center of Earth. If the object were on the moon, a much smaller mass than Earth, the force would be six times less and its weight would be one sixth of its weight on Earth. English physicist, astronomer, and mathematician Isaac Newton was the first person to realize that gravity is a universal force, acting on all objects, and that it explains the movement of planets.

Describing orbits

The shapes of the orbits of the planets were already well-known in Newton's time, based on the three laws of planetary motion introduced by Johannes Kepler. Kepler's first law stated that these orbits were ellipses, with the sun at one focus of each ellipse. The second law described the way that planets moved along their orbits more quickly when they were close to the sun than when they were farther away. The third law

To myself I am only a child playing on the beach, while vast oceans of truth lie undiscovered before me.
Isaac Newton

described the relation between the time taken to complete one orbit and the distance from the sun: the time taken for one orbit, squared, was equal to the cube of the average distance between the planet and the sun. For instance, Earth goes around the sun in one year, while Jupiter is 5.2 times farther away from the sun than Earth. 5.2 cubed equals 140, and the square root of 140 gives the correct figure for one Jupiter year: 11.86 Earth years.

However, although Kepler had correctly discovered the shapes and speeds of planetary

Isaac Newton

Isaac Newton was born on a farm in Woolsthorpe, Lincolnshire, on December 25, 1642. After school in Grantham, he attended Trinity College Cambridge, where he became a Fellow and taught physics and astronomy. His book *Principia* set out the principle of gravity and celestial mechanics.

Newton invented the reflecting telescope; wrote theses on optics, the prism, and the spectrum of white light; was one of the founders of calculus; and studied the cooling of bodies. He also explained why Earth was oblate (a squashed sphere) in shape and

why the equinox moved, and formalized the physics of the speed of sound. He spent much time on biblical chronology and alchemy. Newton was at various times President of the Royal Society, Warden and Master of the Royal Mint, and member of parliament for Cambridge University. He died in 1727.

Key works

1671 *Method of Fluxions*
1687 *Philosphiae Naturalis Principia Mathematica*
1704 *Optics*

See also: Elliptical orbits 50–55 ▪ Halley's comet 74–77 ▪ The discovery of Neptune 106–07 ▪ The theory of relativity 146–53 ▪ Lagrange (Directory) 336

The Great Comet appeared in 1680, then again in 1681. John Flamsteed proposed that it was the same comet. Newton disagreed, but changed his mind after examining Flamsteed's data.

orbits, he did not know why the planets moved as they did. In his 1609 book *Astronomia Nova,* he suggested that Mars was being carried around its orbit by an angel in a chariot. A year later, he had changed his mind, suggesting that the planets were magnets and were being driven around by magnetic "arms" extending from the spinning sun.

Newton's insight

Before Newton, several scientists, including Englishman Robert Hooke and Italian Giovanni Alfonso Borelli, suggested that there was a force of attraction between the sun and the individual planets. They also stated that the force decreased with distance.

On December 9, 1679, Hooke wrote to Newton saying that he thought the force might decrease as the inverse square of distance. However, Hooke did not publish the idea and did not possess the mathematical skills to fully demonstrate his proposition. By contrast, Newton was able to prove rigorously that an inverse square law of attractive force would result in an elliptical planetary orbit.

Newton used mathematics to demonstrate that, if the force of attraction (F) between the sun and the planets varied precisely as an inverse square of the distance (r) between them, this fully explained the planetary orbits and why they follow Kepler's three laws. This is written mathematically as

$F \propto 1/r^2$. It means that doubling the distance between the objects reduces the strength of the attractive force to a quarter of the original force.

The Great Comet

Newton was a shy, reclusive man, and reluctant to publish his breakthrough. Two things forced his hand. The first was the Great Comet of 1680, and the second was the astronomer Edmond Halley.

The Great Comet of 1680 was the brightest comet of the 17th century—so bright that for a short time it was visible in the daytime. Two comets were seen: one that was approaching the sun in November and December 1680; and another that was moving away from the sun between late December 1680 and March 1681. As with all

comets at the time, its orbit was a mystery, and the two sightings were at first not widely recognized as the same object. Astronomer John Flamsteed suggested that the two sightings might be of the same comet, which had come from the outer edge of the solar system, swung around the sun (where it was too close to the sun to be seen), and moved out again.

Halley was fascinated by the mysterious form of cometary orbits, and traveled to Cambridge to discuss the problem with his friend Newton. Using his law that related force to acceleration and his insistence that the strength of the force varied as the inverse square of distance, Newton calculated the parameters of the comet's orbit as it passed through the inner solar » system. This breakthrough intrigued Halley so much that he went on to calculate the orbits of 24 other comets, and to prove that one comet (Halley's comet) returned to the sun around every 76 years. Perhaps more importantly, Halley was so impressed by Newton's work that he strongly encouraged him to publish his findings. This resulted in the book *Philosophiae Naturalis Principia Mathematica*, published in Latin on July 5, 1687, in which Newton describes his laws of motion, his gravitational theory, the proof of Kepler's three laws, and the method he used to calculate a comet's orbit.

The planets' elliptical orbits are explained by an **attractive force** that reduces at a rate of **the square of the distance** between objects.

Gravity explains the motions of the planets, but does not explain what sets them in motion.

This force is universal and applies to all bodies with mass at all distances.

The masses of the two bodies (m_1 and m_2)

The gravitational constant (**G**)

$$F = \frac{Gm_1m_2}{r^2}$$

The force of attraction between the bodies (**F**)

The distance between the bodies (**r**)

Newton's law of universal gravitation shows how the force produced depends on the mass of the two objects and the square of the distance between them.

In his book, Newton stressed that his law was universal—gravity affects everything in the universe, regardless of distance. It explained how an apple fell on his head in the orchard of Woolsthorpe where his mother lived, the tides in the seas, the moon orbiting Earth, Jupiter orbiting the sun, and even the elliptical orbit of a comet. The physical law that made the apple fall in his yard was exactly the same as the one that shaped the solar system, and would later be discovered at work between stars and distant galaxies. Evidence was all around that Newton's law of gravitation worked. It not only explained where planets had been, but also made it possible to predict where they would go in the future.

Constant of proportionality
Newton's law of gravitation states that the size of the gravitational force is proportional to the masses of the two bodies (m_1 and m_2) multiplied together and divided by the square of the distance, r, between them (see left). It always draws masses together and acts along a straight line between them. If the object in question is spherically symmetrical, like Earth, then its gravitational pull can be treated as if it were coming from a point at its center. One final value is needed to calculate the force—the constant of proportionality, a number that gives the strength of the force: the gravitational constant (G).

Measuring G
Gravity is a weak force, and this means that the gravitational constant is rather difficult to measure accurately. The first laboratory test of Newton's theory was made by the English aristocrat scientist Henry Cavendish in 1798, 71 years after Newton's death. He copied an experimental system proposed by the geophysicist John Michell and successfully measured the gravitational force between two lead balls, of diameters 2 and

Nature and Nature's laws lay hid in night: God said, "Let Newton be!" and all was light.
Alexander Pope

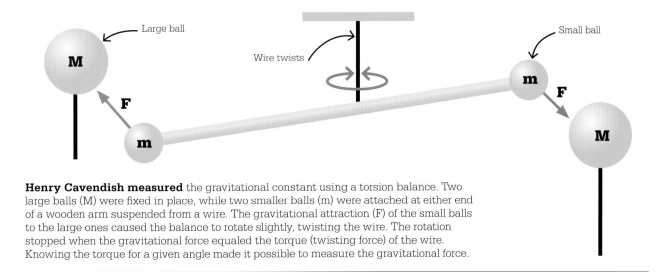

Henry Cavendish measured the gravitational constant using a torsion balance. Two large balls (M) were fixed in place, while two smaller balls (m) were attached at either end of a wooden arm suspended from a wire. The gravitational attraction (F) of the small balls to the large ones caused the balance to rotate slightly, twisting the wire. The rotation stopped when the gravitational force equaled the torque (twisting force) of the wire. Knowing the torque for a given angle made it possible to measure the gravitational force.

12 in (5.1 and 30 cm) (see above). Many have tried to refine and repeat the experiment since. This has led to a slow improvement in the accuracy of G. Some scientists suggested that G changed with time. However, recent analysis of type 1a supernovae has shown that, over the last nine billion years, G has changed by less than one part in 10 billion, if at all. The light from distant supernovae was emitted nine billion years ago, allowing scientists to study the laws of physics as they were in the distant past.

Seeking meaning
Like many of the scientists of his time, Newton was deeply pious and sought a religious meaning behind his observations and laws. The solar system was not regarded as a random collection of planets, and the sizes of the specific orbits were thought to have some specific meaning. For example, Kepler had sought meaning with his notion of "the music of the spheres." Building on ideas first put forward by Pythagoras and Ptolemy, Kepler suggested that each planet was responsible for an inaudible musical note that had a frequency proportional to the velocity of the planet along its orbit. The slower a planet moved, the lower the note that it emitted. The difference between the notes produced by adjacent planets turned out to be well-known musical intervals such as major thirds.

There is some scientific merit behind Kepler's idea. The solar system is about 4.6 billion years old. During its lifetime, the planets and their satellites have exerted gravitational influences on each

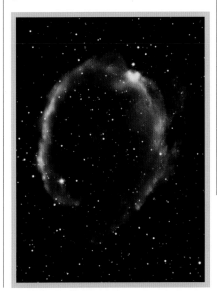

other and have fallen into resonant intervals, similar to the way musical notes resonate. Looking at three of the moons of Jupiter, for every once that Ganymede orbits the planet, Europa goes around twice and Io four times. Over time, they have been gravitationally locked into this resonance.

The three-body problem
The solar system as a whole has fallen into similar resonant proportions to Jupiter's moons. On average, each planet has an orbit that is about 73 percent larger than the planet immediately closer to the sun. Here, however, there appears a difficult mathematical problem, and one that Newton had grappled with. The movement of a low-mass body under the gravitational influence of a large-mass body can be understood, and predicted. But when three bodies are involved, the mathematical problem becomes exceedingly difficult. »

Distant supernovae are seen today as they were billions of years ago. Analysis of their structure shows that the law of gravity operated with the same value of G then as today.

An example of a three-body system is the moon-Earth-sun. Newton thought about this system but the mathematical difficulties were insurmountable, and human knowledge of where the moon will be in the distant future is still very limited. Variations in the orbit of Halley's comet are another indicator of the influence of the gravitational fields of the planets operating in addition to the gravitation of the sun. Recent orbits have taken 76.0,

I have not been able to discover the cause of these properties of gravity from phenomena, and I frame no hypotheses.
Isaac Newton

76.1, 76.3, 76.9, 77.4, 76.1, 76.5, 77.1, 77.8, and 79.1 years respectively due to the combined gravitational influence of the sun, Jupiter, Saturn, and other planets on the comet.

Shaping the planets

While Newton searched for religious meaning in his scientific work, he could find none behind his theory of gravity. He did not discover the hand of God setting the planets in motion, but he had found a formula that shaped the universe.

The action of gravity is key to understanding why the universe looks as it does. For instance, gravity is responsible for the spherical shapes of the planets. If a body has sufficient mass, the gravitational force that it exerts exceeds the strength of the material of the body and it is pulled into a spherical shape. Astronomical rocky bodies, such as the asteroids between the orbits of Mars and Jupiter, are irregular in shape if they have a diameter of less than about 240 miles (380 km) (the Hughes-Cole limit).

Gravitation is also responsible for the size of the deviations from a sphere that can occur on a planet. There are no mountains on Earth higher than the 5.5 miles (8.8 km) of Mount Everest because the gravitational weight of a taller mountain would exceed the strength of the underlying mantle rock, and sink. On planets with lower mass, the weight of objects is less, and so mountains can be bigger. The highest mountain on Mars, for instance, Olympus Mons, is nearly three times as high as Everest. The mass of Mars is about one-tenth that of Earth, and its diameter is about half Earth's. Putting these numbers into Newton's formula for gravitation, this gives a weight on the surface of Mars of just over one-third that on Earth, which explains the size of Olympus Mons.

In his great work *Principia,* Newton plotted the parabolic path of the Great Comet by taking accurate observations and correcting them to allow for the motion of Earth.

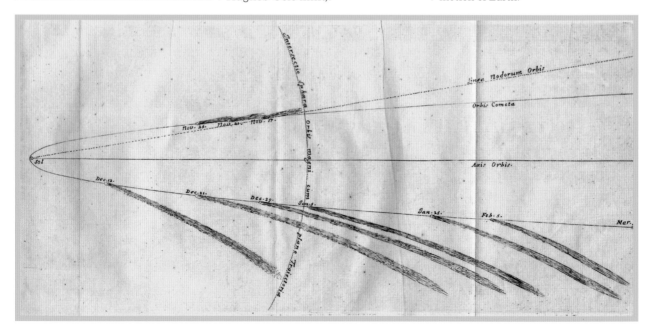

The motions of the comets are exceedingly regular, and they observe the same laws as the motions of the planets.
Isaac Newton

Gravity thus also shapes life on Earth by limiting the size of animals. The largest land animals ever were dinosaurs weighing up to 40 tons. The largest animals of all, whales, are found in the oceans, where the water supports their weight. Gravity is also responsible for the tides, which are produced because water bulges toward the sun and moon on the side of Earth nearer to them, and also bulges away from them on the other side where their gravitational pull is weaker. When the sun and moon are aligned, there is a high spring tide; when they are at right angles, there is a low neap tide.

Escape velocity
Gravity profoundly affects human mobility. The height a person can jump is determined by the gravitational field at ground level. Newton realized that the strength of gravity would affect the ease of travel beyond the atmosphere. To break free from Earth's gravitational pull, it is necessary to travel at 25,020 mph (40,270 km/h). It is much easier to get away from less massive bodies such as the moon and Mars. Turning the problem around, this escape velocity is also the minimum velocity that an incoming asteroid

Newton illustrated escape velocity with a thought experiment of a cannon firing horizontally from a high mountain. At velocities less than orbital velocity at that altitude, the cannonball will fall to earth (A and B). At exactly orbital velocity, it will enter a circular orbit (C). At greater than orbital velocity but less than escape velocity, it will enter an elliptical orbit (D). Only at escape velocity will it fly off into space (E).

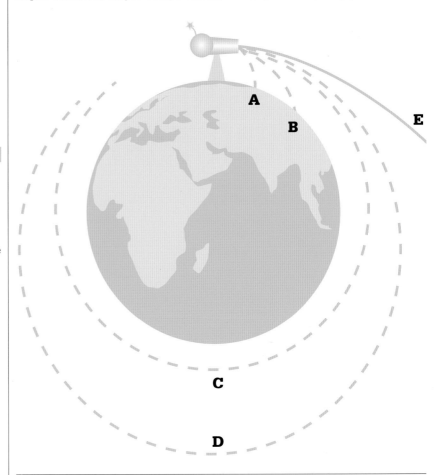

or comet can have when it hits Earth's surface, and this affects the size of the resulting crater.

Today, gravity is held to be most accurately described by the general theory of relativity proposed by Albert Einstein in 1915. This does not describe gravity as a force, but instead as a consequence of the curvature of the continuum of spacetime due to the uneven distribution of mass inside it. This said, Newton's concept of a gravitational force is an excellent approximation in the vast majority of cases. General relativity only needs to be invoked in cases requiring extreme precision or where the gravitational field is very strong, such as close to the sun or in the vicinity of a massive black hole. Massive bodies that are accelerating can produce waves in spacetime, and these propagate out at the speed of light. The first detection of one of these gravitational waves was announced in February 2016 (pp.326–29). ∎

I DARE VENTURE TO FORETELL THAT THE COMET WILL RETURN AGAIN IN THE YEAR 1758

HALLEY'S COMET

In the 16th century and for much of the 17th, advances were made in understanding the motions of planets, but the nature of comets remained a mystery. Up until at least 1500, comets had been feared as harbingers of doom in Europe. Astronomers were familiar with these bright blotches of light and their long, beautiful tails moving slowly across the sky over periods of a few weeks or months, but had no idea where they came from, nor where they disappeared to.

Things changed, however, in 1577, when an exceptionally bright comet lit up the night sky for several months. By studying

See also: The Tychonic model 44–47 ▪ Elliptical orbits 50–55 ▪
Gravitational theory 66–73

observational data from different
parts of Europe, the Danish
astronomer Tycho Brahe calculated
that the comet must be at least four
times farther away than the moon,
and this allowed him to fit comets
into his model of the universe.
He saw them as objects that
could move freely through the
same regions of space as planets.
But what was not agreed on in
Brahe's time, nor for many decades
afterward, was the shape of the
paths that comets carved through
space. Brahe's one-time student
Johannes Kepler believed that they
traveled in straight lines. Polish
astronomer Johannes Hevelius,
however, suggested that a comet
of 1664 had traveled in a curved
orbit around the sun.

Newton tackles comets

From about 1680, stimulated by
the appearance of a particularly
bright comet that year, the great
English scientist Isaac Newton
began studying cometary orbits
while developing his universal

Halley's comet appeared in 1066
and is shown in the Bayeaux Tapestry,
with Anglo-Saxons pointing fearfully
at the sky. Its appearance was taken
by some to foretell the fall of England.

theory of gravitation. Using his
new theory, Newton analyzed and
predicted the future path that the
1680 comet would take. He came
to the conclusion that comets
(like planets) had orbits in the
shapes of ellipses, with the sun
at one focus of the ellipse. These
ellipses were so stretched out,
however, that they could be
approximated to an open-ended
curve called a parabola. If Newton
was right, then once a comet had
visited the inner solar system and
curved around the sun, it would
either never return (if its orbit was
parabolic) or would not return for
thousands of years (if its orbit was
an extremely stretched-out ellipse,
but not a parabola).

In 1684, Newton received a
visit from a young acquaintance
named Edmond Halley, who was »

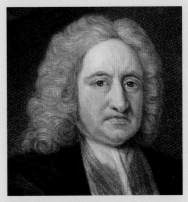

Edmond Halley

Edmond Halley was born
in 1656 in London. In 1676,
he sailed to the island of
St. Helena in the South
Atlantic where he charted
the stars of the southern
hemisphere, publishing a
catalog and star charts after
his return. In 1687, he helped
persuade Isaac Newton to
publish *Principia*, which
included details on how to
calculate cometary orbits.

Halley was appointed
Astronomer Royal in 1720,
and he resided at the Royal
Observatory, Greenwich,
until his death in 1742.
Although remembered mainly
as an astronomer, Halley
did important work in many
fields. He published studies
on variations in Earth's
magnetic field; invented and
tested a diving bell; devised
methods for calculating life
insurance premiums; and
produced oceanic charts of
unprecedented accuracy.

Key works

1679 *Catalogus Stellarum
Australium*
1705 *Astronomiae cometicae
synopsis*
1716 *An Account of Several
Nebulae*

interested in discussing what force might account for the motions of planets and other celestial bodies such as comets. Newton told his astonished visitor that he had been studying the matter himself and had already solved the problem (the answer was gravity), but that he had not yet published his findings. This meeting eventually led to Halley editing and financing the publication in 1687 of Newton's

great book on gravity and the laws of motion, *Philosophiae Naturalis Principia Mathematica*.

Historical records

Halley suggested to Newton that he might apply his new theory to studying the orbits of more comets. However, Newton's mind had turned to other matters so, from the early 1690s, Halley conducted his own detailed study. In all, over a period of more than 10 years, he studied the orbits of 24 comets— some that he had observed himself, and others for which he had obtained data from historical records. He suspected that, while some comets followed paths that are parabolas (open-ended curves) as Newton had proposed, others followed elliptical orbits, meaning that they might pass through the inner solar system, and thus become visible from Earth, more than once in a person's lifetime.

During his studies, Halley had noticed something strange. In general, the orbit of each comet

Even in an age renowned for unusual savants, Halley stands out as a man of extraordinary breadth and depth.
J. Donald Fernie
Professor Emeritus of Astronomy at the University of Toronto

had a few characteristics that clearly distinguished it from the orbits of other comets, such as its orientation in relation to the stars. However, three of the comets he had studied—one he had seen himself in 1682, and others observed by Kepler in 1607, and Petrus Apianus in 1531—seemed to have remarkably similar orbits. He suspected that these were

Three comets of 1531, 1607, and 1682 had very **similar orbits**.

↓

The **small differences** in their orbits can be accounted for in terms of **the gravitational pull of Jupiter and Saturn**.

↓

The three comets are therefore **the same comet**, which reappears every 75–76 years.

↓

The comet will reappear around 1758.

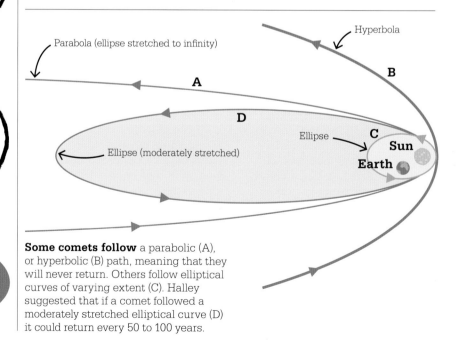

Some comets follow a parabolic (A), or hyperbolic (B) path, meaning that they will never return. Others follow elliptical curves of varying extent (C). Halley suggested that if a comet followed a moderately stretched elliptical curve (D) it could return every 50 to 100 years.

On its last appearance in 1986, Halley's comet passed to within 0.42 astronomical units (AU) of Earth. It has passed much closer. In 1066, for instance, it came within 0.1 AU.

successive reappearances, which occurred about once every 75 to 76 years, of the same comet, which was traveling on a closed, elliptical orbit. In 1705, Halley outlined his ideas in a paper called *Astronomiae cometicae synopsis (A synopsis of the astronomy of comets)*. He wrote: "Many considerations incline me to believe the Comet of 1531 observed by Apianus to have been the same as that described by Kepler and Longomontanus in 1607 and which I again observed when it returned in 1682. All the elements agree. Whence I would venture confidently to predict its return, namely in the year 1758."

One uncertainty still worried Halley. The time intervals between the three appearances were not precisely the same—they differed by about a year. Remembering research he had done some years earlier on Jupiter and Saturn, Halley suspected that the gravitational pull from these two giant planets might slightly throw the comet off its course and delay its timing. Halley asked Newton to reflect on this problem, and Newton came up with gravitational calculations by which Halley was able to refine his forecast. His revised prediction was that the comet would reappear either in late 1758 or in early 1759.

Halley is proved right

Interest in Halley's prediction spread throughout Europe. As the year of the comet's predicted return approached, three French mathematicians—Alexis Clairaut, Joseph Lalande, and Nicole-Reine Lepaute—spent several arduous months recalculating when it might reappear, and where it might first be seen in the night sky. Amateur and professional astronomers alike began watching for the comet's return as early as 1757. On December 25, 1758, it was finally spotted by Johann Palitzsch, a farmer and amateur astronomer from Germany.

The comet passed closest to the sun in March 1759, only a couple of months later than Halley had

Aristotle's opinion that comets were nothing else than sublunary vapors prevailed so far that this sublimest part of astronomy lay altogether neglected.
Edmond Halley

predicted. By then, Halley had been dead for 17 years, but the comet's reappearance brought him posthumous fame. The comet was named Halley's comet in his honor by the French astronomer Nicolas-Louis de Lacaille.

Halley's comet was the first object other than a planet that had been proven to orbit the sun. It also provided one of the earliest proofs of Newton's theory of gravity, demonstrating that the theory could be applied to all celestial bodies. Comets themselves, once feared as unpredictable omens of ill fortune, were now understood.

Subsequent research found that the comet had made regular appearances going back to at least 240 BCE, including some particularly bright apparitions in 87 BCE, 12 BCE, 837 CE, 1066, 1301, and 1456. In 1986, the comet was closely approached by spacecraft, which provided data on the structure of its nucleus (solid part) and its tail. It is the only known short-period comet (comet with an orbit of less than 200 years) that may be seen with a naked eye and appear twice in a human lifetime. ◼

THESE DISCOVERIES ARE THE MOST BRILLIANT AND USEFUL OF THE CENTURY

STELLAR ABERRATION

IN CONTEXT

KEY ASTRONOMER
James Bradley (1693–1762)

BEFORE
17th century The general acceptance of a sun-centered cosmos leads astronomers to search for stellar parallax—the apparent movement of stars caused by Earth's movement.

1676 Danish astronomer Ole Rømer estimates the speed of light using observations of the Jovian satellites.

1748 Swiss mathematician Leonhard Euler outlines the physical cause of nutation.

AFTER
1820 German optician Joseph von Fraunhofer builds a new type of heliometer (a device for measuring the sun's diameter) for the study of stellar parallax.

1838 Friedrich Bessel measures the parallax of the star 61 Cygni. He finds that it is 600,000 times farther away from Earth than the sun.

In the 1720s, while seeking proof that Earth was moving by tracking changes in the apparent positions of stars, Oxford astronomer James Bradley found another phenomenon that also provided proof—stellar aberration. The aberration of light causes objects to appear to be angled toward the direction of a moving observer (in this case, Earth as it moves through space). Aberration angles are tiny: no more than the speed of Earth perpendicular to the star's direction divided by the speed of light, which is 20 arcseconds at most. Earth moves at about 20 miles/s (30 km/s), but both its speed and direction of travel change as it orbits the sun. As a result, a star's observed position follows a small ellipse around its real position. Bradley observed this in the case of the star Gamma Draconis—the first irrefutable proof that Earth moves.

He also discovered another periodic variation in star positions, called nutation. Like aberration, the effect is small. Earth's spin axis gradually changes its orientation in space. The greatest change is precession, and a full cycle takes 26,000 years to complete. Nutation is a small wobble in precession with an 18.6-year cycle. Precession and nutation are both caused by gravitational interactions between the moon, Earth, and sun. Bradley made his discovery public in 1748, after 20 years of observations. ∎

Observed position

Real position

Stellar aberration is caused by Earth's movement. Changes in Earth's velocity can be detected through changes in the position of the stars.

Earth

Movement of Earth

See also: Shifting stars 22 ▪ Stellar parallax 102 ▪ Rømer (Directory) 335

A CATALOG OF THE SOUTHERN SKY

MAPPING SOUTHERN STARS

French astronomer and mathematician Nicolas-Louis de Lacaille had the idea to use trigonometry to measure the distance to the planets after observing them from different places. To provide the longest possible baseline for his calculations, Lacaille needed simultaneous observations in Paris and at the Cape of Good Hope. To this end, he traveled to South Africa in 1750 and set up an observatory at Cape Town. There, he not only observed the planets, but also measured the positions of 10,000 southern stars. His results were published posthumously in 1763 in *Coelum Australe Stelliferum*. They proved to be his greatest legacy to astronomy.

Southern stars
Parts of the sky surveyed by Lacaille are too far south to be visible from Europe and many of the stars he observed had not been allocated to constellations. To give designations to the stars in his catalog, Lacaille introduced 14 new constellations

Lacaille laid the foundation of exact sidereal astronomy in the southern hemisphere.
Sir David Gill

that are still recognized and used today, and he defined the boundaries of existing southern constellations. Before leaving South Africa, he also carried out a major surveying project with the aim of better understanding the shape of the Earth.

Lacaille was a zealous and highly skilled observer who appreciated the value of accurate measurements. He demonstrated an exceptional ability and energy to pioneer a thorough survey of the southernmost sky. ■

See also: Consolidating knowledge 24–25 ▪ The southern hemisphere 100–01

URANU
NEPTU
1750–1850

S TO
NE

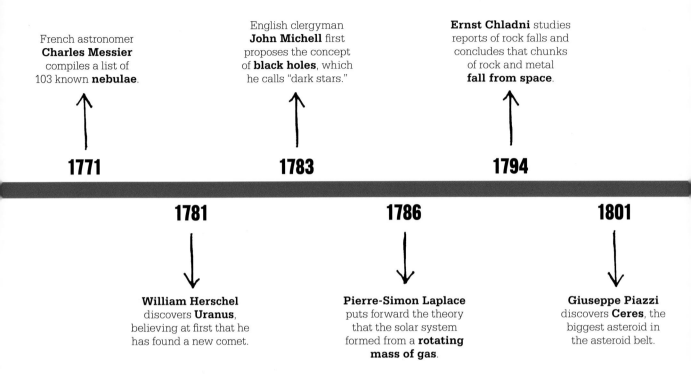

French astronomer **Charles Messier** compiles a list of 103 known **nebulae**.

1771

English clergyman **John Michell** first proposes the concept of **black holes**, which he calls "dark stars."

1783

Ernst Chladni studies reports of rock falls and concludes that chunks of rock and metal **fall from space**.

1794

1781

William Herschel discovers **Uranus**, believing at first that he has found a new comet.

1786

Pierre-Simon Laplace puts forward the theory that the solar system formed from a **rotating mass of gas**.

1801

Giuseppe Piazzi discovers **Ceres**, the biggest asteroid in the asteroid belt.

In the space of 75 years across the 18th and 19th centuries, two new planets were discovered, bringing the number of known major planets to eight (including Earth). However, the circumstances under which Neptune was found in 1846 were very different from those that resulted in the accidental identification of Uranus in 1781. In between these two discoveries, many other bodies were found in the solar system, showing that it contains a far greater number and variety of objects than had previously been imagined.

Powers of observation

Briton William Herschel is considered by many to have been the greatest visual astronomer of all time. He built better telescopes than any of his contemporaries and was an obsessive observer of apparently boundless stamina and enthusiasm. In addition, he persuaded members of his family to help his enterprises, notably his sister Caroline, who gained recognition as an astronomer in her own right.

William was not looking for a planet when he noticed Uranus, but his discovery was a consequence of his skill at telescope-making and systematic approach to observing, which enabled him to spot the movement of the planet over time. Herschel also studied double and multiple stars, cataloged nebulae and star clusters, and attempted to map the structure of the Milky Way. Always alert to the unexpected, he discovered infrared radiation by accident when studying the spectrum of the sun in 1800. Better telescopes led to far more detailed surveys of the sky. William's son, John Herschel, inherited his father's aptitude for astronomy and spent five years in South Africa completing his father's surveys.

All the effects of Nature are only the mathematical consequences of a small number of immutable laws.
Pierre-Simon Laplace

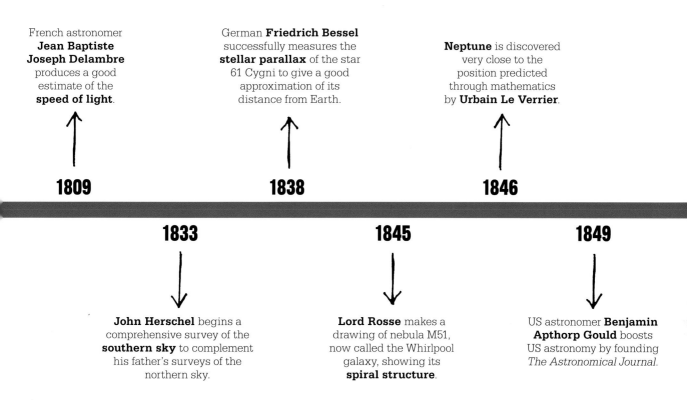

French astronomer **Jean Baptiste Joseph Delambre** produces a good estimate of the **speed of light**.

German **Friedrich Bessel** successfully measures the **stellar parallax** of the star 61 Cygni to give a good approximation of its distance from Earth.

Neptune is discovered very close to the position predicted through mathematics by **Urbain Le Verrier**.

1809

1838

1846

1833

1845

1849

John Herschel begins a comprehensive survey of the **southern sky** to complement his father's surveys of the northern sky.

Lord Rosse makes a drawing of nebula M51, now called the Whirlpool galaxy, showing its **spiral structure**.

US astronomer **Benjamin Apthorp Gould** boosts US astronomy by founding *The Astronomical Journal*.

William Parsons, 3rd Earl of Rosse, took the next big step in the investigation of nebulae. In the 1840s, he set himself the ambitious task of constructing the largest telescope in the world. With it, he discovered that some nebulae (which we now know to be galaxies) have a spiral structure.

More planets

Herschel's discovery of Uranus aroused fresh interest in the wide gap in the solar system between the orbits of Mars and Jupiter. The regular spacing of the other planets suggested that there was an unknown planet in the gap. It turned out to be occupied not by a single major planet but by numerous minor planets, which William Herschel dubbed "asteroids." Italian Giuseppe Piazzi discovered the first asteroid, Ceres, in 1801, as he was observing for a new star catalog. Three more were located in the following six years. The next was not found until 1845, after which the pace of discovery increased.

Meanwhile, German Ernst Chladni had correctly concluded that meteorites reaching Earth are chunks of rock and metal from space. Clearly, the solar system contained a great variety of bodies.

The might of mathematics

By contrast to the fortuitous discovery of Uranus, the discovery of Neptune was a demonstration of the power of mathematics. While astronomers were working with better telescopes, mathematicians grappled with the practical difficulties of applying Newton's theory of gravitation to the complex interplay of gravitational forces between the larger bodies of the solar system. The calculations of German mathematician Carl Friedrich Gauss in 1801 allowed Ceres to be relocated, while between 1799 and 1825, Frenchman Pierre-Simon Laplace produced a monumental definitive work on celestial mechanics.

It soon became evident that Uranus was not following its predicted course. The pull of an unknown planet was suspected. Building on Laplace's work, compatriot Urbain Le Verrier tackled the problem of predicting the undiscovered planet's possible position. Neptune was duly found close to where Le Verrier thought it would be. For the first time, astronomers now had an idea of the true extent of the solar system. ∎

I FOUND THAT IT IS A COMET, FOR IT HAS CHANGED ITS PLACE

OBSERVING URANUS

Uranus has been observed but **not recognized** as a planet.

Observations a few days apart show that **it has moved**, meaning that it **might be a comet**.

Calculations show that **its orbit is almost circular**, so it **must be a planet**.

Irregularities in its orbit indicate that there may be **an eighth planet** in the solar system.

Uranus, the seventh planet from the sun, is visible to the naked eye, and it is believed that the ancient Greek Hipparchus observed it in 128 BCE. The development of telescopes in the 17th century led to further sightings, such as one by English astronomer John Flamsteed in 1690 when it was recorded as 34 Tauri,

a star. It was also observed by the French astronomer Pierre Lemonier several times between 1750 and 1769. However, none of the observers figured out that it was a planet.

William Herschel observed Uranus on March 13, 1781, while looking for multiple star systems. He spotted it again four nights later, and on this second occasion

See also: Shifting stars 22 ▪ Gravitational theory 66–73 ▪ The discovery of Neptune 106–07

> I compared it to H. Geminorum and the small star in the quartile between Auriga and Gemini, finding it so much larger than either of them.
> **William Herschel**

he noticed that its position had changed in relation to the stars around it. He also noted that if he increased the power of the telescope he was using the new object increased in size more than the fixed stars. These two observations indicated that it was not a star, and when he presented his discovery to the Royal Society he announced that he had found a new comet. The Astronomer Royal, Nevil Maskelyne, looked at Herschel's discovery and decided that the new object was as likely to be a planet as a comet. Swedish-Russian Anders Johan Lexell and German Johann Elert Bode independently computed the orbit of Herschel's object and concluded that this was a planet in a near-circular orbit, roughly twice as far away as Saturn.

Naming the planet

Herschel's discovery was praised by King George III, who appointed Herschel "The King's Astronomer." Maskelyne asked Herschel to name the new planet, and he chose Georgium Sidus (George's Star) in honor of his patron. Other names, including Neptune, were proposed, and Bode suggested Uranus. His suggestion became universal in 1850, when the UK's Greenwich Observatory finally abandoned the name Georgium Sidus.

The detailed study of the orbit of Uranus by subsequent astronomers showed that there were discrepancies between its observed orbit and the orbit predicted according to Newton's laws—irregularities that could only be explained by the gravitational influence of an eighth, even more distant planet. This led to the discovery of Neptune by Urbain Le Verrier in 1846. ▪

Herschel observed Uranus using a 7-ft (2.1-m) reflector telescope. He would later construct a 40-ft (12-m) telescope, which was the largest telescope in the world for half a century.

William Herschel

Born in Hanover, Germany, Frederick William Herschel emigrated to Britain at the age of 19 to make a career in music. His studies of harmonics and mathematics led to an interest in optics and astronomy, and he set out to make his own telescopes.

Following his discovery of Uranus, Herschel detected two new moons of Saturn and the largest two moons of Uranus. He also showed that the solar system is in motion relative to the rest of the galaxy and identified numerous nebulae. While studying the sun in 1800, Herschel discovered a new form of radiation, now known as infrared radiation.

Herschel's sister Caroline (1750–1848) acted as his assistant, polishing mirrors and recording and organizing his observations. She began to make observations of her own in 1782, and went on to discover a number of comets.

Key works

1781 *Account of a Comet*
1786 *Catalogue of 1,000 New Nebulae and Clusters of Stars*

THE BRIGHTNESS OF THE STAR WAS ALTERED

VARIABLE STARS

IN CONTEXT

KEY ASTRONOMER
John Goodricke (1764–1786)

BEFORE
130 BCE Hipparchus defines a magnitude scale for the apparent brightness of stars, which is popularized by Ptolemy in the *Almagest*.

1596 David Fabricius discovers that the star Mira Ceti varies in brightness with periodic regularity.

AFTER
1912 Henrietta Swan Leavitt discovers that the period of some variable stars is related to their absolute (true) brightness.

1913 Ejnar Hertzsprung calibrates this variation in brightness, allowing Cepheid variables to be used as "standard candles" to calculate the distance to galaxies.

1929 Edwin Hubble identifies the link between the velocity of a galaxy and its distance.

Ancient Greek astronomers were the first to classify stars by their apparent brightness—that is, their brightness as observed from Earth. In the 18th century, British amateur astronomer John Goodricke grew interested in changes in apparent brightness after his neighbor, astronomer Edward Pigott, provided him with a list of stars known to vary. In the course of his observations, he discovered more.

In 1782, Goodricke observed the variation in brightness of Algol, a bright star in the constellation Perseus. He was the first person to propose a reason for this change in brightness, suggesting that Algol was in fact a pair of stars orbiting one another, with one brighter than the other. When the dimmer of the two stars passed in front of the brighter one, the eclipse would reduce the brightness detected by observers. Today this is referred to as an eclipsing binary system (it is now known that Algol is actually a three-star system).

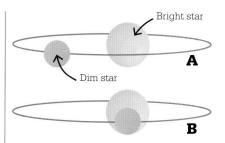

In an eclipsing binary system, maximum brightness occurs when both stars are visible (A); minimum brightness occurs when the dim star is eclipsing the bright star (B).

Goodricke also discovered that the star Delta Cephei in the constellation Cepheus varies in brightness with a regular period. It is now known that Delta Cephei is one of a class of stars whose apparent brightness varies because the star itself changes. Stars such as this are called Cepheid variables, and they are key to calculating the distance to other galaxies.

Goodricke presented his findings to the Royal Society in 1783. He died shortly after from pneumonia, at just 21 years old. ∎

See also: A new kind of star 48–49 ▪ Measuring the universe 130–37 ▪ Beyond the Milky Way 172–77

OUR MILKY WAY IS THE DWELLING, THE NEBULAE ARE THE CITIES
MESSIER OBJECTS

IN CONTEXT

KEY ASTRONOMER
Charles Messier (1730–1817)

BEFORE
150 CE Ptolemy records five stars that appear nebulous and one nebula not linked to a star.

964 Persian astronomer Abd al-Rahman al-Sufi notes several nebulae in his *Book of Fixed Stars*.

1714 Edmond Halley publishes a list of six nebulae.

1715 Nicolas Louis de Lacaille identifies 42 nebulae.

AFTER
1845 Lord Rosse observes that some nebulae have a spiral structure.

1864 William Huggins examines the spectra of 70 nebulae, finding that a third of them are clouds of gas, while the rest are masses of stars.

1917 Vesto Slipher identifies spiral nebulae as distant galaxies.

By the 18th century, large telescopes that could magnify images by several hundred times were being produced. This allowed astronomers to identify various fuzzy patches of light, which were called nebulae, after the Latin word for "cloud."

French astronomer Charles Messier was primarily interested in finding comets, which often look just like nebulae. A fuzzy object could only be identified as a comet if it changed position against the stars over a period of weeks or months. Messier therefore compiled a list of known nebulae to eliminate them as potential comets. His initial list was published in 1774 and identified 45 nebulae. The final 1784 version listed 80 objects. These nebulae are now known as Messier objects. Other astronomers added further nebulae that were observed by Messier but not recorded by him in his catalog, bringing the total to 110.

With more powerful telescopes, it has been possible to determine the nature of the Messier objects. Some are galaxies beyond the Milky Way, some are clouds of gas where stars are forming, and others are the remains of supernova explosions or the gas thrown off by dying stars the size of our sun. ∎

Messier 31 is also known as the Andromeda Galaxy. It is the nearest major galaxy to the Milky Way.

See also: Halley's comet 74–77 ▪ Mapping southern stars 79 ▪ Examining nebulae 104–05 ▪ Properties of nebulae 114–15 ▪ Spiral galaxies 156–61

ON THE CONSTRUCTION OF THE HEAVENS

THE MILKY WAY

IN CONTEXT

KEY ASTRONOMER
William Herschel (1738–1822)

BEFORE
1725 English astronomer
John Flamsteed's catalog of
3,000 stars is issued, followed
by his star atlas in 1729.

1750 Thomas Wright suggests
that the solar system is part
of a disk of stars.

1784 Charles Messier
produces his final catalog
of nebulae.

AFTER
1833 John Herschel continues
his father's work, publishing a
systematic mapping of the sky
including observations made
from the southern hemisphere.

1845 Lord Rosse observes
that some nebulae have a
spiral structure.

1864 William Huggins uses
emission spectra to determine
that some nebulae are masses
of stars.

One of the most spectacular features in the sky visible to the naked eye is the dense band of light called the Milky Way. This light from billions of stars is not seen by many people today because of light pollution, but was a common sight before street lighting.

In the 1780s, British astronomer William Herschel attempted to determine the shape of the Milky Way and the sun's position within it by observing the stars. In this endeavor, Herschel built upon the work of his compatriot Thomas Wright, who, in 1750, had argued that the stars appeared as a band of light because they were not

From Earth, the Milky Way appears as a band of light whose individual stars cannot be seen with the naked eye. The band is the galaxy's disk-shaped structure viewed from within.

randomly scattered but formed a vast ring around Earth, held together by gravity.

The Milky Way appeared to circle Earth and so Herschel concluded that the galaxy was disklike. He observed the numbers of stars of different magnitudes (brightness) and discovered that these were equally distributed within the band of the Milky Way in all directions. This led him to assume that the

See also: Messier objects 87 ▪ The southern hemisphere 100–01 ▪ Properties of nebulae 114–15 ▪ Spiral galaxies 156–61 ▪ The shape of the Milky Way 164–65

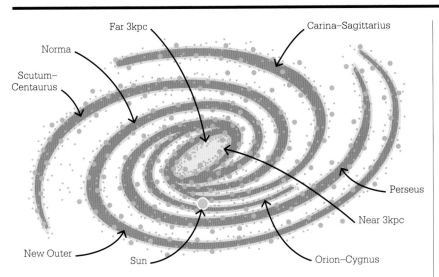

The Milky Way comprises stellar arms spiraling out from the bulging "bar" at the center. The arms are labeled here. The sun is located in the Orion–Cygnus arm, 26,000 light-years from the center.

I have observed stars of which the light, it can be proved, must take two million years to reach the Earth.
William Herschel

brightness of a star indicated its distance from Earth, with dimmer stars being more distant. The even distribution, he believed, must mean that the solar system was close to the center of the galaxy. Herschel's model was refined by other astronomers, but was not replaced until the early 20th century.

New nebulae

Herschel did not limit himself to the study of stars in his investigation into a galaxy's shape. He also observed nebulae, the fuzzy patches of light that dotted the sky. Herschel was a skilled telescope-maker as well as an astronomer, and used two large, powerful telescopes with 49½-in (126-cm) and 18½-in (47-cm) apertures. From 1782, he used these instruments to conduct systematic observations of the "deep sky," searching for objects that were not stars. He listed these as nebulae or as clusters of stars, and published details of 1,000

new objects in 1786, with further catalogs appearing in 1789 and 1802. Herschel classified the objects he listed into eight categories, depending on their brightness, size, or whether they appeared to consist of dense or scattered clusters of stars. He also conjectured that most nebulae were similar in

nature and size to the Milky Way, decades before it was confirmed that nebulae were in fact galaxies in their own right.

The current model of the Milky Way is a barred spiral galaxy. Around two-thirds of spiral galaxies have central bars like the Milky Way's. The early idea of a disk of stars is broadly correct, but the stars within the disk are arranged in a series of spiral arms, with the sun in a sparse area of the Orion–Cygnus arm. ▪

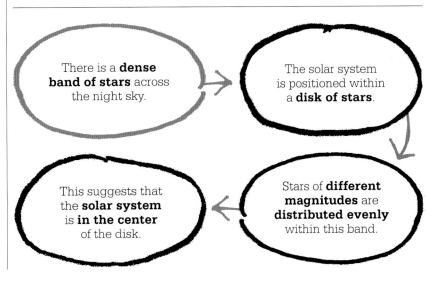

There is a **dense band of stars** across the night sky.

The solar system is positioned within a **disk of stars**.

Stars of **different magnitudes** are **distributed evenly** within this band.

This suggests that the **solar system** is **in the center** of the disk.

ROCKS FALL FROM SPACE
ASTEROIDS AND METEORITES

IN CONTEXT

KEY ASTRONOMER
Ernst Chladni (1756–1827)

BEFORE
1718 Isaac Newton proposes that nothing can exist between the planets.

1771 A spectacular fireball is recorded passing over Sussex in southern England and continuing to be seen over northern France.

AFTER
1798 British chemist Edward Howard and French mineralogist Jacques-Louis de Bournon analyze stones and irons from falls in Italy, England, and India. They find similar proportions of nickel in the stones, indicating a link between them.

1801 Giuseppe Piazzi discovers Ceres, the largest object in the asteroid belt, now classed as a dwarf planet.

In the 18th century, the real source and nature of what are now called meteorites was not known. Interplanetary space was thought to be empty, and the fiery lumps of rock and iron that fell from the sky were believed to originate either in volcanoes on Earth that had thrown them up, or from dust in the atmosphere, perhaps by the action of lightning.

This idea can be traced back to Isaac Newton, who wrote that it was "necessary to empty the Heavens of all Matter" in order for the planets and comets to move unimpeded in their regular orbits.

In the early 1790s, a German physicist named Ernst Chladni attempted to solve the mystery of these "fallen stones" by examining historical records. One that he studied had landed in 1768 in France, where it had been subjected to chemical analysis. The results showed that it had formed from a lump of sandstone that had been struck by lightning and flung into the air. Chladni then examined an object found in 1772 that had a mass of more than 1,500 lb (700 kg). It had a rough surface, was filled with cavities, and was totally unlike the rock of the landscape where it was found. It had also very clearly been melted.

Falling from space

Chladni realized that neither lightning nor a forest fire could have produced enough heat to melt bedrock (the solid rock that underlies loose deposits). Yet the rock he examined had become a mass of metallic iron. This "iron," he concluded, could only have come from space. It had melted on its passage through the atmosphere.

This iron-nickel meteorite was found on an Arctic ice sheet. The odd shape of the meteorite is due to spinning and tumbling at a high temperature on entry into the atmosphere.

See also: Gravitational theory 66–73 ▪ The discovery of Ceres 94–99 ▪ Investigating craters 212

> **Reports** of rock-falls from the sky are **all very similar**.

> These are **reliable reports**.

> The rocks **do not resemble local rocks.**

> The rocks **melted as they fell** through the atmosphere.

> The rocks show the effects of **extreme heating**.

> **Rocks fall from space.**

Ernst Chladni

Ernst Chladni was born in Saxony to a family of prominent academics. Chladni's father disapproved of his son's interest in science and insisted he study law and philosophy. He obtained a degree in these subjects from the University of Leipzig in 1782. However, when his father died that year, he turned to physics.

Initially, Chladni applied his physics knowledge to work in acoustics, which brought him renown. He identified the way rigid surfaces vibrate, and his observations were applied to the design of violins. His later work on meteorites drew less favorable attention from the scientists of the day, and might have vanished into obscurity had it not been for the popular writing of Jean-Baptist Biot, whose findings supported Chladni's ideas.

Key works

1794 *On the Origin of the Iron Masses Found by Pallas and Others Similar to it, and on Some Associated Natural Phenomena*
1819 *Igneous Meteors and the Substances that have Fallen from them*

Chladni published his findings in a book in 1794, which set out his main conclusions: that masses of iron or stone fall from the sky; and that friction in the atmosphere causes them to heat up, creating visible fireballs ("shooting stars"); that the masses do not originate in Earth's atmosphere but far beyond it; and that they are fragments of bodies that never joined together to make planets.

Chladni's conclusions were correct, but at the time he was ridiculed—until some chance rock-falls helped to change opinion. The first of these took place within two months of the publication of Chladni's book, when a large fall of stones came down on the outskirts of Siena, Italy. Analysis of them showed they were very different from anything found on Earth. Then, in 1803, nearly 3,000 stones fell in fields around L'Aigle in Normandy. French physicist Jean-Baptiste Biot investigated this fall. He concluded that they could not have originated anywhere nearby.

Solar system fragments

Thanks to Chladni's work, scientists know that shooting stars are lumps of rock or metal from space heated to glowing point as they pass through the atmosphere. The object that causes the glowing trail is called a meteor. If any of it survives to reach the ground, it is termed a meteorite. Meteorites can originate in the asteroid belt between Jupiter and Mars, or they can be rocks thrown up from Mars or the moon. Many meteorites contain small particles called chondrules, which are thought to be material from the asteroid belt that never formed into larger bodies. These are some of the oldest materials in the solar system, and can tell scientists much about its early composition. ▪

THE MECHANISM OF THE HEAVENS
GRAVITATIONAL DISTURBANCES

IN CONTEXT

KEY ASTRONOMER
Pierre-Simon Laplace
(1749–1827)

BEFORE
1609 Johannes Kepler determines that the planets move in elliptical orbits.

1687 Isaac Newton publishes *Principia Mathematica*, which includes his law of universal gravitation and a mathematical derivation of Kepler's laws of planetary motion.

1734 Swedish philosopher Emanuel Swedenborg outlines the nebular theory of the formation of the solar system.

AFTER
1831 Mary Somerville translates Laplace's *Méchanique céleste* to English.

1889 French mathematician Henri Poincaré shows that it is not possible to prove that the solar system is stable, laying the foundations for chaos theory.

There are **disturbances** in **the mechanism** of the heavens.

Without divine intervention these disturbances look like they should make the **orbits** of the planets **unstable**.

But the disturbances **continually self-correct** over time.

The self-correction is made by **the force of gravity** that caused the disturbance itself.

By the end of the 18th century, the structure of the solar system was well-known. The planets moved in elliptical orbits around the sun, held in place by gravity. Isaac Newton's laws allowed a mathematical basis for this model of the solar system to be developed, but there were still problems. Newton himself tested his ideas against observations, but noted "perturbations" to the planets' orbits. By this he meant a disturbance to the orbits caused by an additional force, which would make the orbits unstable if not corrected. As a result, Newton decided that the hand of God was occasionally required to maintain the solar system in a stable state.

Orbital resonance
French mathematician Pierre-Simon Laplace rejected the notion of divine intervention, however. In

See also: Elliptical orbits 50–55 ▪ Galileo's telescope 56–63 ▪ Gravitational theory 66–73 ▪
The theory of relativity 146–53 ▪ Delambre (Directory) 336

1784, he turned his attention to a long-standing question known as the "great Jupiter-Saturn inequality." Laplace showed that perturbations in the orbits of these two planets were due to the orbital resonance of their motions. This refers to the situation in which the orbits of two bodies relate to each other in a ratio of whole numbers. In the case of Jupiter and Saturn, Jupiter orbits the sun almost exactly five times for every two orbits of Saturn. This means that their gravitational fields have a greater effect on each other than they would for orbits that are not in resonance.

The nebular hypothesis

Laplace published his work on the solar system in two influential books—a popular account called *Exposition du système du monde* and the mathematical *Méchanique celeste*. In his *Exposition*, Laplace explored the idea that the solar system developed from a primeval nebula. Laplace described a rotating mass of hot gases that

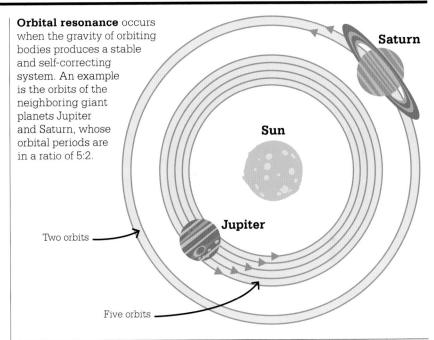

Orbital resonance occurs when the gravity of orbiting bodies produces a stable and self-correcting system. An example is the orbits of the neighboring giant planets Jupiter and Saturn, whose orbital periods are in a ratio of 5:2.

Saturn

Sun

Jupiter

Two orbits

Five orbits

cooled and contracted, breaking off rings from its outer edge. The core material formed the sun and the matter in the rings cooled to form the planets.

Shortly after Laplace's death, his work was translated into English by the Scottish mathematician Mary Somerville, and this led to a wide dissemination of his ideas. Using Laplace's new theorems, his fellow countryman Jean Baptiste Joseph Delambre was able to produce far more accurate tables predicting the motions of Jupiter and Saturn. ▪

Pierre-Simon Laplace

Pierre-Simon Laplace was born in Normandy, France, to a minor landowner. His father intended him for the Church, and he studied theology at the University of Caen, but it was there that he developed his interest in mathematics. He gave up any intention of entering the priesthood and moved to Paris, where he obtained a teaching position in the École Militaire. Here, he taught a young Napoleon Bonaparte. The post allowed him time to devote himself to research, and through the 1780s, he produced a string of influential mathematical papers.

When Napoleon seized power in 1799, Laplace became a member of the senate and served on many scientific commissions. He continued to work on the mathematics of astronomy until his death, publishing five volumes on celestial mechanics.

Key works

1784 *Théorie du movement et de la figure elliptique des planètes*
1786 *Exposition du système du monde*
1799–1825 *Méchanique céleste*

I SURMISE THAT
IT COULD BE SOMETHING
BETTER
THAN A COMET
THE DISCOVERY OF CERES

IN CONTEXT

KEY ASTRONOMER
Giuseppe Piazzi (1746–1826)

BEFORE
1596 Johannes Kepler suggests there are unobserved planets in the solar system.

1766 Johann Titius predicts the gap between Mars and Jupiter contains a planet.

1781 William Herschel's discovery of Uranus confirms the pattern of orbits proposed by Johann Bode.

1794 Ernst Chladni suggests meteorites are rocks that were once in orbit.

AFTER
1906 Trojan asteroids are found in the orbit of Jupiter.

1920 Hidalgo, the first "centaur" asteroid (an asteroid with an unstable orbit), is found between Jupiter and Neptune.

2006 Ceres is classified as a dwarf planet.

The **orbits of the planets** appear to follow a **mathematical formula**.

The formula predicts that the **gap between Mars and Jupiter** should contain an orbiting body.

Ceres, a body seen in the gap, is too small to be a planet, but it **does not have the orbit of a comet**.

Ceres is a **minor planet**, or **asteroid**—one of thousands in that region of space.

F or centuries, the number of known "wandering stars," or planets, that trailed through the night sky was five. Together with the sun and moon, that brought the total of major celestial bodies visible from Earth to seven—a number imbued with mystical significance. Then, in 1781, William Herschel spotted Uranus out beyond the orbit of Saturn, which forced astronomers to rethink this number. However, when the new planet's orbit was placed in an updated plan of the solar system, it revealed another number conundrum.

Finding a gap

In 1766, a German astronomer named Johann Titius discovered a mathematical link between the orbital distances of the planets. He divided the orbital distance of Saturn by 100 to create a unit to measure all the other orbits. Mercury's orbit was 4 units from the sun, and every other planet's position from there was linked to a doubling of 3, or the number sequence 0, 3, 6, 12, 24, 48, and 96. So Mercury was located at 4 + 0 units from the sun, Venus at 4 + 3,

---

Guiseppe Piazzi

As was common for younger sons in wealthy Italian families, Guiseppe Piazzi's career began in the Catholic Church. By his mid-20s, it was obvious that his abilities lay in academia. In 1781, he was appointed math professor at a newly founded academy in Palermo, Sicily, but soon switched to astronomy. His first task in this role was to build a new observatory, which he equipped with the Palermo Circle, a telescope built in London with a 5-ft (1.5-m) wide altitude scale. It was the most accurate telescope in the world at the

time. Piazzi was famed for his diligence, and would take measurements on at least four consecutive nights to average out errors. In 1806, he recorded the large proper motion of the star 61 Cygni. This prompted several astronomers to use the parallax of that star to measure the distance between stars.

Key works

1803 *Præcipuarum stellarum inerrantium (Star catalog)*
1806 *Royal Observatory of Palermo (Book 6)*

See also: Elliptical orbits 50–55 ▪ Observing Uranus 84–85 ▪ Asteroids and meteorites 90–91

> From Mars there follows a space of 4 + 24 = 28 such parts, but so far no planet was sighted there. But should the Lord Architect have left that space empty? Not at all.
> **Johann Titius**

Earth at 4 + 6, and Mars was at 4 + 12. Jupiter was at 4 + 48 and Saturn was at 4 + 96. There was no known planet in the sequence at 4 + 24 = 28, so there appeared to be a gap in the solar system between Mars and Jupiter. Titius proposed that the gap must be occupied by an unknown body. However, his findings seemed too good to be true—and the results for Mars and Saturn were slightly out, so few astronomers paid them much heed.

A few years later, in 1772, a fellow German named Johann Bode published a slightly modified version of Titius's work, which met with greater acclaim. As a result, the theory is best remembered as Bode's law. When Uranus was discovered, Bode's law predicted that it would be 196 units from the sun. It was finally shown to be nearer to 192 units, but that seemed close enough. Surely, it meant the 28-unit gap must also contain a planet.

In 1800, a group of German-based astronomers led by Franz Xaver von Zach, Heinrich Olbers, and Johann

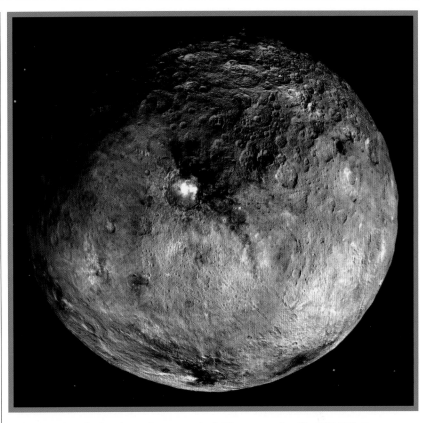

Photographed by NASA's *Dawn* spacecraft in 2015, Ceres is the largest object in the asteroid belt, and the only object large enough to have been made spherical by its own gravity.

Schröter decided to launch a search of the gap. Their plan was to divide up the zodiac—the strip of sky in which all the planets move—and ask Europe's top 24 astronomers to patrol one zone each, searching for planetlike motion. The team they put together was dubbed the Celestial Police. But in the end it was straightforward luck, not efficiency, that filled the gap.

Surveying telescope
One of the astronomers among the Celestial Police was Giuseppe Piazzi, who was based in Palermo, Sicily. Like most astronomers at the time, Piazzi was mainly concerned with creating accurate star maps. To that end, he had acquired a surveying telescope now known

as the Palermo Circle. Although it was not the most powerful telescope of its day, its altazimuth mounting could move both vertically and horizontally, enabling it to make very accurate measurements of stellar positions, a feature that would pay rich dividends.

On the evening of New Year's Day 1801, the instructions from the Celestial Police were still en route to Piazzi so he spent the evening surveying stars and recorded a new, faint object (with a magnitude of eight) in the » constellation of Taurus. The following night, Piazzi checked his measurements and found that the object had moved slightly. This meant it was definitely not a star.

Piazzi watched the object for 24 days before informing Bode. He thought at first that it was a comet—a relatively common discovery—but his observations soon suggested otherwise. He could see no fuzzy coma, or tail, and while comets sped up as they approached the sun, Piazzi's discovery took a more stable, circular path. In his letter to Bode, Piazzi made his suspicions clear: this could be the missing planet everyone was looking for.

On hearing the news in late March, Bode wasted no time in announcing the discovery of a new planet, which he named Juno. (He had recently chosen the name for Uranus, and clearly felt confident in his right to do so.) Other astronomers preferred the name Hera, but Piazzi, still the only person who had actually seen the object, had opted for Ceres, after the Roman goddess of agriculture.

By June, Ceres's orbit had taken it into the glare of the sun. Piazzi had been sick in the interim and so had not had the chance to map anything but the simplest orbital arc. He calculated that his discovery would be visible again in the fall. But, try as they might, neither Piazzi nor anyone else could find Ceres.

Mathematical hunch

Von Zach decided to follow a hunch, and sent the details of Ceres's orbit to the mathematician Carl Friedrich Gauss. In less than six weeks, Gauss calculated all the places Ceres was likely to be. It took von Zach most of December to search through Gauss's predictions, but on the evening of New Year's Eve 1801, almost exactly a year to the day after it was first seen, he found Ceres once again.

The orbital distance of Ceres was 27.7 Bode units, a remarkably close fit to the predicted location. However, the orbital data showed that this new member of the solar system was far smaller than the known planets. William Herschel's early estimate put Ceres at just 160 miles (260 km) across. A few years later, Schröter proposed a diameter of 1,624 miles (2,613 km). The actual figure is 588 miles (946 km), which means it would be a comfortable fit over the Iberian Peninsula or Texas.

Piazzi's telescope, the Palermo Circle, was built by Jesse Ramsden. Its precision mounting allowed it to measure stellar positions with an accuracy of a few seconds of an arc.

The evening of the third, my suspicion was converted into certainty, being assured it was not a fixed star. I waited till the evening of the fourth, when I had the satisfaction to see it had moved at the same rate as on the preceding days.
Giuseppe Piazzi

The Celestial Police kept up the search and, in March 1802, Olbers discovered a second body like Ceres located at the same distance from the sun, calling it Pallas. In 1804, Karl Harding found a third, named Juno, while it was Olbers again who spotted the fourth, Vesta, in 1807. All these bodies were later shown to be smaller than Ceres—Vesta and Pallas were slightly more than 300 miles (500 km) wide and Juno was half that size.

Asteroid belt

The Celestial Police called their discoveries minor planets, but William Herschel chose another name—asteroid, which means starlike. Herschel reasoned that, unlike true planets, these small objects had no discernible features, or at least none that could be made out with the telescopes of the day, so would be indistinguishable from starlight were it not for the fact that they moved. Perhaps still smarting from his failure to name the planet he had found 20 years earlier,

Herschel qualified his suggestion by reserving for himself "the liberty of changing that name, if another, more expressive of their nature, should occur."

Nothing more expressive did occur, and after the Celestial Police was disbanded in 1815, a steady trickle of asteroid discoveries continued. By 1868, their number stood at 100; by 1985, it was 3,000. The advent of digital photography and image analysis has now boosted the number of recorded asteroids to more than 50,000, spread around the 28-Bode-unit gap. Olbers and Herschel had discussed the possibility that the asteroids were the remains of a planet that once orbited in the gap before being smashed by an astronomical cataclysm. Today, it is thought that the gravitational disruption of nearby Jupiter prevented the asteroids from accreting into a planet in the first place, as similar disks had done elsewhere in the primordial solar system.

Under constant influence from the cumulative gravity of other asteroids, about 80 percent of known asteroids have unstable orbits. The 13,000 or so bodies

They resemble small stars so much as hardly to be distinguished from them. From this, their asteroidal appearance, if I take my name, and call them asteroids.
William Herschel

The asteroid Vesta was visited by the *Dawn* spacecraft from 2011–12. Its orbit lies within that of Ceres, and it is the brightest asteroid as seen from Earth.

that come particularly close to Earth—the Near Earth Asteroids (NEAs)—are monitored in the hope of predicting and preventing devastating future impacts.

Trojans

There are also asteroids known as trojans, which travel in the same orbits as planets, gathering far from their host in gravitationally stable "libration points." Most of these are in the Jupiter system, where they form two clusters: the "Trojan Camp" and "Greek Camp." Mars and Neptune have trojans, and the first Earth trojan was discovered in 2011.

In 2006, the International Astronomical Union gave Ceres the status of dwarf planet, the only one in the asteroid belt. At the same time, Pluto was reclassified as a dwarf planet. The orbits of neither Neptune nor Pluto match the predictions of Bode's law. Despite the fact that it was instrumental in the discovery of Ceres, Bode's law is now viewed as a mathematical coincidence, and not a key to unlocking the formation of the solar system. ∎

A SURVEY OF THE WHOLE SURFACE OF THE HEAVENS

THE SOUTHERN HEMISPHERE

IN CONTEXT

KEY ASTRONOMER
John Herschel (1792–1871)

BEFORE
1784 Charles Messier publishes a list of 80 known nebulae.

AFTER
1887 The Cartes du Ciel, an ambitious project to survey the entire sky photographically, is initiated by the director of the Paris Observatory, Amédée Mouchez.

1918 The Henry Draper Catalog, covering most of the sky, is published by Harvard College Observatory.

1948–58 The Palomar Observatory in California completes its major sky survey, which includes data from nearly 2,000 photographic plates.

1989–93 The Hipparcos satellite gathers data that allow more than 2.5 million stars to be cataloged.

Between 1786 and 1802, William Herschel published catalogs listing more than 1,000 new objects in the night sky. Following his death in 1822, William's son John continued his work, but expanded its scope and ambition to carry out a complete survey of the night sky. William's observations had all been made from southern England, and so were limited to objects down to around 33° below the celestial equator. To survey the rest of the sky, his son's observations would have to be made from somewhere in the southern hemisphere.

Herschel settled on South Africa, then a part of the British Empire. He moved there in 1833, taking with him his wife and young family, an assistant, and his father's 20-ft (6-m) focal length telescope. This was the same instrument that had been used to survey the northern skies, and Herschel chose it to ensure that the new information gathered from the southern hemisphere was comparable to that already produced. The family set themselves up in a house near the base of Table Mountain, far enough away to avoid the clouds that often gathered on its summit, and Herschel spent the next four years completing his survey.

The southern skies

The Magellanic clouds are two dwarf galaxies close to the Milky Way, and are only visible from the southern hemisphere. They can be

The Milky Way's core is clearest from the southern hemisphere. The dark regions are where starlight is blocked by interstellar dust.

See also: Messier objects 87 ▪ The Milky Way 88–89 ▪ Examining nebulae 104–05

From **each hemisphere**, part of the celestial sphere is **always hidden**.

↓

A survey taken **from Britain** misses everything **33° below** the celestial equator.

↓

Adding **observations from South Africa** would make a **complete survey**.

↓

Combining observations from both hemispheres produces a survey of the whole surface of the heavens.

John Herschel

John Herschel left Cambridge University in 1816, already a renowned mathematician. He worked with his father, William, and continued his work after William's death in 1822. Herschel became one of the founders of the Royal Astronomical Society and served as president for three separate terms. He married in 1826 and fathered 12 children. Herschel had numerous interests in addition to astronomy. While in South Africa, Herschel and his wife produced a portfolio of botanical illustrations. He also made important contributions to photography, experimenting with color reproduction, and published papers on meteorology, telescopy, and other subjects.

Key works

1831 *A Preliminary Discourse on the Study of Natural Philosophy*
1847 *Results of Astronomical Observations Made at the Cape of Good Hope*
1864 *General Catalog of Nebulae and Clusters of Stars*
1874 *General Catalog of 10,300 Multiple and Double Stars*

seen by the naked eye, but Herschel's telescopic surveys provided the first detailed observations available to astronomers. He compiled a list of more than 1,000 stars, star clusters, and nebulae within these galaxies.

Herschel also made careful observations of the distributions of stars within the Milky Way.

The stars are the landmarks of the universe.
John Herschel

Due to the orientation of the solar system within the Milky Way, the brightest section of it, which is now known to be the core of the galaxy, is only visible low on the horizon from the northern hemisphere during the summer when nights are short. From the southern hemisphere, the brighter core is visible higher in the sky and during the darker months of the year, allowing easier and more detailed observations.

The end result of Herschel's labors, *The General Catalog of Nebulae and Clusters of Stars*, listed more than 5,000 objects in total. These included all the objects observed by John and his father, and also many discovered by others such as Charles Messier, since it was intended to be a complete catalog of the stars. ■

AN APPARENT MOVEMENT OF THE STARS

STELLAR PARALLAX

arallax is the apparent movement of a nearby object against distant objects due to the changing position of the observer. According to this phenomenon, nearby stars should appear to change position against the background of more distant stars as Earth moves around its orbit. The idea that it might be possible to measure the distance to nearby stars using parallax dates back to ancient Greece. However, it was not achieved until the 19th century, due to the distances involved being far greater than anyone supposed.

Much of German astronomer Friedrich Bessel's career had been dedicated to the accurate determination of the positions of stars and finding their proper motion (changes in position due to the motion of the star, rather than changes in apparent position due to the time of night or the season). By the 1830s, with improvements in the power of telescopes, there was a race to carry out the first accurate measurement of stellar parallax. In

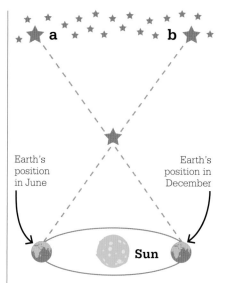

Due to the effects of parallax, a nearby star's apparent position against distant background stars moves from **b** in June to **a** in December.

1838, Bessel measured parallax with an angle of 0.314 arc seconds for the star 61 Cygni, which indicated that it was 10.3 light-years away. The current estimate is 11.4 light-years, giving Bessel's measurement an error of just under 10 percent. ∎

See also: The Tychonic model 44–47 ∙ Measuring the universe 130–37 ∙ Beyond the Milky Way 172–77

SUNSPOTS APPEAR IN CYCLES
THE SURFACE OF THE SUN

unspots are cooler areas on the sun's surface caused by changes in its magnetic field. The first written observations of sunspots date from about 800 BCE, in China, but it was not until 1801 that the British astronomer William Herschel made the connection between sunspots and changes in Earth's climate.

Samuel Schwabe, a German astronomer, started observing sunspots in 1826. He was looking for a new planet thought to orbit closer to the sun than Mercury, provisionally named Vulcan. It would have been very difficult to observe such a planet directly, but Schwabe thought he might see it as a dark spot moving in front of the sun. He did not find Vulcan, but he did discover that the number of sunspots varied over 11-year cycles.

Swiss astronomer Rudolf Wolf studied Schwabe's and other observations, including some from as far back as Galileo, and numbered the cycles starting at 1 for the 1755–66 cycle. Eventually, he saw that there are long periods in each cycle when the number of sunspots is low. Herschel had not noticed the pattern because he was observing during what is now called the Dalton Minimum, when overall numbers of sunspots were low. ■

Sunspots can last from a few days to several months. The largest can be the size of Jupiter.

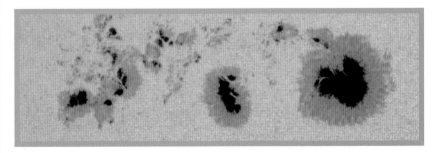

See also: Observing Uranus 84–85 ▪ The properties of sunspots 129 ▪ Carrington (Directory) 336

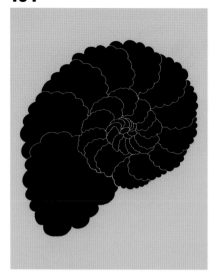

A SPIRAL FORM OF ARRANGEMENT WAS DETECTED

EXAMINING NEBULAE

IN CONTEXT

KEY ASTRONOMER
Lord Rosse (1800–1867)

BEFORE
1784 Charles Messier
publishes a catalog
of the visible nebulae.

1785 William Herschel
publishes catalogs of
nebulae and speculates that
many are similar in shape
and size to the Milky Way.

1833 John Herschel expands
his father's catalogs by
surveying objects from the
southern hemisphere.

1864 William Huggins
discovers that some nebulae
are clouds of luminous gas,
not aggregations of stars.

AFTER
1917 Vesto Slipher concludes
that spiral galaxies are "island
universes" and that the Milky
Way is one such galaxy seen
by us from within.

To the **naked eye**, nebulae are **fuzzy patches** of light that could comprise gas or stars.

Telescopes show some nebulae to be **clusters of stars**.

Larger telescopes reveal a spiral form of arrangement.

In the 1840s, a British aristocrat named William Parsons, Lord Rosse, decided to commit some of his considerable wealth to building the world's largest reflecting telescope. Rosse was curious to reexamine some of the nebulae listed by John Herschel in the early 19th century, in particular those nebulae that did not appear to be clusters of stars.

To re-observe these nebulae, Rosse needed to build a larger and better telescope than that used by Herschel. He experimented for many years with methods for casting a 36-inch (0.9-m) mirror. Mirrors at the time were made from a metal called speculum, an alloy of copper and tin—a brittle material that was prone to cracking as it cooled.

Despite this difficulty, by 1845 Rosse had succeeded in casting a mirror that was 72 in (1.8 m) in diameter. He mounted it in his telescope at Birr Castle, near Parsonstown in Ireland, where it became known as the Leviathan of Parsonstown. This telescope remained the world's largest reflecting type until the 100-in (2.5-m) reflector was built at Mount Wilson in California in 1917.

Central Ireland proved a far from ideal place to build a telescope, as overcast or windy conditions often prevented viewing. The telescope itself had limited mobility, meaning that only a small area of the sky could be examined. Nonetheless, when the weather was clear, Rosse was able to use

See also: Messier objects 87 ▪ The Milky Way 88–89 ▪ The southern hemisphere 100–01 ▪ Properties of nebulae 114–15 ▪ Spiral galaxies 156–61

the huge instrument to observe and record the spiral nature of some nebulae—now called spiral galaxies—for the first time. The first of these spirals that Rosse identified was M51, later known as the Whirlpool galaxy. Today, about three-quarters of all the galaxies that have been observed are spiral galaxies. However, these are thought ultimately to transform into elliptical galaxies. Formed of older stars, elliptical galaxies are dimmer and much harder to spot, but astronomers believe that they are probably the most common galaxy type in the universe.

The nebular hypothesis

In the mid-19th century, astronomers debated whether nebulae consisted of gas or stars. In 1846, Rosse found numerous

The Leviathan Telescope at Parsonstown held a mirror weighing 3.3 tons (3 metric tons), inside a 54 ft- (16.5 m-) long tube. The whole structure weighed about 13 tons (12 metric tons).

The light by which we recognize the nebulae now must be merely that which left their surfaces a vast number of years ago … phantoms of processes completed long in the Past.
Edgar Allen Poe

stars in the Orion nebula, and so for a time, the idea of gaseous nebulae was rejected. However, although the stars were real, their presence did not mean that there was no gas. The gaseous nature of some nebulae was not demonstrated until spectroscopic analysis was used by William Huggins in 1864. ▪

Lord Rosse

William Parsons was born in Yorkshire in 1800 and became Third Earl of Rosse on the death of his father in 1841. He was educated at Trinity College, Dublin, and at Oxford University, where he was awarded a first-class degree in mathematics. He married in 1836, but only four of his 13 children survived to adulthood. Lord Rosse's estates were in Ireland, and this is where he built his telescopes.

In 1845, after he made public his findings on nebulae, Rosse was criticized by John Herschel, who was convinced that nebulae were gaseous in nature. Both men accused each other of using flawed instruments. Ultimately, however, neither succeeded in demonstrating sufficient scientific evidence to resolve conclusively the question of whether nebulae were composed of gas or stars.

Key works

1844 *On the construction of large reflecting telescopes*
1844 *Observations on some of the Nebulae*
1850 *Observations on the Nebulae*

THE PLANET WHOSE POSITION YOU HAVE POINTED OUT ACTUALLY EXISTS

THE DISCOVERY OF NEPTUNE

IN CONTEXT

KEY ASTRONOMER
Urbain Le Verrier (1811–1877)

BEFORE
March 1781 William Herschel discovers Uranus.

August 1781 Finnish–Swedish astronomer Anders Lexell finds irregularities in Uranus's orbit and suggests that they are due to other, undiscovered planets.

1799–1825 Pierre-Simon Laplace explains perturbations mathematically.

1821 French astronomer Alexis Bouvard publishes predictions of future positions of Uranus. Subsequent observations deviate from his predictions.

AFTER
1846 Briton William Lassell discovers Triton, Neptune's largest moon, only 17 days after the discovery of the planet.

1915 Albert Einstein explains perturbations in the orbit of Mercury using relativity.

In the months following William Herschel's discovery of Uranus in 1781, astronomers found irregularities, or perturbations, in its orbit. Most perturbations in orbits are caused by the gravitational effects of other large bodies, but with Uranus there were no known planets that could cause the observed motion. This led some astronomers to suggest that there must be a planet orbiting beyond Uranus.

Searching for the invisible

Frenchman Urbain Le Verrier tackled the problem of the perturbations of Uranus by assuming the location of an undiscovered planet and using Newton's law of gravity to work out what its effect might be on Uranus. This prediction was compared to observations of Uranus, and the position was revised according to the planet's movements. After many repetitions of this process, Le Verrier established the likely position of an unknown planet. He presented his ideas before the Académie des Sciences in 1846, and he also sent his predictions to Johann Galle (1812–1910) at the Berlin Observatory.

Galle received Le Verrier's letter on the morning of September 23, 1846, and obtained permission to

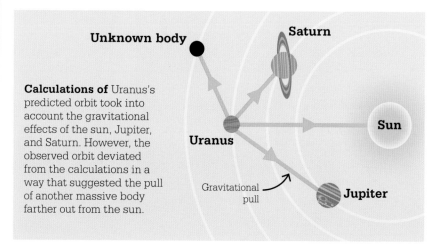

Calculations of Uranus's predicted orbit took into account the gravitational effects of the sun, Jupiter, and Saturn. However, the observed orbit deviated from the calculations in a way that suggested the pull of another massive body farther out from the sun.

See also: The Milky Way 88–89 ▪ Gravitational disturbances 92–93 ▪ The theory of relativity 146–53

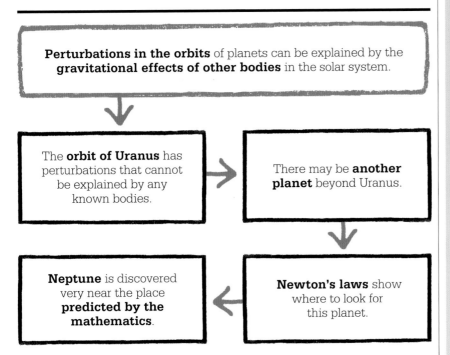

> **Perturbations in the orbits** of planets can be explained by the **gravitational effects of other bodies** in the solar system.

The **orbit of Uranus** has perturbations that cannot be explained by any known bodies.

There may be **another planet** beyond Uranus.

Newton's laws show where to look for this planet.

Neptune is discovered very near the place **predicted by the mathematics**.

Urbain Le Verrier

Urbain Le Verrier studied at the École Polytechnique, near Paris. After graduating, his initial interests were in chemistry, before he switched to astronomy. His astronomical work was focused on celestial mechanics—the description of the movements of the bodies in the solar system using mathematics. Le Verrier obtained a position at the Paris Observatory and spent most of his life there, acting as director from 1854. However, his management style was not popular and he was replaced in 1870. He took up the position again in 1873 after his successor drowned, and held it until his own death in 1877.

Le Verrier spent his early career building on Pierre-Simon Laplace's work on the stability of the solar system. He later went on to study periodic comets before turning his attention to the puzzle of Uranus's orbit.

Key work

1846 *Recherches sur les Mouvements de la Planète Herschel (Research on the Movements of the Planet Herschel)*

look for the planet. Working with his assistant Heinrich D'Arrest, he located an unknown object within 1° of the predicted position that same night. Observations on subsequent nights showed that the object was moving against the background of stars and was, indeed, a planet—one that would later be named Neptune at Le Verrier's suggestion. Galle later gave the credit for the discovery to Le Verrier.

Independent discovery

At the same time as Le Verrier was calculating the position of the unknown planet, British astronomer John Couch Adams (1819–92) was also looking at the cause of the perturbations in the orbit of Uranus. He arrived at a similar conclusion to Le Verrier, completely independently, but his results were not published until after Galle had observed the new

planet. There was some controversy over who should have the credit for the discovery, but Adams always acknowledged that Le Verrier had the better claim.

Galle was not the first person to observe Neptune. Once the orbit of Neptune had been worked out, it was possible to go through old records and find that others had already observed it without realizing it was a planet, including both Galileo and John Herschel. Later, Le Verrier used a similar technique to analyze the orbit of Mercury and found that perturbations in its orbit could not be explained by Newtonian mechanics. He suggested that this might be due to the influence of another planet even closer to the sun, provisionally named Vulcan. This speculation ended when Einstein explained the perturbations using his general theory of relativity. ▪

THE RIS

ASTROP

1850–1915

E OF
PHYSICS

Germans **Gustav Kirchhoff** and **Robert Bunsen** investigate the physics behind **spectral lines**.

Italian priest **Angelo Secchi** starts a project to **classify stars** according to their spectra.

Pioneering American astrophotographer **Henry Draper** takes the first photograph of the **Orion nebula**.

1854

1863

1880

1862

1868

1888

Scottish physicist **James Clerk Maxwell** produces a set of equations that describe the **wavelike behavior** of light.

British astronomer **Joseph Norman Lockyer** discovers a new element in the sun, which he calls **helium**.

Using long-exposure photography, British amateur astronomer **Isaac Roberts** reveals the structure of the **Andromeda nebula**.

In the early 19th century, astronomy was mainly concerned with cataloging the positions of stars and planets, and understanding and predicting the movements of the planets. New comets continued to be discovered, and there was a growing awareness of assorted distant phenomena, such as variable stars, binary stars, and nebulous objects. However, there seemed little scope for learning more about the nature of these remote objects—their chemical composition or temperature, for example. The key that unlocked these mysteries was the analysis of light using spectroscopy.

Decoding starlight
A glowing object gives out light over a range of wavelengths, which we perceive as a rainbow of colors from the longest wavelength (red) to the shortest (violet). When a spectrum is examined in close detail, a multitude of fine variations can emerge. A typical star spectrum appears crossed by numerous dark lines, some fine and faint, some broad and black.

Light is to us the sole evidence of the existence of these distant worlds.
James Clerk Maxwell

Such lines in the sun's spectrum had been noted as early as 1802, but the first physicists to examine the physics behind particular kinds of spectra were Gustav Kirchhoff and Robert Bunsen. Importantly, in about 1860, Kirchhoff showed that different patterns of dark lines are the spectral fingerprints of different chemical elements. Here was a way to investigate the composition of the sun and stars. It even led to the discovery of the previously unknown element, helium.

This new branch of astronomy was enthusiastically taken up by the British astronomer William Huggins and his wife, Margaret, who also pioneered photography as a way of recording observations. They did not restrict themselves to stars, but studied the spectra of nebulae as well.

The **Harvard College Observatory** produces the first *Draper Catalog of Star Spectra*.

1890

While investigating X-rays, French physicist **Henri Becquerel** demonstrates the effects of the **radioactive decay** of uranium.

1896

Harvard computer **Henrietta Swan Leavitt** shows how stars called **Cepheid variables** can be used to measure distances in the universe.

1907

1895

In experiments with cathode ray tubes, German physicist **Wilhelm Röntgen** discovers **X-rays**.

1900

Max Planck lays the foundation for **quantum mechanics** by suggesting that energy can only exist in distinct sizes of "quanta."

1912

Austrian physicist **Victor Hess** shows that powerful rays, now called **cosmic rays**, come from space.

By the end of the 19th century, it seemed that, in order to fully understand the nature of stars, it was necessary to systematically record their spectra and classify them into different types.

Star classification

This immense task was undertaken at Harvard College Observatory, where the director Edward Pickering employed a large team of women to carry out the exacting work. Here, Annie Jump Cannon devised the stellar classification system still used today, based on a temperature sequence. Cannon personally classified some 500,000 stellar spectra. The star catalog included not only their position but also precise information about their magnitude (apparent brightness) and spectrum. This information quickly paid dividends as astronomers analyzed the new data. Cannon's colleague at Harvard, Antonia Maury, realized that the simple temperature sequence did not take account of subtle variations within each star type. Ejnar Hertzsprung and Henry Norris Russell independently followed this up, leading to the discovery that stars of the same color could be giants or dwarfs, and the identification of the first known white dwarf star.

The physics of the stars

In an interval of some 50 years, cutting-edge astronomy had changed its focus. By the early 20th century, physics—the study of matter, forces, and energy, and how they are related—could be applied to the sun and stars, and would strongly influence the future direction of astronomy. Significant developments in basic physics impacted on astronomy. For example, Briton James Clerk Maxwell published his theory of electromagnetism in 1873, describing electromagnetic radiation such as light in terms of its wavelike properties. X-rays were discovered in 1895 and radioactivity in 1896. In 1900, German physicist Max Planck prepared the ground for quantum physics with a leap of inspiration, postulating that electromagnetic energy comes in "packets" of a particular size, called "quanta." These discoveries would lead to new ways of looking at the skies, and shed new light on the processes taking place within stars. Physics and astronomy would be inseparable from this point on. ∎

SODIUM IS TO BE FOUND IN THE SOLAR ATMOSPHERE
THE SUN'S SPECTRUM

IN CONTEXT

KEY ASTRONOMER
Gustav Kirchhoff
(1824–1887)

BEFORE
1802 After creating an image of the sun's spectrum by passing sunlight through a narrow slit and prism, English chemist William Hyde Wollaston notices seven dark lines in the spectrum.

1814 Joseph von Fraunhofer, the German inventor of the spectroscope, discovers 574 of the same dark lines in the sun's spectrum. He maps these in detail.

AFTER
1912 Danish physicist Niels Bohr introduces a model of the atom in which movements of electrons switching between different energy levels cause radiation to be emitted or absorbed at particular wavelengths.

I n 1814, a German maker of optical instruments named Joseph von Fraunhofer invented the spectroscope (see diagram on p.113). This allowed the spectrum of the sun, or any other star, to be displayed and measured with high precision. Fraunhofer noticed that there were more than 500 dark lines crossing the sun's spectrum, each located at a precise wavelength (color). These came to be known as Fraunhofer lines.

By the 1850s, German scientists Gustav Kirchhoff and Robert Bunsen had discovered that, if different chemical elements are heated in a flame, they emit light at one or more wavelengths that are characteristic for that element, acting like a fingerprint to indicate its presence. Kirchhoff noticed that the wavelengths of light given off by some elements corresponded to the wavelengths of some Fraunhofer lines. In particular, sodium's emissions at wavelengths of 589.0 and 589.6 nanometers exactly matched two Fraunhofer lines. Kirchhoff suggested that a

hot, dense gas, such as the sun, will emit light at all wavelengths and thus produce a continuous spectrum. However, if the light passes through a cooler, lower-density gas, such as the sun's atmosphere, some of that light might be absorbed by an element (sodium, for example), at the same wavelengths at which the element emits light when heated. The absorption of the light causes gaps in the spectrum, which are now known as absorption lines. ∎

The path is opened for the determination of the chemical composition of the sun and the fixed stars.
Robert Bunsen

See also: Analyzing starlight 113 ▪ The characteristics of stars 122–27 ▪ Refining star classification 138–39 ▪ Stellar composition 162–63

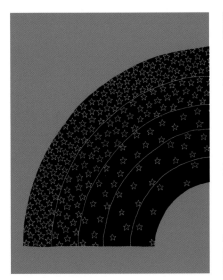

STARS CAN BE GROUPED BY THEIR SPECTRA
ANALYZING STARLIGHT

IN CONTEXT

KEY ASTRONOMER
Angelo Secchi (1818–1878)

BEFORE
1802 William Hyde Wollaston notices that there are dark gaps in the sun's spectrum.

1814 German lensmaker Joseph von Fraunhofer measures the wavelengths of these dark lines.

1860 Gustav Kirchhoff and Robert Bunsen use a gas burner to make systematic recordings of the wavelengths produced by burning elements.

AFTER
1868 English scientist Norman Lockyer identifies a new element, helium, from emission lines in the sun's light.

1901 The Harvard System for the classification of stellar spectra, devised by Williamina Fleming and Annie Jump Cannon, supersedes Secchi's system.

Angelo Secchi was one of the pioneers of astrophysics, an arm of science that focuses on the properties of a star, rather than merely its position in the sky. He was the first to group stars according to their spectra, or the particular colors of light they emit.

A Jesuit priest as well as a noted physicist, Secchi founded a new observatory at the order's Collegio Romano in Rome. There he became a pioneer of the technique of spectroscopy, a new way to measure and analyze starlight.

Gustav Kirchhoff had shown that gaps in a stellar spectrum were caused by the presence of specific elements (see opposite). Armed with this knowledge, Secchi began to class stars according to their spectra. At first he used three classes: Class I were white or blue stars that showed large amounts of hydrogen in their spectra; Class II were yellow stars, with metallic spectral lines (for astronomers, "metallic" refers to any element heavier than helium); and Class III

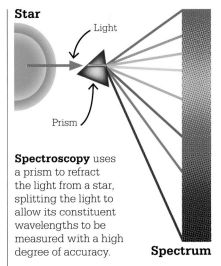

Star

Light

Prism

Spectroscopy uses a prism to refract the light from a star, splitting the light to allow its constituent wavelengths to be measured with a high degree of accuracy.

Spectrum

were orange, with a complex array of elements present. In 1868, Secchi added Class IV for redder stars with carbon present, and finally in 1877 came Class V for stars that showed emission lines (not absorption lines, as in the other four).

Secchi's stellar classes were later amended by other scientists, and in 1880 became the foundation of the Harvard System, which is used to classify stars to this day. ∎

See also: The sun's spectrum 112 ▪ The sun's emissions 116 ▪ The star catalog 120–21 ▪ The characteristics of stars 122–27

ENORMOUS MASSES OF LUMINOUS GAS
PROPERTIES OF NEBULAE

In the 1860s, a pioneering British astronomer named William Huggins made key discoveries by studying the composition of stars and nebulae using a spectroscope. This instrument, a glass prism attached to a telescope, splits white light into its constituent light wavelengths, producing a spectrum of color. Gustav Kirchhoff and Robert Bunsen had already noted the chemical composition of the sun by studying the dark absorption lines that occur in its spectrum. These lines are caused by the atoms of different chemical elements absorbing radiation at certain precise wavelengths. Huggins, encouraged by his astronomer wife, Margaret, turned his attention deeper into space, toward nebulae, the fuzzy patches of light that had long mystified astronomers. He used spectroscopy to divide these patches into two distinct types.

The spectra of nebulae
Huggins observed that nebulae, such as the Andromeda nebula, had a spectrum of light similar to that of the sun and other stars—a broad band of color with dark absorption lines. The reason for this (not discovered until the 1920s, after Huggins's death) was that such

Spectroscopes allow astronomers to measure **a nebula's spectrum of light**.	Some nebulae are found to have spectra **similar to those of stars**.
These nebulae are enormous masses of luminous gas.	Others have spectra that **emit energy** at a **single wavelength**.

See also: Observing Uranus 84–85 ▪ Messier objects 87 ▪ The sun's spectrum 112

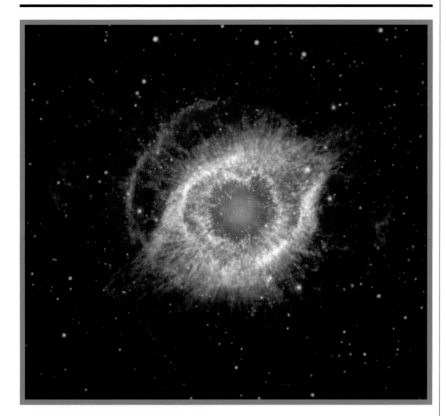

William Huggins

After selling the family drapery business when he was 30, William Huggins ran a private observatory in Tulse Hill in South London. He used his new wealth to buy a powerful 8-in (20-cm) refracting telescope.

In 1875, age 51, Huggins married 27-year-old Irish astronomy enthusiast Margaret Lindsey, who encouraged him to adopt photography to record his spectra and was an active partner in his later research, co-authoring many papers. Huggins was a pioneer in the use of photography to record astronomical objects. He also developed a technique to study the radial velocity of stars using the Doppler shift of their spectral lines.

As a pioneer astronomical spectroscopist, Huggins was elected to serve as president of the Royal Society from 1900 to 1905. He died at his home in Tulse Hill in 1910, age 86.

Key works

1870 *Spectrum analysis in its application to the heavenly bodies*
1909 *Scientific Papers*

nebulae are indeed composed of stars, and are galaxies in their own right. The second type of nebula he observed was entirely different. Its light spectrum was made of single-wavelength emission lines—energy was being emitted as one color; there were no absorption lines.

Huggins realized that these second kind of nebula were huge clouds of hot, low-density gas. Some of this gas could be in the process of forming new stars; other gas clouds, like the planetary nebulae, could have been ejected from evolving stars.

Huggins' 1864 observations of the Cat's Eye planetary nebula in the constellation Draco revealed a spectrum with a single absorption line, produced by hot hydrogen gas. However, the nebula also emitted

Huggins was the first to analyze the spectrum of a planetary nebula (the Cat's Eye nebula), confirming that it was gaseous and not composed of stars.

energy in two strong green lines, which did not correspond to any known chemical element. Some astronomers suggested that they were produced by a new element, dubbed nebulium.

Huggins concluded from his spectroscopic observations that all of the heavenly bodies he had studied were made of exactly the same elements as Earth. However, the mystery of nebulium was not solved until after his death. In 1927, it was found to be simply doubly ionized oxygen—oxygen atoms that have lost electrons and have a double positive charge. ▪

THE SUN'S YELLOW PROMINENCE DIFFERS FROM ANY TERRESTRIAL FLAME
THE SUN'S EMISSIONS

IN CONTEXT

KEY ASTRONOMERS
Jules Janssen (1824–1907)
Joseph Norman Lockyer
(1836–1920)

BEFORE
1863 Gustav Kirchhoff develops spectroscopy, showing how light can be used to identify hot substances.

1864 William and Margaret Huggins find that the spectra of nebulae contain different emission lines, showing that they are largely clouds of gas.

AFTER
1920 Arthur Eddington states that stars are fueled by the fusion of hydrogen into helium.

1925 Cecilia Payne-Gaposchin shows that stars are largely made from the elements hydrogen and helium.

1946 US cosmologist Ralph Alpher calculates that most of the universe's helium was formed in the first few minutes after the Big Bang.

In August 1868, the French astronomer Pierre Jules César Janssen traveled to India to observe a solar eclipse. The eclipse covered the sun's bright disk, leaving only a narrow ring of light. This was the chromosphere, the middle of three layers in the sun's atmosphere, which was normally hidden by the glare. Janssen found that the spectrum of the chromosphere's light contained numerous bright emission lines. Using discoveries made by Gustav Kirchhoff, Janssen was able to confirm that the chromosphere was a layer of gas. He also noticed a previously unseen yellow emission line in the sun's spectrum. He assumed this unknown light was produced by sodium, helping to give the sun its yellow hue.

In October that year, English astronomer Joseph Norman Lockyer developed a spectroscope for observing the chromosphere directly. He also detected its curious light and also assumed it was produced by sodium, but after consulting the chemist Edward Frankland,

A total eclipse of the sun reveals the chromosphere. This image of an eclipse was captured in 1919 by British astronomer Arthur Eddington.

he changed his mind—the light was not from sodium but from a hitherto unknown element, which he named helium, after *helios*, the Greek word for the sun. For some years, it was thought that helium only existed on the sun, but in 1895, Scottish chemist William Ramsay succeeded in isolating a sample from a radioactive uranium mineral. ∎

See also: The sun's spectrum 112 ▪ Nuclear fusion within stars 166–67 ▪ The primeval atom 196–97

MARS IS TRAVERSED BY A DENSE NETWORK OF CHANNELS
MAPPING MARS'S SURFACE

IN CONTEXT

KEY ASTRONOMER
Giovanni Schiaparelli
(1835–1910)

BEFORE
1858 Angelo Secchi first uses the word *canali* (channels) in connection with Mars.

AFTER
1897 Italian astronomer Vincenzo Cerulli theorizes that the Martian canals are just an optical illusion.

1906 A book by American astronomer Percival Lowell, *Mars and Its Canals*, promotes the idea that there may be artificial canals on Mars, built by intelligent beings.

1909 Photographs of Mars taken at the new Baillaud dome at the Pic du Midi observatory in France discredit the Martian canals theory.

1960s NASA's Mariner flyby missions fail to capture any images of the canals or find any evidence of them.

By the mid-19th century, scientists were increasingly speculating about the possibility of life on Mars, which had been found to have certain similarities to Earth, including ice caps, a similar length of day, and an axial tilt that meant it experienced seasons. However, it had also been found that it did not rain on Mars.

Between 1877 and 1890, Italian astronomer Giovanni Schiaparelli carried out a series of detailed observations of Mars to produce a map of the planet's surface.

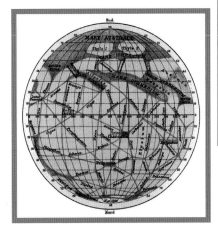

Schiaparelli described various dark areas as "seas" and lighter areas as "continents." He also portrayed what seemed to him to be a network of long, dark, straight lines, or streaks, crisscrossing Mars's equatorial regions. In his book *Life on Mars*, Schiaparelli suggested that, in the absence of rain, these channels might be the mechanism by which water was transported across the dry surface of the planet to allow life to exist there.

In the following years, many eminent scientists, including American astronomer Percival Lowell, speculated that these dark lines were irrigation canals constructed by intelligent beings on Mars. However, others could not see the channels at all when they looked for them—and by 1909, observations with telescopes of higher resolution had confirmed that the Martian canals did not exist. ∎

Schiaparelli's 1888 atlas of Mars shows land, seas, and a network of straight channels. Here, the south pole is shown at the top.

See also: Observing Saturn's rings 65 ∎ Analyzing starlight 113 ∎
Life on other planets 228–35

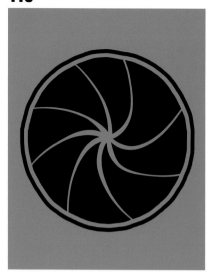

PHOTOGRAPHING THE STARS
ASTROPHOTOGRAPHY

IN CONTEXT

KEY ASTRONOMER
David Gill (1843–1914)

BEFORE
1840 The first clear photograph of the moon is taken by American John Draper, using a 20-minute exposure.

1880 John Draper's son Henry takes a 51-minute exposure of the Orion nebula. He also takes the first wide-angle photograph of the tail of a comet.

AFTER
1930 US astronomer Clyde Tombaugh discovers Pluto by spotting a moving object on photographic plates.

1970s Charge-coupled devices replace photographic plates and film with digital photographs.

1998 The Sloan Digital Sky Survey begins to make a 3-D map of galaxies.

Photographs of the stars can be used to make very accurate **star maps**.

Photographing the stars requires **long exposures**.

However, **Earth's rotation** makes images blurry. **A precise tracking mechanism** is needed to move the camera.

Accurate maps reveal that stars are moving at **different speeds** and in **different directions**.

Isaac Newton's theory of gravitation, like many advances of the Scientific Revolution (pp.42–43), was based on a belief that the universe worked like clockwork. In the 1880s, David Gill, a master clockmaker from Aberdeen, Scotland, applied his precision clockmaking machinery to astronomical telescopes—and, ironically enough, offered a way of showing that the stars were not all moving in clocklike unison.

Gill was a pioneer in the field of astrophotography. In the mid-1860s, then still an amateur astronomer working in his father's backyard, he built a tracking mount for his 12-in (30-cm) reflecting telescope, and used it to photograph the moon with a clarity that had never been seen before. The photographs earned Gill a fellowship in the Royal Astronomical Society, and, by 1872, his first job as a professional astronomer at the Dunecht Observatory in Aberdeen.

Gill applied clockwork tracking mechanisms to telescope mounts so that the telescope could move in near-perfect harmony with the rotation of Earth. This allowed the instrument to remain fixed and focused on a single patch of sky. Gill was not the first to attempt to photograph the heavens with telescopes, but imaging faint celestial light required exposures of several minutes at least, and poor tracking meant that early star photographs were mostly incomprehensible blurs.

Southern sky

In 1879, Gill became the chief astronomer at the Cape Observatory in South Africa. By now, he was using the latest dry-plate system (a photographic plate pre-coated in light-sensitive chemicals), which he employedto capture the "Great Comet" that appeared over the southern hemisphere in 1882.

Working in partnership with the Dutch astronomer Jacobus Kapteyn, Gill spent the best part of the next two decades creating a

Frank McClean, an astronomer friend of David Gill, donated the McClean Telescope to the Cape Observatory in 1897. David Gill used it extensively.

photographic record of the southern sky. The result was the *Cape Photographic Durchmusterung* (catalog), showing the position and magnitude of nearly half a million stars. Gill also became a key figure in the *Carte du Ciel* ("Map of the Sky") project, a global collaboration of observatories begun in 1887 with the goal of making a definitive photographic map of the stars. This ambitious, expensive, and decades-long project involved teams of human computers who would measure the plates by hand. It was, however, superseded by new methods and technologies before it was finished.

The accurate maps produced by Gill's photographic techniques may seem fairly unremarkable today, but at the turn of the 20th century they were the first reliable means for showing the proper

motion of nearby stars relative to more distant ones. This information was invaluable for measuring stellar distances on a large scale, and it began to reveal to astronomers the true scale of the galaxy and the universe beyond. ▪

David Gill

The eldest son of a successful clockmaker, David Gill was destined to take over the family business. However, while at the University of Aberdeen, he became a student of the great physicist James Clerk Maxwell, whose lectures gave Gill a passion for astronomy. When offered a job as a professional astronomer in 1872, Gill sold the family business and began work at the Dunecht Observatory, Aberdeen.

In addition to his pioneering work in astrophotography, Gill developed the use of the heliometer, a device for

measuring stellar parallax (p.102). His measurements, used in conjunction with his star maps, did much to reveal the distances between stars. By the time he left the Cape Observatory in 1906, Gill was a renowned astronomer. In one of his final jobs, the government consulted him on the implementation of daylight saving hours.

Key work

1896–1900 *Cape Photographic Durchmusterung* (with Jacobus Kapteyn)

A PRECISE MEASUREMENT OF THE STARS
THE STAR CATALOG

Edward C. Pickering, in his role as director of the Harvard College Observatory from 1877 to 1906, laid the foundations for precise stellar astronomy. His team carried out star surveys that broke new ground in understanding the scale of the universe. Pickering combined the latest techniques in astrophotography with spectroscopy (splitting light into its constituent wavelengths) and photometry (measuring the brightness of stars) to create a catalog that listed a star's location, magnitude, and spectral type. He did this with the help of the Harvard "computers"— a team of mathematically minded women upon whom Pickering relied to process the huge amounts of data required to create the catalog.

More than 80 computers, known in those less enlightened days as "Pickering's Harem," worked at the Harvard Observatory. The first of them was Williamina Fleming, who had been Pickering's maid. Upon taking over the observatory, Pickering fired his male assistant, deeming him "inefficient," and hired Fleming in his place. Other notable names among the computers included Antonia Maury, Henrietta Swan Leavitt, and Annie Jump Cannon.

Color and brightness
Pickering's individual contributions to the star catalog were twofold. In 1882, he developed a method of photographing multiple star spectra simultaneously by transmitting the stars' light through a large prism and onto photographic plates. In 1886, he designed a wedge photometer, a device for measuring the apparent magnitude of a star. Magnitudes had previously been recorded psychometrically—using

A woman had no chance at anything in astronomy except at Harvard in the 1880s and 1890s. And even there, things were rough.
William Wilson Morgan
US astronomer

the naked eye as a means of comparing one star's brightness with that of another. The wedge photometer was much more objective; the observer viewed a target star alongside one of several stars with an accepted brightness and then edged a wedge of calcite in front of the known source, diminishing its magnitude in increments until the two sources looked to have the same brightness.

In 1886, Mary Draper, the widow of spectral photography pioneer Henry Draper, agreed to fund Pickering's work in her husband's name. In 1890, the first *Draper Catalogue of Stellar Spectra* was published. Pickering then opened an observatory in Arequipa,

Many of the Harvard computers were trained in astronomy, but as women, they were excluded from academic positions. Their wages were similar to those of unskilled workers.

I do not know if God is a mathematician, but mathematics is the loom on which God weaves the universe.
Edward C. Pickering

Peru, to survey the southern sky and produce the first all-sky photographic map.

In combination with the work of the Harvard computers, Pickering's data was the basis for the *Henry Draper Catalogue* published in 1918, which contained spectral classifications for 225,300 stars across the entire sky. ▪

Edward C. Pickering

Edward C. Pickering was the dominant figure of American astronomy at the turn of the 20th century. Many of the first steps in the development of today's astrophysics and cosmology were made by people he employed at the Harvard College Observatory. Known as a progressive for his attitudes to the education of women and their role in research, Pickering nonetheless asserted a rigid authority over his team. On more than one occasion, he forced out researchers with whom he did not agree, only for them later to be proved right; one example of this is Antonia Maury, whose work on stellar spectra Pickering dismissed.

Pickering spent his whole career in academia, but was also an avid outdoorsman, and was a founding member of the Appalachian Mountain Club. The club became a leading voice in the movement to preserve wilderness areas.

Key works

1886 *An Investigation in Stellar Photography*
1890 *Draper Catalogue of Stellar Spectra*
1918 *Henry Draper Catalogue*

CLASSIFYING THE STARS

ACCORDING TO THEIR SPECTRA REVEALS THEIR

AGE AND SIZE

THE CHARACTERISTICS OF STARS

IN CONTEXT

KEY ASTRONOMER
Annie Jump Cannon
(1863–1941)

BEFORE
1860 Gustav Kirchhoff shows that spectroscopy can be used to identify elements in starlight.

1863 Angelo Secchi classifies stars using their spectra.

1868 Jules Janssen and Joseph Norman Lockyer discover helium in the solar spectrum.

1886 Edward Pickering begins compiling the Henry Draper Catalogue using a photometer.

AFTER
1910 The Hertzsprung–Russell diagram reveals the different sizes of stars.

1914 US astronomer Walter Adams records a white dwarf.

1925 Cecilia Payne-Gaposchkin finds that stars are composed almost entirely of hydrogen and helium.

American astronomer Annie Jump Cannon was the early 20th-century's leading authority on the spectra of stars. When she died in 1941, Cannon was described as "the world's most notable woman astronomer." Her great contribution was to create the basis of the system for classifying the spectra of stars that is still in use today.

Cannon worked at the Harvard College Observatory, as part of the team of "Harvard Computers," a group of women employed by the director Edward C. Pickering to help compile a new stellar catalog. The college's catalog, begun in the 1880s with funding from the widow of astrophotographer Henry Draper, used new techniques to collect data on every star in the sky brighter than a certain magnitude, including obtaining the spectra of as many stars as possible. In the 1860s, Angelo Secchi had set out a provisional system for classifying

The seven main classes of star, categorized according to spectra and temperature, are, from left to right: O, B, A, F, G, K, and M, with O the hottest and M the coolest.

Each substance sends out its own vibrations of particular wavelengths, which may be likened to singing its own song.
Annie Jump Cannon

stars according to their spectra. Pickering's team modified this system. By 1924, the catalog contained 225,000 stars.

Early approaches
Williamina Fleming, the first of Pickering's female computers, made the earliest attempt at a more detailed classification system, by subdividing Secchi's classes into 13 groups, which she labeled with the letters A to N (excluding I), then adding O, P, and Q. In the next phase of the work, fellow computer Antonia Maury, working with better data received from observatories

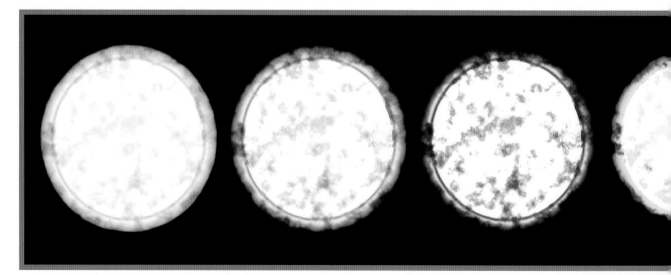

See also: The sun's spectrum 112 ▪ Analyzing starlight 113 ▪ The sun's emissions 116 ▪ The star catalog 120–21 ▪ Analyzing absorption lines 128 ▪ Refining star classification 138–39 ▪ Stellar composition 162–63

around the the world, noticed more variety in the detail. She devised a more complex system of 22 groups designated by Roman numerals, each divided into three subgroups. Pickering was concerned that applying such a detailed system would delay the task of compiling the catalog. However, Maury's approach to stellar classification proved a crucial step toward the creation of the Hertzsprung–Russell diagram in 1910, and consequent discoveries about stellar evolution.

Cannon joined the Harvard College Observatory staff in 1896 and began working on the next part of the catalog, which was published in 1901. With Pickering's approval, to make classification clearer and easier, she reverted to Fleming's spectral classes designated by letters, but she changed the order.

Maury had realized that stars of similar colors have the same characteristic absorption lines in the spectra. She had also deduced that a star's temperature is the main factor affecting the

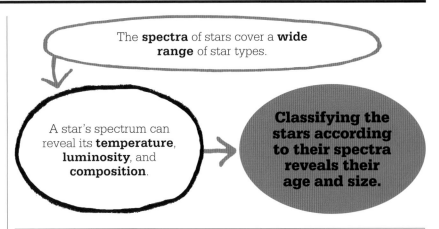

The **spectra** of stars cover a **wide range** of star types.

A star's spectrum can reveal its **temperature**, **luminosity**, and **composition**.

Classifying the stars according to their spectra reveals their age and size.

appearance of its spectrum and made her classes a temperature sequence from hotter to cooler. On this, Cannon followed Maury's lead. Some of Fleming's letters were dropped because they were unnecessary, so the final sequence became O, B, A, F, G, K, M, based on the presence and strength of certain spectral lines, especially those due to hydrogen and helium. Students of astronomy still learn it by remembering the mnemonic, "Oh Be A Fine Girl, Kiss Me," attributed to Henry Norris Russell.

Harvard system

Cannon's 1901 system laid the foundations for the Harvard Spectral Classification system. By 1912, she had extended it to introduce a range of more precise subclasses, adding 0 to 9 after the letter, with 0 the hottest in the class and 9 the coolest. A few new classes have been added since.

The Harvard system essentially classifies stars by temperature and takes no account of the luminosity or size of the star. In 1943, however, luminosity was added as an »

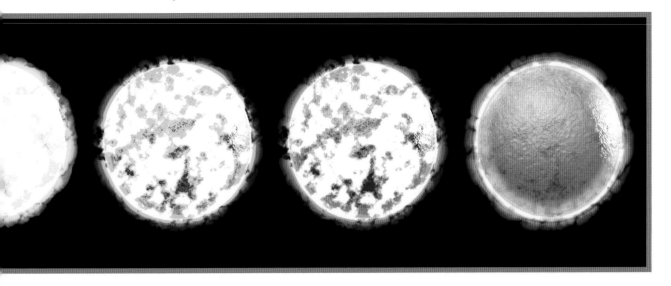

additional dimension, creating the Yerkes classification system, otherwise called the MKK system after William Morgan, Philip Keenan, and Edith Kellman, the astronomers based at the Yerkes Observatory in Wisconsin who formulated it. This system denotes luminosity with Roman numerals, although a few letters are also used.

The advantage of the MKK system is that it gives a star a size as well as a temperature, so that stars can be described in colloquial terms such as white dwarf, red giant, or blue supergiant. The main sequence stars, including the sun, are all small enough to be called dwarfs. The sun is a G2V star, which indicates that it is a yellow dwarf with a surface temperature of about 5,800 K.

Classes and characteristics

The hottest class of star, O-types have a surface temperature in excess of 30,000 K. Most of the radiation these stars emit is in the ultraviolet part of the spectrum and appear blue when viewed in visible light. O stars are mainly giants, typically 20 times as massive as the sun and 10 times as wide. Only 0.00003 percent of main sequence stars

> The prism has revealed to us something of the nature of the heavenly bodies, and the photographic plate has made a permanent record of the condition of the sky.
> **Williamina Fleming**

The strengths of the absorption lines of different elements vary according to the surface temperature of the star. Lines of heavier elements are more prominent in the spectra of cooler stars.

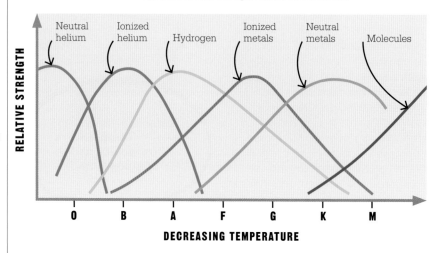

are this hot. O-type stars burn their fuel very quickly and release huge amounts of energy. As a result, they have a short life expectancy, which is measured in tens of millions of years, compared to billions for cooler stars. Members of this class have weak lines of hydrogen in their spectra, and strong evidence of ionized helium, which is present because of the high temperature.

With a surface temperature of between 10,000 and 30,000 K, B-type stars are brighter in visible light than O-types, despite being cooler. This is because more of the radiation is emitted as visible light, making them "blue-white." Again, B-type dwarfs are rare, making up less than 0.1 percent of main sequence stars. When they do occur, they are perhaps 15 times more massive than the sun. B-type stars have non-ionized helium in their spectra and more evidence of hydrogen. Because they live for only a short time, B-type stars are found in molecular clouds or star-forming regions, since they have not had time to move far from the location in which they formed. About

twice as large as the sun, main sequence A-type stars have a surface temperature of between 7,500 and 10,000 K. They have strong hydrogen lines in their spectra and emit a wide spectrum of visible light, which makes them look white (with a blueish tinge). As a result, they are some of the most easily seen stars in the night sky, and include Vega (in Lyra), Gamma Ursae Majoris (in the Big Dipper), and Deneb (in Cygnus). However, only 0.625 percent of main sequence stars are A-type stars.

Cooling stars

As dwarf stars cool, the hydrogen in their spectra becomes less intense. They also exhibit more absorption lines due to metals. (To an astronomer, everything heavier than helium is a metal.) This is not because their composition is different from that of hotter stars but because the gas near the surface is cooler. In hotter stars, the atoms are too ionized to create absorption lines. F-type stars have a surface temperature of between 6,000 and 7,500 K. Called yellow-white dwarfs, they make up 3 percent of the main sequence and are a little larger than the sun. The spectra of these stars contain medium-intensity hydrogen lines and strengthening lines for iron and calcium.

The sun's class

Type-G yellow dwarfs, of which the sun is one, make up 8 percent of the main sequence. They are between 5,200 and 6,000 K on the surface and have weak hydrogen lines in their spectra, with more prominent metal lines. Type-K dwarfs are orange and make up 12 percent of the main sequence. They are between 3,700 and 5,200 K on the surface and have very weak hydrogen absorption lines but strong metallic ones, including manganese, iron, and silicon. Type-M are red dwarfs. These are by far the most common

A white dwarf sits at the heart of the Helix planetary nebula. When its fuel ends, the sun will become a white dwarf.

main sequence stars, making up 76 percent of the total, although no red dwarf is visible to the naked eye. They are just 2,400–3,700 K on the surface and their spectra contain absorption bands for oxide compounds. The majority of the yellow, orange, and red dwarfs are believed to have planetary systems.

Extended classification

Stellar spectral classes now cover even more types of stars. Class W are thought to be dying supergiant stars. Class C, or carbon stars, are declining red giants. Classes L, Y, and T are a diminishing scale of colder objects, from the coolest red dwarfs to the brown dwarfs, which are not quite large or hot enough to be classed as stars. Finally, white dwarfs are class D. These are the hot cores of red giant stars that no longer burn with fusion and are gradually cooling. Eventually they should fade to black dwarfs, but it is estimated it will take a quadrillion years for that to happen. ■

Annie Jump Cannon

Born in Delaware, Annie Jump Cannon was the daughter of a state senator, and was introduced to astronomy by her mother. She studied physics and astronomy at Wellesley College, an all-women's college. Graduating in 1884, Cannon returned to her family home for the next 10 years. On the death of her mother, in 1894, she began to teach at Wellesley and joined Edward C. Pickering's Harvard Computers two years later.

Cannon suffered from deafness, and the ensuing difficulties in socializing led her to immerse herself in scientific work. She remained at Harvard for her entire career, and is said to have classified 350,000 stars over 44 years. Subject to many restrictions over her career due to her gender, she was finally appointed a member of the Harvard faculty in 1938. In 1925, she became the first woman to be awarded an honorary degree by Oxford University.

Key work

1918–24 *The Henry Draper Catalogue*

THERE ARE TWO KINDS OF RED STAR
ANALYZING ABSORPTION LINES

IN CONTEXT

KEY ASTRONOMER
Ejnar Hertzsprung
(1873–1967)

BEFORE
1866 Angelo Secchi creates the first classification of stars according to their spectral characteristics.

1880s At the Harvard College Observatory, Edward Pickering and Williamina Fleming establish a more detailed classification system.

1890s Antonia Maury develops her own system for classifying star spectra, taking into account differences in width and sharpness of spectral lines.

AFTER
1913 Henry Norris Russell creates a diagram, similar to one made by Hertzsprung, that plots the absolute magnitude (intrinsic brightness) of stars against spectral class. This later becomes known as a Hertzsprung–Russell diagram.

In the late 19th and early 20th centuries, Edward Pickering and his assistants carried out extensive work classifying star spectra. They cataloged the range of light wavelengths coming from a star, which, among other information, contains dark absorption lines. These lines indicate the presence of particular elements in the star's atmosphere that are absorbing those wavelengths.

One of Pickering's assistants, Antonia Maury, developed her own classification system, taking into account differences in the width of absorption lines in star spectra. She noticed that some spectra, which she denoted as "c," had sharp, narrow lines. Using Maury's system, Danish astronomer Ejnar Hertzsprung saw that stars with "c-type" spectra were far more luminous than other stars.

Bright and dim red stars

Hertzsprung figured out that what Maury had identified as "c-type" stars were radically different from other types in the same category. For example, within the M-class

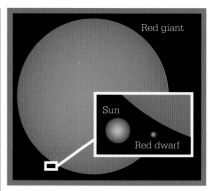

A typical red giant has a diameter about 50 times that of the sun, and 150 times that of a typical red dwarf. However, a red giant has only about 8–10 times the mass of a red dwarf.

category, or red-colored stars, he noticed that the "c-types" were highly luminous, high-mass, comparatively rare stars—today, these are called red giants or red supergiants, depending on their size. The remaining majority of non-"c-type" M-class stars were low-mass, faint stars that are now known as red dwarfs. A similar distinction of two main kinds also applied to K-class (orange) stars. ∎

See also: The sun's spectrum 112 ▪ Analyzing starlight 113 ▪ The star catalog 120–21 ▪ The characteristics of stars 122–27 ▪ Refining star classification 138–39

SUNSPOTS ARE MAGNETIC
THE PROPERTIES OF SUNSPOTS

IN CONTEXT

KEY ASTRONOMER
George Ellery Hale
(1868–1938)

BEFORE
800 BCE The appearance of dark spots on the sun is recorded in the Chinese *Book of Changes*.

1600 English physicist William Gilbert discovers that Earth has a magnetic field.

1613 Galileo demonstrates that sunspots are features on the surface of the sun.

1838 Samuel Heinrich Schwabe notes a cycle in the numbers of sunspots seen each year.

1904 British astronomers Edward and Annie Maunder publish evidence of an 11-year sunspot cycle.

AFTER
1960 US physicist Robert Leighton introduces the field of helioseismology, a study of the motion of the solar surface.

American George Hale was just 14 when his wealthy father bought him his first telescope, and 20 when his father built him an observatory on the family property. Two years later, while at MIT, Hale developed a new design for a spectroheliograph—a device for viewing the surface of the sun one wavelength of light at a time. He used this device to study the spectral lines of sunspots.

Some years later, Hale organized the building of some of the largest telescopes in the world at the time,

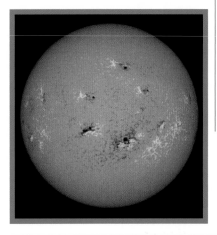

including the 60-inch (150-cm) Hale Telescope, built at California's Mount Wilson Observatory in 1908, paid for by a bequest from his father. Working at Mount Wilson that same year, Hale was able to take clear images of sunspots in a deep red wavelength emitted by hydrogen. The speckled images reminded Hale of the way iron filings mapped the force field around a magnet. This led him to look for signs of the Zeeman effect in light coming from the sunspots.

The Zeeman effect is a split in spectral lines caused by the presence of a magnetic field, first observed by the Dutch physicist Pieter Zeeman in 1896. The spectral lines in light coming from sunspots had indeed been split, which suggested to Hale that sunspots were swirling magnetic storms on the surface of the sun. ∎

The variations in the strength of the sun's magnetic field are shown in this magnetogram, produced using the Zeeman effect. The marks correspond to the locations of sunspots.

See also: Galileo's telescope 56–63 ▪ The surface of the sun 103 ▪ The sun's vibrations 213 ▪ Maunder (Directory) 337

THE KEY

TO A DISTANCE

SCALE

OF THE UNIVERSE

MEASURING THE UNIVERSE

IN CONTEXT

KEY ASTRONOMER
Henrietta Swan Leavitt
(1868–1921)

BEFORE
1609 German pastor David Fabricius discovers the periodically variable star Mira.

1638 Dutch astronomer Johannes Holwarda observes Mira's variation in brightness over a regular 11-month cycle.

1784 John Goodricke discovers a periodic variation in the star Delta Cephei: the prototypic example of a Cepheid variable.

1838 Friedrich Bessel measures the distance to the star 61 Cygni using the parallax method.

AFTER
1916 Arthur Eddington studies why Cepheids pulsate.

1924 Edwin Hubble uses observations of a Cepheid in the Andromeda nebula to calculate its distance.

Some of the most important, but often most challenging, measurements for astronomers to make have been the distances to extremely remote objects—which includes most celestial objects aside from the moon, sun, and other planets of the inner solar system. Nothing in the light coming from distant stars and galaxies gives any direct indication of how far that light has traveled through space to reach Earth.

For several hundred years, scientists realized that it should be possible to measure the distances to relatively nearby stars by a method called parallax. This is based on comparing the position of a nearby star against the background of more distant stars from two perspectives—usually Earth's different positions in space six months apart in its orbit around the sun. Although many others had tried (and failed) before him, the first astronomer to measure a star's distance accurately using this method was Friedrich Bessel, in 1838. However, even with increasingly powerful telescopes, measuring star distances by parallax proved difficult and,

A remarkable relation between the brightness of these (Cepheid) variables and the length of their periods will be noticed.
Henrietta Swan Leavitt

by the year 1900, the distances to only about 60 stars had been measured. Furthermore, the parallax method could be applied only to nearby stars. The difference in perspective for more distant stars over the course of a year was too small to be accurately determined. New methods were therefore needed to measure large distances in space.

Measuring brightness
In the 1890s and early 1900s, the Harvard College Observatory in Massachusetts was one of the world's leading astronomical research

---

Henrietta Swan Leavitt

Henrietta Swan Leavitt developed an interest in astronomy while studying at Radcliffe College, Cambridge, Massachusetts. After graduation, she suffered a serious illness that caused her to become increasingly deaf for the rest of her life. From 1894 to 1896 and then again from 1902, she worked at Harvard College Observatory. Leavitt discovered more than 2,400 variable stars and four novae. In addition to her work on Cepheid variables, Leavitt also developed a standard of photographic measurements, now called the Harvard Standard.

Due to the prejudices of the day, Leavitt did not have the chance to use her intellect to the fullest, but she was described by a colleague as "possessing the best mind at the Observatory." She was remembered as hardworking and serious-minded, "little given to frivolous pursuits." Leavitt worked at the Observatory until her death from cancer in 1921.

Key work

1908 *1777 Variables in the Magellanic Clouds*

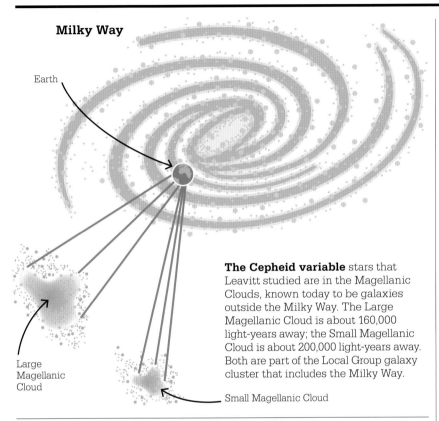

Milky Way

Earth

Large Magellanic Cloud

Small Magellanic Cloud

The Cepheid variable stars that Leavitt studied are in the Magellanic Clouds, known today to be galaxies outside the Milky Way. The Large Magellanic Cloud is about 160,000 light-years away; the Small Magellanic Cloud is about 200,000 light-years away. Both are part of the Local Group galaxy cluster that includes the Milky Way.

At the time, the SMC and LMC were thought to be very large star clusters within the Milky Way, which itself was assumed to comprise the entire universe. Today, they are known to be relatively small, separate galaxies that lie outside the Milky Way. The Magellanic Clouds are visible to the naked eye in the night sky of the southern hemisphere, but are never visible from Massachusetts, where Leavitt lived and worked. Therefore, although she examined numerous photographic plates of the LMC and SMC obtained by astronomers at an observatory in Peru, it is highly unlikely that she ever physically observed them in the sky.

After several years' work, Leavitt had found 1,777 variables in the SMC and LMC. One particular kind that caught Leavitt's attention, representing a small fraction of all the variables she had found (47 out of 1,777), was of a type called a Cepheid variable. Leavitt called them "cluster variables"—the term Cepheid variable was introduced »

institutions. Under the supervision of its director, Edward C. Pickering, the Observatory employed many men to build equipment and take photographs of the night sky, and several women to examine photographic plates taken from telescopes throughout the world, measure their brightness, and perform computations based on their assessment of the plates. These women had little chance to do theoretical work at the Observatory, but several of them, including Williamina Fleming, Henrietta Swan Leavitt, Antonia Maury, and Annie Jump Cannon, nevertheless left a lasting legacy.

Henrietta Swan Leavitt, who had originally joined the Observatory as an unpaid volunteer in 1894,

eventually became the head of the photographic photometry department. This mainly involved measuring the brightness of stars, but a specific aspect of Leavitt's work was to identify stars that fluctuate in brightness—known as variable stars. To do this, she would do a comparison of photographic plates of the same part of the sky, made on different dates. Occasionally she would find a star that was brighter on different dates, indicating that it was a variable.

Cluster variables

A specific task that Leavitt took on was to examine some of the photographic plates of stars in the Small Magellanic Cloud (SMC) and the Large Magellanic Cloud (LMC).

One of the most striking accomplishments of Miss Leavitt was the discovery of 1,777 variable stars in the Magellanic Clouds.
Solon I. Bailey
Colleague of Leavitt

The **period** of the fluctuation in brightness of a **Cepheid variable** is closely related to its **intrinsic brightness**.

Measuring its period gives a value for its **intrinsic brightness**.

Comparing its **intrinsic brightness** to its **apparent brightness** from Earth gives a value for its distance from Earth.

Cepheid variables can be used as **"standard candles"** to **measure distances** in the universe.

later. These are stars that regularly vary in brightness with a period (cycle length) that could be anything from one to more than 120 days. Cepheid variables are reasonably easy to recognize because they are among the brightest variable stars, and they have a characteristic light curve, showing fairly rapid increases

in brightness followed by a slower tailing off. Today, they are known to be giant yellow stars that "pulsate"—varying in diameter as well as brightness over their cycles—and are very rare. As a class of stars, they also have an exceptionally high average brightness, which means they stand out even in other

galaxies. In examining her records of Cepheid variables in either the LMC or SMC, Leavitt noticed something that seemed significant. Cepheids with longer periods seemed to be brighter on average than those with shorter periods. In other words, there was a relationship between the rate at which Cepheids "blinked" and their brightness. Furthermore, Leavitt correctly inferred that, since the Cepheids she was comparing were all in the same distant nebula (either the LMC or the SMC), they were all at much the same distance from Earth. It followed that any difference in their brightness as viewed from Earth (their apparent magnitude) was directly related to differences in their true or intrinsic brightness (their absolute magnitude). This meant there was a definite relationship between the periods of Cepheid variables and their average intrinsic brightness or their optical luminosity (the rate at which they emit light energy).

Leavitt published her initial findings in a paper that first appeared in the *Annals of the*

A straight line can readily be drawn among each of the two series of points corresponding to maxima and minima, thus showing that there is a simple relation between the brightness of the variables and their periods.
Henrietta Swan Leavitt

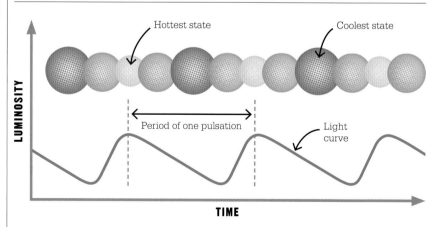

A Cepheid variable belongs to a class of star called a pulsating variable. These stars expand and contract over a regular cycle, at the same time regularly varying in brightness. They are hottest and brightest shortly after reaching their most contracted phase. The graph of the star's luminosity (light output) against time is called its light curve.

Brightness and magnitudes of stars

Apparent magnitude is the brightness of a star as viewed from Earth.

Absolute visual magnitude is the brightness of a star as viewed from a set distance and indicates the true or intrinsic brightness of a star.

Optical luminosity is the rate at which a star emits light energy from its surface and is closely related to absolute visual magnitude.

Astronomical Observatory of Harvard College in 1908. Then, in 1912, after further study, which included plotting graphs of the periods of Cepheid variables in the SMC against values for their minimum and maximum brightness, she confirmed her discovery in more detail. It became known as the "period–luminosity" relationship. Formally, it stated that the logarithm of the period of a Cepheid variable is linearly (i.e., directly) related to the star's average measured brightness.

Building on Leavitt's work

Although it is possible that Leavitt did not realize the full implications right away, she had discovered an extremely valuable tool for measuring distances in the universe, far beyond the limitations of parallax measurements. Cepheid variables were to become the first "standard candles"—a class of celestial objects that have a known luminosity, allowing them to be used as tools to measure vast distances in space.

One of the first people to appreciate the significance of Leavitt's discovery was Danish astronomer Ejnar Hertzsprung. Due to the period–luminosity relationship discovered by Leavitt, Hertzsprung realized that by measuring the period of any Cepheid variable it should be possible to determine its luminosity and intrinsic brightness. Then by comparing its intrinsic brightness to its apparent magnitude (measured average brightness from Earth), it should be possible to calculate the distance to the Cepheid variable. In this way, it should also be possible to determine the distance to any object that contained one or more Cepheid variable star.

However, there was still a problem to be solved: although Leavitt had established the important period–luminosity relationship, initially all this promised was a system for measuring the distance to remote objects relative to the distance to the SMC. The reason for this is that Leavitt had no accurate information about the distance to the SMC, nor indeed any accurate data about the intrinsic brightness of any Cepheid variable.

Calibrating the variables

To turn Leavitt's finding into a system that could be used to determine absolute distances, not just relative distances, it needed calibrating in some way. In order to do this, it would be necessary to measure accurately the distances to and intrinsic brightness of a few Cepheid variables. Hertzsprung therefore set about determining the distances to a handful of Cepheids in the Milky Way galaxy, using an alternative complex method called statistical parallax, which involves calculating the average movement of a set of stars assumed to be at a similar distance from the sun.

Having obtained the stars' distances, it was a straightforward step to figure out the intrinsic brightness of each of the nearby Cepheids. Hertzsprung used these values to calibrate a scale, which allowed him to calculate the distance to the SMC and the intrinsic brightness of each of Leavitt's Cepheids in the SMC. »

I should be willing to pay thirty cents an hour in view of the quality of your work, although our usual price, in such cases, is twenty five cents an hour.
Edward C. Pickering

Leavitt left behind a legacy of a great astronomical discovery.
Solon I. Bailey

Following these calibrations, Hertzsprung was able to establish a system for determining the distance to any Cepheid variable from just two items of data—its period and its apparent magnitude.

Further applications

It was not long before Leavitt's findings, tuned by the work of Hertzsprung, led to further important results in terms of helping to understand the scale of the universe. From 1914 to 1918, the American astronomer Harlow Shapley (who was also the first person to show that Cepheid variables are pulsating stars) was one of the first to use the newly developed concept that the distances of variable stars could be found from knowing their periods and apparent brightness. Shapley found that objects called globular star clusters—all part of the Milky Way—were distributed roughly in a sphere whose center lay in the direction of the constellation of Sagittarius. He was able to conclude from this that the center of the Milky Way galaxy is at a considerable distance (tens of thousands of light-years) in the direction of Sagittarius and that the sun is not, as had previously been supposed, at the center of the galaxy. Shapley's work, which led

to the first realistic estimate of the true size of the Milky Way, was an important milestone in galactic astronomy.

Right up to the 1920s, many scientists (including Harlow Shapley) maintained that the Milky Way galaxy was the whole universe. Although there were those that believed otherwise, neither side could conclusively prove their argument one way or another. In 1923, however, the American astronomer Edwin Hubble, using the latest in telescopic technology, found a Cepheid variable in the

The star RS Puppis is one of the brightest Cepheid variables in the Milky Way. It is about 6,500 light-years from Earth and has a cycle of variability lasting 41.4 days.

Andromeda nebula, allowing its distance to be measured. This led directly to the confirmation that the Andromeda nebula is a separate large galaxy (and is now called the Andromeda galaxy) outside the Milky Way. Later, Cepheids were similarly used to show that the Milky Way is just one of a vast number of galaxies in the universe. The study of Cepheids was also employed by Hubble in his discovery of the relationship between the distance and recessional velocity of galaxies, leading to confirmation that the universe is expanding.

Revising the scale

In the 1940s, the German astronomer Walter Baade was working at the Mount Wilson Observatory in California. Baade made observations of the stars at the center of the Andromeda galaxy during the enhanced viewing conditions afforded by the wartime blackout. He distinguished two separate populations, or groups, of Cepheid variables that have different period–luminosity relationships. This led to a dramatic revision in the extragalactic distance scale— for example, the Andromeda galaxy was found to be double the distance from the Milky Way that Hubble had calculated. Baade announced his findings at the International Astronomical Union in 1952. The two groups of Cepheids became known as

classical and Type II Cepheids, and started to be used for different purposes in distance measuring.

Today, classical Cepheids are used to measure the distance of galaxies out to about 100 million light-years—well beyond the local group of galaxies. Classical Cepheids have also been used to clarify many characteristics of the Milky Way galaxy, such as its local spiral structure and the sun's distance from the plane of the galaxy. Type II Cepheids have been used to measure distances to the galactic center and globular clusters.

The measurement of distances to Cepheid variables for more accurate calibration of period–luminosity relationships is still considered extremely important, and it was one of the primary missions of the Hubble Space Telescope project when it was launched in 1990. A better calibration is crucial, among other things, to calculate the age of the universe. Leavitt's findings from over a century ago are still having significant repercussions in terms of truly understanding the scale of the cosmos. ∎

Hubble's underwhelming acknowledgment of Leavitt is an example of the ongoing denial and lack of professional and public recognition that she suffers from, despite her landmark discovery.
Pangratios Papacosta
Science historian

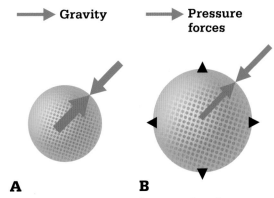

Gravity **Pressure forces**

A simplified version of the mechanisms that cause Cepheid variables to fluctuate in size is shown here. The pressure forces inside a star include gas pressure, maintained by heat output from the star's core, and radiation pressure. Another mechanism that may be involved is a cyclical change in the opacity (resistance to the transmission of radiation) in gas within the star's outer layers.

A
Pressure forces exceed gravity. The star begins to expand.

B
Pressure and gravity are now in balance but inertia causes the star to expand further.

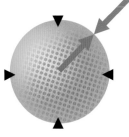

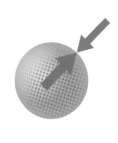

C
With continued expansion, the pressure forces decrease, as does the gravity, though to a lesser extent. Eventually, gravity exceeds the pressure from the pressure forces, and the star stops expanding and begins to shrink.

D
Pressure and gravity are in balance again but inertia causes the star to shrink further.

E
As the star contracts, the pressure forces increase until they exceed the inward-pulling gravity. The star stops shrinking and begins to expand again, starting a new pulsation cycle.

STARS ARE GIANTS OR DWARFS
REFINING STAR CLASSIFICATION

IN CONTEXT

KEY ASTRONOMER
Henry Norris Russell
(1877–1957)

BEFORE
1901 Annie Jump Cannon, working at the Harvard College Observatory, introduces the star spectral classes O, B, A, F, G, K, and M (based on surface temperature of stars).

1905 Based on analyses of star spectra, Ejnar Hertzsprung states that there are two fundamentally different kinds of star within some spectral classes, one of which is far more luminous.

AFTER
1914 Walter Adams discovers white dwarf stars—white-hot but relatively faint.

1933 Danish astronomer Bengt Strömgren introduces the term "Hertzsprung–Russell diagram" to denote a plot of the absolute magnitudes of stars against spectral class.

Among **most stars**, blue stars are brighter than yellow stars, which are brighter than orange/red stars. These are **dwarf stars**.

However, a few **exceptionally bright stars** do not follow this rule. These **are giant stars**.

Stars fall into two **distinct groups** when plotted on a diagram showing **luminosity and temperature**.

Stars are either giants or dwarfs.

A round 1912, American Henry Russell began comparing stars' absolute magnitude (or true brightness) and their color, or spectral class. Before the early 20th century, no one had figured out how different star types might be related in some overall scheme, but it had long been recognized that they differ in certain properties, such as color. While some stars shine with a pure white light, others have distinct colors: many have reddish or bluish hues, while the sun is yellow. In 1900, German physicist Max Planck worked out the precise mathematics to describe how the mix of wavelengths of light given off by hot objects, and hence their color, varies according to their temperature. Thus, star colors are related to surface temperature—red stars have the coolest surfaces, and blue stars the hottest. By around 1910, stars were considered to fit into spectral classes related to their colors and surface temperatures.

The other obvious way in which stars differ is in their brightness. Since ancient times, stars have been classified into brightness classes. This developed into the apparent magnitude scale, which rated stars according to how bright

See also: Analyzing starlight 113 ▪ The characteristics of stars 122–27 ▪ Analyzing absorption lines 128 ▪ Measuring the universe 130–37 ▪ Discovering white dwarfs 141 ▪ Stellar composition 162–63

they look from Earth. However, it was realized that, in order to know a star's absolute brightness, it would be necessary to correct for its distance from Earth: the farther away a star is, the dimmer it will appear. From the mid-19th century, reasonably precise distances to some stars began to be calculated, and the absolute brightness of these stars could be established.

Russell's discovery

Among the majority of stars, Russell found a definite relationship—hot blue-white stars (spectral classes B and A) tend to have higher absolute magnitudes than cooler white and yellow stars (classes F and G), while white and yellow stars have higher absolute magnitudes than orange and red stars (classes K and M). However, some exceptionally bright red, orange, and yellow stars departed from this rule. These were the "giant" stars.

Russell plotted the absolute magnitudes of stars against their spectral classes on a scatter diagram, which he published in

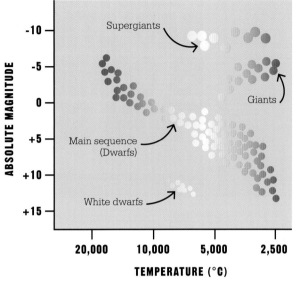

The Hertzsprung–Russell diagram shows the distribution of stars by absolute magnitude and spectral class. The diagram formed the basis for developing theories about how stars evolve. (In the absolute magnitude scale, the lower the number, the higher the magnitude.)

1913. However, unknown to him, Danish chemist and astronomer Ejnar Hertzsprung had performed a similar exercise a couple of years earlier, and the diagram is now known as the Hertzsprung–Russell diagram. The diagram shows stars divided into a group of bright giant stars and a much larger group of ordinary stars running diagonally.

Russell called these ordinary stars "dwarfs"; Hertzsprung referred to them as "main sequence." The newly discovered hot but faint white dwarfs were later added to the diagram, forming a third group. Today, it is known that most stars spend most of their lives on the main sequence, some later evolving into giants or supergiants. ▪

Henry Norris Russell

Henry Norris Russell was born in Oyster Bay, Long Island, in 1877. At age 5, his parents encouraged him to observe a transit of Venus across the sun's disc, which inspired an interest in astronomy. He was awarded a doctoral degree by the astronomy department of Princeton University for an analysis of the way that Mars perturbs the orbit of the asteroid Eros. From 1903 to 1905, he worked at the Cambridge Observatory, England, on star photography, binary stars, and stellar parallax. In 1905, he was appointed as an instructor in astronomy at Princeton University, and in 1911, he became professor of astronomy there. He was also director of Princeton University Observatory from 1912 to 1947.

Key works

1927 *Astronomy: A Revision of Young's Manual of Astronomy; Volume 1: The Solar System; Volume 2: Astrophysics and Stellar Astronomy*
1929 *On the Composition of the sun's Atmosphere*

PENETRATING RADIATION IS COMING FROM SPACE

COSMIC RAYS

IN CONTEXT

KEY ASTRONOMER
Victor Hess (1883–1964)

BEFORE
1896 French physicist Henri Becquerel detects radioactivity.

1909 German scientist Theodor Wulf measures air ionization near the top of the Eiffel Tower. Levels are higher than expected.

AFTER
1920s American physicist Robert Millikan coins the term "cosmic ray."

1932 American physicist Carl Anderson discovers the positron (antiparticle of the electron) in cosmic rays.

1934 Walter Baade and Fritz Zwicky propose the idea that cosmic rays come from supernova explosions.

2013 Data from the Fermi Space Telescope suggest that some cosmic rays come from supernova explosions.

Austrian-born physicist Victor Hess made a series of dangerous high-altitude ascents over eastern Germany in a hydrogen balloon in the years 1911 and 1912. His goal was to measure air ionization at a height of 3 miles (5 km).

Ionization is the process by which electrons are stripped from atoms. In the early years of the 20th century, scientists were puzzled by the levels of ionization in Earth's atmosphere. After the discovery of radioactivity in 1896, it was suggested that ionization was caused by radiation emitted by substances in the ground, meaning that air ionization should decrease with altitude. However, measurements made at the top of the Eiffel Tower in Paris in 1909 indicated a higher level of ionization than expected.

Hess's results showed that ionization decreased up to an altitude of about half a mile (1 km), and then increased above that point. He concluded that powerful radiation from space was penetrating and ionizing the atmosphere. This radiation later became known as cosmic rays.

In 1950, scientists found that cosmic rays consisted of charged particles, some possessing very high energies. They smash into atoms in the atmosphere, creating new subatomic particles that may themselves create collisions, which in turn cause a cascade of collisions called a cosmic ray shower. ∎

In 1951, the Crab nebula was found to be a major source of cosmic rays. Since then, supernovae and quasars have also been identified as sources.

See also: Supernovae 180–81

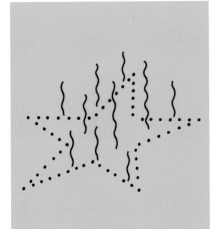

A WHITE HOT STAR THAT IS TOO FAINT

DISCOVERING WHITE DWARFS

IN CONTEXT

KEY ASTRONOMER
Walter Adams (1876–1956)

BEFORE
1783 William Herschel discovers 40 Eridani B and C.

1910 Williamina Fleming answers an inquiry from Henry Norris Russell about the spectrum of 40 Eridani B, confirming that it is a Type A star.

AFTER
1926 British astronomer Ralph Fowler applies new ideas in quantum physics to explain the nature of the extremely dense material in white dwarfs.

1931 Subrahmanyan Chandrasekhar calculates that white dwarfs cannot be more massive than 1.4 times the mass of the sun.

1934 Walter Baade and Fritz Zwicky suggest that stars too massive to become white dwarfs form neutron stars.

In the first decade of the 20th century, American astronomer Walter Adams developed a method for calculating the absolute magnitude of stars from the relative intensities of particular wavelengths in their spectra. One of the original team members at the Mount Wilson Observatory in California, Adams used his method to investigate the triple-star system 40 Eridani, which contained a mysterious star that seemed very dim but also very hot.

White dwarf

The brightest of the three stars, 40 Eridani A, was being orbited by a much dimmer binary pair, 40 Eridani B and C. Stars as faint as 40 Eridani B and C were expected to be of spectral class M, meaning that their starlight is red, indicating a relative coolness. 40 Eridani C fitted this profile, but 40 Eridani B was one of the whitest and hottest types of star. When Adams published the data in 1914, astronomers were presented with a puzzle: a star that hot had to be getting its energy from somewhere.

The answer could only be that, although it was small (about the size of Earth), its density must be immense—about 25,000 times that of the sun. 40 Eridani B was the first white dwarf star to be discovered. White dwarfs were later shown to be the hot stellar cores left behind when main sequence stars run out of fuel for nuclear fusion. ∎

Composed of material 3,000 times denser than anything you have ever come across, a ton of [this] material would be a little nugget that you could put in a matchbox.
Arthur Eddington
describing white dwarfs

See also: Observing Uranus 84–85 ▪ Refining star classification 138–39 ▪ The life cycles of stars 178 ▪ Energy generation 182–83

ATOMS, AND GA

1915—1950

STARS, AXIES

Albert Einstein publishes his **general theory of relativity**, which explains gravity as a warping of spacetime.

1916

Observing a solar eclipse, **Arthur Eddington** shows that **light from stars is bent** by the sun's gravity, just as relativity predicts.

1919

Edwin Hubble finds a relation between the redshift and distance of nebulae, showing that **spiral nebulae are galaxies**.

1924

1917

Vesto Slipher shows that many nebulae show large **redshift**, meaning that they are moving away from us rapidly.

1920

At the Smithsonian museum, a **"Great Debate"** takes place over whether or not **spiral nebulae are galaxies**.

1926

Austrian physicist **Erwin Schrödinger** formalizes the equation describing **quantum mechanics**, which describes strange behavior at the quantum level.

Despite the vast difference in scale, atoms, stars, and galaxies share a property in common: each in its own size domain is a fundamental construction unit of the universe. Galaxies define the distribution of matter in the universe on the grandest scale; stars are a defining constituent of those galaxies (although galaxies may harbor quantities of gas, dust, and mysterious dark matter as well); atoms are the units of matter that make up the hot gas of stars (with some simple molecules in cooler stars). If we think of galaxies as cities, stars are like individual buildings within the city, and atoms are the bricks.

In a mere 30-year period in the first half of the 20th century, astronomy took huge leaps in understanding how the hierarchy of matter in the universe is organized. Underpinning these developments was Einstein's general theory of relativity, in which the concepts of mass and energy are inseparable in a unified fabric of space and time.

Looking inside a star
Between 1916 and 1925, Briton Arthur Eddington worked on the physical nature of ordinary stars such as the sun. He pieced together a detailed physical description of a sphere of hot gas, in which energy makes its way from a central source to the surface, from where it then radiates into space. Eddington also did much to convince astronomers that stars are fueled by subatomic processes—what we would now call nuclear energy.

In 1919, the New Zealand physicist Ernest Rutherford was able to transmute atoms of nitrogen into oxygen by firing particles at them from a radioactive element. There was now ample evidence that nuclear processes could produce new elements and release unimaginable quantities of energy. For any remaining doubters, Eddington reflected on the experiments conducted at Cambridge University by pointing out that "what is possible in the Cavendish Laboratory may not be too difficult in the sun."

When British astronomer Cecilia Payne-Gaposchkin, working in the US, concluded in 1925 that stars are overwhelmingly made of hydrogen atoms, astronomers at last had a real grasp on the true nature of "ordinary" stars.

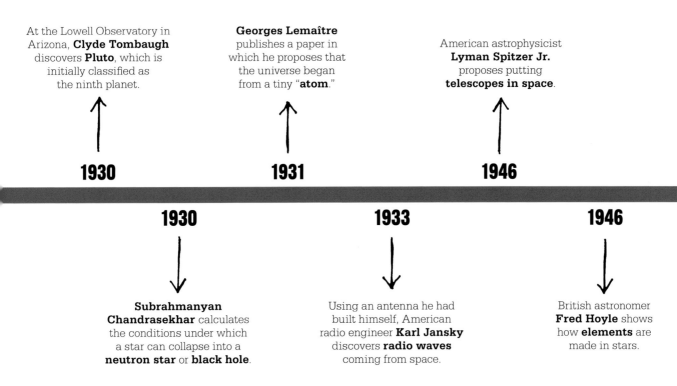

At the Lowell Observatory in Arizona, **Clyde Tombaugh** discovers **Pluto**, which is initially classified as the ninth planet.

1930

Georges Lemaître publishes a paper in which he proposes that the universe began from a tiny "**atom**."

1931

American astrophysicist **Lyman Spitzer Jr.** proposes putting **telescopes in space**.

1946

1930

Subrahmanyan Chandrasekhar calculates the conditions under which a star can collapse into a **neutron star** or **black hole**.

1933

Using an antenna he had built himself, American radio engineer **Karl Jansky** discovers **radio waves** coming from space.

1946

British astronomer **Fred Hoyle** shows how **elements** are made in stars.

However, not all stars are quite so ordinary. White dwarfs, for example, are clearly extraordinarily dense. In the 1930s, the tools of the new quantum physics were used to explain how a star could become so compacted and predicted even more exotic types of collapsed star. It turned out that 1.46 solar masses would be the upper limit to make a white dwarf, but there was nothing to stop more massive stars from collapsing into a much denser neutron star or a black hole.

Black holes may be real

Walter Baade and Fritz Zwicky speculated that the central remnant of a supernova explosion would be a neutron star, and with the work of Indian Subrahmanyan Chandrasekhar and others, the theoretical concept of black

holes was born, although many astronomers found it hard to believe they could really exist. In any event, it was four decades before the first neutron stars and candidate black holes were identified.

The universe of galaxies

Meanwhile, the whole concept of the nature of the universe was changing rapidly. In 1917, American Vesto Slipher recognized that many so-called "nebulae" were galaxies, akin to our own Milky Way, and in rapid motion. Some 10 years later, Belgian priest Georges Lemaître realized that an expanding universe was consistent with Einstein's theory of relativity. American Edwin Hubble discovered that the more distant a galaxy, the faster it is receding from us, and Lemaître suggested that the universe began by exploding from a

tiny "primeval atom" like a firework. In just a handful of years, astronomers had learned that the universe was far larger and more complex than they had ever imagined. ■

We used to think that if we knew one, we knew two, because one and one are two. We are finding that we must learn a great deal more about "and."
Arthur Eddington

TIME AND SPACE AND GRAVITATION HAVE NO SEPARATE EXISTENCE FROM MATTER

THE THEORY OF RELATIVITY

IN CONTEXT

KEY ASTRONOMER
Albert Einstein (1879–1955)

BEFORE
1676 Ole Rømer shows that light speed is not infinite.

1687 Isaac Newton publishes his laws of motion and universal law of gravitation.

1865 James Clerk Maxwell shows that light is a wave moving though an electromagnetic field at a constant speed.

AFTER
1916 Karl Schwarzschild uses Einstein's equations to show how much matter warps space.

1919 Arthur Eddington provides evidence for the warping of spacetime.

1927 Georges Lemaître shows that a relativistic universe can be dynamic and constantly changing, and proposes the Big Bang theory.

The **speed of light** is **always constant** even when observers are moving.

↓

This must mean **that moving** through space makes the flow of **time slower**.

↓

The **slowing of time** makes an object's **mass increase**.

A person undergoing **acceleration** cannot tell if this is due to **gravity** or **another force**. Their body could be thought of as moving, or the universe around it could be thought of as changing.

↓

Mass exists not just in space but in **spacetime**. Mass itself distorts spacetime.

↓

Gravity is best described as the result of **spacetime** being **warped by mass**.

↓ ↓

Time and space and gravitation have no separate existence from matter.

Albert Einstein's general theory of relativity has been called the greatest act of thought about nature ever to take place in a person's head. It explains gravity, motion, matter, energy, space and time, the formation of black holes, the Big Bang, and possibly dark energy. Einstein developed the theory over more than a decade at the start of the 20th century. It went on to inspire Georges Lemaître, Stephen Hawking, and the LIGO team, which searched for the gravitational waves predicted by the theory.

The theory of relativity arose from a contradiction between the laws of motion described by Isaac Newton and the laws of electromagnetism defined by Scottish physicist James Clerk Maxwell. Newton described nature in terms of matter in motion governed by forces that act between objects. Maxwell's theories concerned the behavior of electric and magnetic fields. Light, he said, was an oscillation through these fields, and he predicted that the speed of light was always constant, regardless of how fast the source was moving.

Measuring the speed of light is not an easy thing to do. Danish astronomer Ole Rømer tried in 1676 by measuring the time delay in the light arriving from Jupiter's moons. His answer was 25 percent too slow, but he did show that light's speed was finite. By the 1850s, more accurate measurements had been made. However, in a Newtonian universe, there must also be changes in the speed of light to account for the relative motion of its source and observer. Try as researchers might, no such differences could be measured.

See also: Gravitational theory 66–73 ▪ Curves in spacetime 154–55 ▪ The birth of the universe ▪ Dark energy 298–303 ▪ Gravitational waves 328–31

At the end of the 19th century, many believed that physicists had fully figured out the laws of the universe. All that was now needed were more precise measurements. However, even as a child, Einstein was not convinced that physics had been solved. At the age of 16, he asked himself a question: "What would I see if I were sitting on a beam of light?" In the Newtonian context, young Albert would be traveling at the speed of light. Light coming from in front would reach his eyes at twice the speed of light. When looking back, he would see nothing at all. Even though light from behind was traveling at the speed of light, it could never catch up.

Annus mirabilis

Einstein's first job was working as a patent clerk in Bern, Switzerland. It afforded him a lot of spare time to devote to private study. The fruit of this solitary work was the Annus Mirabilis (miracle year) of 1905, when he presented four papers. These included two linked discoveries: special relativity and the equivalence of mass and energy, summed up by the equation $E=mc^2$ (p.150).

Special relativity

Einstein used thought experiments to develop his ideas, the most significant of which involved two men—one on a speeding train and the other standing on the platform. In one version (below), inside the train, Bob shines a flashlight at a mirror directly above him on the ceiling. He measures the time the light takes to travel to the mirror and back. At the same time, the train is passing the platform at close to the speed of light. From the platform, the stationary observer Pat sees the light beam shine to the mirror and back, but in the time it takes for the beam to travel, the train has moved, meaning that, rather than

> If you can't explain it to a six year old, you don't understand it yourself.
> **Albert Einstein**

traveling straight up and down, the beam travels diagonally. To Pat on the platform the light beam has traveled farther, so, since light always travels at the same speed, more time must have passed.

Einstein's explanation for this took an enormous leap of imagination, which became the basis of special relativity. Speed is a measure of units **»**

Inside the speeding train, Bob shines a light beam directly up and down. Bob measures the time it takes for the light to be reflected back to him as the distance straight up and down divided by c (the speed of light).

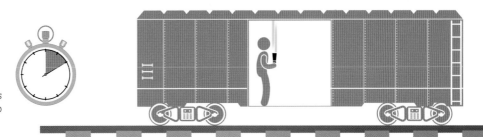

On the platform, Pat observes the beam traveling diagonally. It is still traveling at the same speed c, so more time must have passed than for Bob as the light has traveled a longer distance.

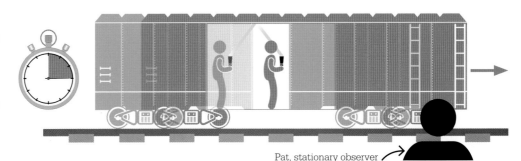

Pat, stationary observer

As an object's velocity (*v*) approaches light speed (*c*), the object becomes increasingly squashed in the direction of travel when viewed by a stationary observer. This is not merely an illusion. In the observer's frame of reference, the object's shape really does change.

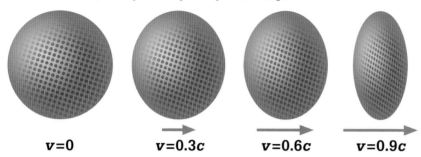

$v=0$ $v=0.3c$ $v=0.6c$ $v=0.9c$

of distance per units of time. Therefore, the constancy of the speed of light must be due to an inconstancy in the flow of time. Objects observed to be traveling faster through space are moving more slowly through time. Clocks on the station and on the train are ticking at different rates, depending on the frame of reference from which they are observed. On the moving train, Bob sees his clock ticking away as normal, but to the observer Pat on the platform, the train's clock is moving very slowly.

The passenger on the speeding train will not notice any slowing of time. The mechanisms by which time is measured—such as the swing of a pendulum, the vibration of a quartz crystal, or the behavior of an atom—are physical phenomena obeying universal laws. According to special relativity, laws remain unchanged within the reference frame—the moving train, or any other set of objects moving together.

Energy is mass

The impact of this dilation of time has far-reaching effects, which Einstein gradually pieced together into a single general theory of relativity in 1915. One early breakthrough was the discovery of $E=mc^2$, which states that E (energy) is equal to mass (m) multiplied by the square of the speed of light (c). c^2 is a very large number—about 90 million billion—and so a small amount of mass contains a huge quantity of energy. This is evident in a nuclear explosion when mass is converted to free energy.

Returning to the train thought experiment, the two observers now throw tennis balls at each other. The balls collide and bounce back to each person (both Pat and Bob have very good aim). If both observers were in the same reference frame, the described motion of the balls would occur because the balls had the same mass and were thrown with the same force. But in this experiment the balls are in different reference frames—one stationary, the other moving at close to the speed of light. Pat would see Bob's ball moving much more slowly than his own due to the time dilation, yet when they collide, both balls are knocked back to their owners. The only way this could work is if Bob's slow tennis ball is heavier, or contains more mass, than Pat's tennis ball.

Albert Einstein

Einstein was born in Germany but spent his formative years in Switzerland. He was an average student, and then struggled to find teaching work, ending up at the patent office in Bern. After the success of his 1905 papers, Einstein took university posts in Bern, Zurich, and then in Berlin, where he presented his general theory in 1915. With the rise of Nazism in 1933, Einstein moved to the United States, where he settled at Princeton University. There he spent the rest of his days trying to link relativity with quantum mechanics.

He failed to do so, and no one else has succeeded yet either. A leading pacifist voice for many years, in 1939 Einstein was instrumental in alerting Allies to the dangers that Germany might build a nuclear weapon. He declined to be involved in the Manhattan Project that built the first atomic bombs. An avid violinist, Einstein stated that he often thought in music.

Key work

1915 *Relativity: the Special and the General Theory*

> Each ray of light moves in the coordinate system "at rest" with a definite, constant velocity independent of whether this ray of light is emitted by a body at rest or a body in motion.
> **Albert Einstein**

Therefore, according to special relativity, when matter moves, it becomes more massive. These mass increases can be measured on the everyday, human scale, but are negligible. However, they have a marked effect when objects are moving very quickly. For example, the protons accelerated by the Large Hadron Collider (LHC) particle accelerator travel very close to the speed of light—within 99.999 percent. Additional energy does very little to this speed, and instead boosts mass. At full power, the protons in the LHC are nearly 7,500 times more massive than they were when stationary.

Speed limit

With the relationship between speed and mass, relativity highlights another basic principle: the speed of light is the upper limit of motion through space. It is impossible for an object with mass—a nuclear particle, spaceship, planet, or star—to travel at the speed of light. As it approaches light speed, its mass becomes almost infinite, time slows nearly to a stop, and it would take an infinite amount of energy to push it to light speed.

To generalize his theory, Einstein linked gravity to his ideas about energy and motion. Taking an object in space and removing all reference points, it is not possible to tell if it is moving. There is no test that can be done to prove that it is. Therefore, from the point of view of any object, or reference frame, it stays still while the rest of the universe moves around it.

Einstein's happiest thought

This is easiest to picture if everything is moving at a constant speed. According to Newton's first law of motion, an object maintains its motion unless a force acts to accelerate it (change its speed or direction). When Einstein included the effects of acceleration in his theory, it led to an insight that he called his "happiest thought": it was not possible to differentiate why an object accelerated—it could be because of gravity, or it could be another force. The effect of both was the same and could be described by the way the rest of the universe moved around the reference frame.

> The theory of relativity cannot but be regarded as a magnificent work of art.
> **Ernest Rutherford**
> *New Zealand physicist*

Einstein had described motion in terms of the links between mass, energy, and time. For a general theory, he needed to add space. It was not possible to understand the path of an object through space without considering its path through time. The result was that mass moves through spacetime, which has a four-dimensional geometry, as opposed to the usual three dimensions (up, down, and side to side) of the everyday concept of space. When an object »

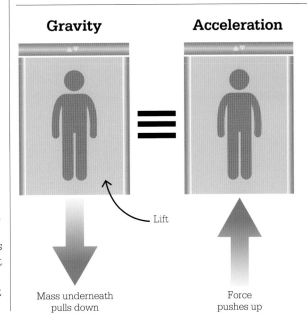

Gravity **Acceleration**

Lift

Mass underneath pulls down

Force pushes up

From inside an elevator, a person cannot tell whether they are being accelerated upward by a force pushing the elevator from below or pulled downward by the gravity of a mass underneath the elevator. Either way, they feel a sense of weight as the floor pushes against them, and objects dropped from a height accelerate down to the floor. This is Einstein's equivalence principle, which he described as his "happiest thought."

moves through spacetime, the time dimension dilates, and the space dimensions contract. From the point of view of Pat back at the station, the speeding train's length is compressed, making it look very squashed and stubby. However, it is all normal to Bob; anything he measures on board will have the same length as when the train was stationary. This is because his means of measurement, such as a ruler, has contracted along with space.

Warping spacetime

In Einstein's universe, gravity is recast not as a force but rather the effect of warps in the geometry of spacetime caused by the presence of mass. A large mass, such as a planet, bends space, and so a smaller object, such as a meteor, moving in a straight line through space nearby, will curve toward the planet. The meteor has not changed course—it is still moving along the same line in space; it is just that the planet has bent that line into a curve.

Warps in spacetime can be visualized as balls deforming a rubber sheet, making depressions or "gravity wells." A large "planet" ball makes a well, and a smaller "meteor" ball will roll into the

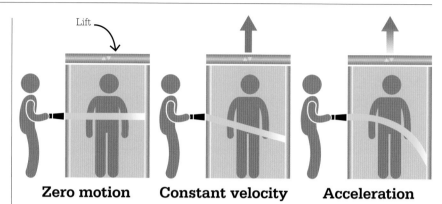

Lift

Zero motion **Constant velocity** **Acceleration**

A light beam shines into an elevator from an observer with a flashlight standing outside. The paths of the light beam are shown as they will be observed from inside the elevator. If the elevator is accelerating, the beam will curve downward. Light is similarly curved toward a source of gravity.

well. Depending on its trajectory, speed, and mass, the meteor might collide with the planet or roll back up the other side of the well and escape. If the trajectory is just right, the meteor will circle around the planet in an orbit.

The warps created by matter also bend time. Two distant objects—for this explanation, a red star and a blue star—are not moving in relation to one another. They are in different points of space, but at the same point in time, the same "now." However, if the red star moves directly away from the blue, its passage through time slows compared to the blue star's. That means the red star shares a "now" with the blue star in the past. If the red star travels directly toward the blue one, its "now" is angled toward the blue star's future. Consequently, events that are observed simultaneously from one reference frame may appear to occur at different times in another.

Relativity solved the puzzle of perturbations in the orbit of the planet Mercury (pictured) that could not be explained by Newtonian physics, which had first been noticed in 1859.

Proof of relativity

Einstein's physics were initially met with bafflement from most of the scientific community. However, in 1919, the English astronomer Arthur Eddington demonstrated that this new way of describing the universe was indeed accurate. He traveled to the Atlantic island of Principe to observe a full solar eclipse and specifically to look at the background of stars near to the sun. Light from stars travels to Earth along the most direct route, known as the geodesic. In Euclidean geometry (the geometry of Newtonian physics), that is a straight line, but in the geometry

Everything must be made as simple as possible. But not simpler.
Albert Einstein

of spacetime, a geodesic can be curved. So, starlight shining very close to the edge of the sun passes into the warp created by the star's mass and follows a bending path. Eddington photographed the stars revealed by the absences of the solar glare. These images showed that the apparent position of the stars had indeed been shifted due to the warping of space, an effect now known as gravitational lensing. Einstein was proved right.

Einstein's general theory of relativity allows astronomers to make sense of what they observe, everywhere from the very edge of the visible universe to the event horizon of a black hole. Today, the time dilations of relativity are taken into account in GPS

Time is an illusion.
Albert Einstein

technology, while the wavelike contractions of space predicted by relativity have recently been discovered in the LIGO experiment. Other ideas from relativity are also being used in the search for possible answers to the mystery of dark energy. ∎

Real   Observed

Mass creates a gravity well that causes an effect called gravitational lensing, first observed in 1919 by Arthur Eddington. The observed position of a star is changed by the effect of the sun's gravity, which causes light from the star to travel past the sun along a curved path.

The twin paradox

A result known as the "twin paradox" is illustrated using a pair of newborn twins. One stays on Earth, while another is taken on a rocket on a journey to a star 4 light-years away. The rocket travels at an average velocity of 0.8c, meaning that it returns from its 8-light-year journey on the 10th birthday of the twin who stayed on Earth. However, to the clock on the rocket, it is only the other twin's 6th birthday. The clock has been in a moving time frame, so has been ticking more slowly.

Relativity insists that the twin on the rocket is also entitled to consider herself at rest, which seems to lead to a paradox—from her point of view, the twin on Earth had been the one moving. The paradox is resolved by the fact that only the twin in the rocket has undergone acceleration, with its consequent time dilation, both on the way out and to change direction and come back. The twin on Earth has remained in one frame of reference, while the twin on the rocket has been in two— one on the way out and another on the way back. Thus, the twins' situations are not symmetrical, and the twin who stayed at home really is now four years older than her sister.

The twin paradox has been a popular theme in science fiction. In the film *The Planet of the Apes*, astronauts return to Earth to find that thousands of years have elapsed, and the planet is now ruled by apes. In the film *Interstellar*, physics consultants were employed to ensure that the time elapsed for each character was correct according to relativity.

AN EXACT SOLUTION TO RELATIVITY PREDICTS BLACK HOLES

CURVES IN SPACETIME

IN CONTEXT

KEY ASTRONOMER
Karl Schwarzschild
(1873–1916)

BEFORE
1799 Pierre-Simon Laplace develops a theory about black holes, which he calls "*corps obscures*," or "dark bodies."

1915 Albert Einstein's general theory of relativity shows that the force of gravity is caused by a warping of space and time.

AFTER
1931 Subrahmanyan Chandrasekhar calculates the mass of stellar cores that become neutron stars and black holes.

1979 Stephen Hawking proposes that black holes do actually emit radiation as a result of quantum fluctuations.

1998 Andrea Ghez shows that a supermassive black hole sits at the center of the Milky Way.

The **gravitational field** of a mass is **a warping of spacetime**.

This warping can be described mathematically using the **Schwarzschild solution**.

The Schwarzschild solution is an exact solution to relativity that predicts black holes.

Black holes are surrounded by an **event horizon**, a boundary beyond which nothing can be observed.

In 1916, German mathematician Karl Schwarzschild managed something that even Albert Einstein had failed to do—he provided a solution to the field equations of general relativity that could yield precise answers. The Einstein field equations are a complex set of formulae that link space and time (or spacetime) with the action of gravity. Schwarzschild's achievement, known as the Schwarzschild solution, was to solve the equations to show exactly how spacetime curved in the presence of mass. This solution showed how the gravity of objects like the sun and Earth was warping spacetime in accordance with the theories of relativity. A generation later, Schwarzschild's mathematics were used to throw light on the darkest of all objects, the black hole.

No escape

In the early days of relativity, black holes were purely theoretical objects, although they had been

See also: Gravitational disturbances 92–93 ▪ The theory of relativity 146–53 ▪
The life cycles of stars 178 ▪ Hawking radiation 255 ▪ The heart of the Milky Way 297

predicted a century before. The French astronomer Pierre-Simon Laplace had theorized *corps obscures*, objects so dense that the velocity required to escape their gravity exceeded the speed of light. The modern definition of black holes is similar: objects in space with such enormous gravity nothing can escape them, not even light.

Event horizon

The Schwarzschild solution can be used to calculate the size of a black hole for a given mass. To create a black hole, mass must be compressed into a volume with a smaller radius than that predicted by the Schwarzschild solution. An object so dense that its radius is smaller than the Schwarzschild radius for its mass will warp spacetime to such an extent that its gravitational pull will be impossible to resist—it will create a black hole. Any mass or light that comes closer than the Schwarzschild radius is doomed to be pulled into the black hole. The points in space that surround a black hole at a distance of the Schwarzschild radius form its "event horizon," so-called because it is impossible to observe the events that take place beyond it. Nothing comes out of a black hole—no mass, no light, and no information about what is inside.

The Schwarzschild solution allows astronomers to estimate the masses of actual black holes, although it is not possible to be exact because black holes rotate and carry an electric charge, and these factors are not accounted for by the mathematics. If the sun became a black hole, its event horizon would be 2 miles (3 km) from the center. A black hole with Earth's mass would have a ⅓-in (9-mm) radius. However, it is not possible to make black holes from bodies this small; it is thought that black holes form from collapsed stars that are at least three solar masses. ▪

Karl Schwarzschild

Karl Schwarzschild's prodigious mathematical abilities were obvious from an early age. By the age of 16, he had published his first scientific paper concerning the mechanics of binary orbits, and by 28, he was a professor at the University of Göttingen in Lower Saxony.

Schwarzschild made contributions to the most significant sciences of the age: radioactivity, atomic theory, and spectroscopy. In 1914, he joined up to fight in World War I, but still found time for mathematics. In late 1915, he sent Albert Einstein some early calculations, saying: "As you see, the war treated me kindly enough, in spite of the heavy gunfire, to allow me to get away from it all and take this walk in the land of your ideas." The following year, Schwarzschild presented the full solution that bears his name. He developed an autoimmune disease while serving on the Russian Front and died in May 1916.

Key work

1916 *On the Gravitational Field of a Mass Point after Einstein's Theory*

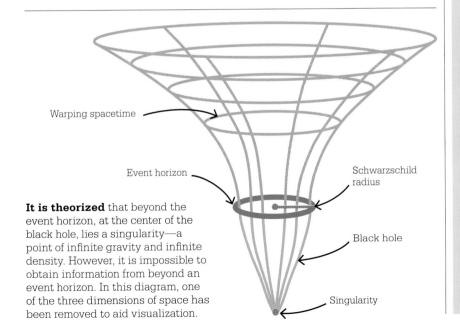

It is theorized that beyond the event horizon, at the center of the black hole, lies a singularity—a point of infinite gravity and infinite density. However, it is impossible to obtain information from beyond an event horizon. In this diagram, one of the three dimensions of space has been removed to aid visualization.

Warping spacetime

Event horizon

Schwarzschild radius

Black hole

Singularity

THE SPIRAL NEBULAE ARE STELLAR SYSTEMS

SPIRAL GALAXIES

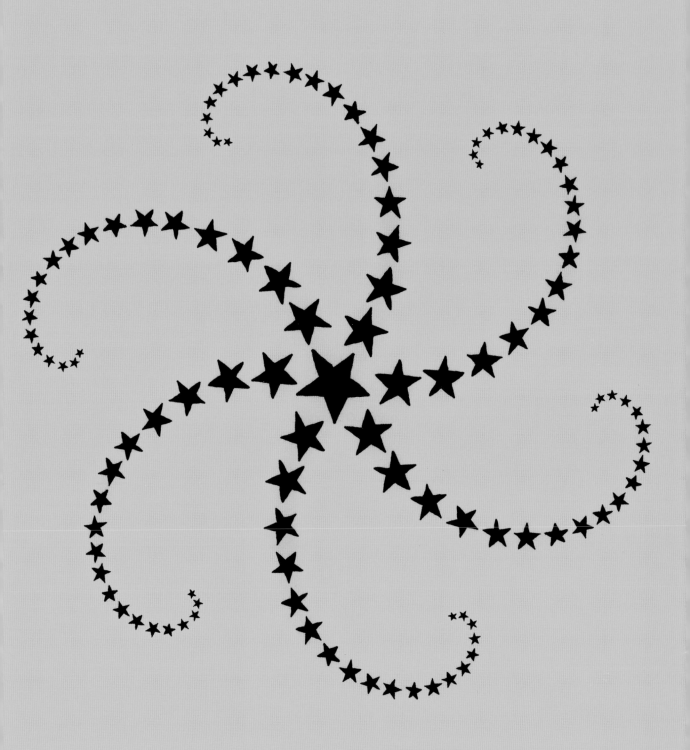

IN CONTEXT

KEY ASTRONOMER
Vesto Slipher (1875–1969)

BEFORE
1842 Austrian physicist Christian Doppler proposes the Doppler effect—a change in the perceived frequency of waves coming from an object moving relative to the viewer.

1868 William Huggins determines the velocity of a star moving away from Earth using the Doppler effect.

AFTER
1929 Edwin Hubble finds a link between the recessional velocities of spiral galaxies and their distances.

1998 Saul Perlmutter and his colleagues discover that the expansion of the universe has been accelerating for the last 5 billion years.

In the 1780s and 1790s, British astronomer William Herschel cataloged large numbers of nebulae and speculated that some of these might be comparable in size and nature to the Milky Way. In his conjectures, Herschel was following an earlier suggestion by German philosopher Immanuel Kant that nebulae might be large disks of stars—"island universes" independent of the Milky Way and separated from it by vast distances. In the 19th century, using improved telescopes, British astronomer Lord Rosse discovered that some nebulae had "arms" arranged in a spiral, while his compatriot William Huggins found that many nebulae consisted of a mass of stars. However, aside from the fact that they might contain stars, nebulae were still little understood by the turn of the 20th century, when a young scientist from Indiana named Vesto Slipher began to study them.

The Lowell Observatory

From 1901, Slipher worked at the Lowell Observatory in Flagstaff, Arizona. The observatory had been founded by American astronomer Percival Lowell in 1894. Lowell had

> It seems to me that with this discovery, the great question, if the spirals belong to the system of the Milky Way or not, is answered with great certainty: they do not.
> **Ejnar Hertzsprung**
> *in a letter to Vesto Slipher*

selected the site because its high altitude, at over 6,900 ft (2,100 m), with few clouds, and its distance from city lights, meant it guaranteed good visibility almost every night. Lowell's venture marked the first time an observatory had been built deliberately in a remote, high place for optimal observations.

Lowell initially hired him for a short-term position, but Slipher would remain for his entire career. Lowell and Slipher worked well together, with the unassuming new recruit content to leave the limelight to his flamboyant employer. Slipher was a talented mathematician and had practical mechanical skills, which he put to use installing new spectrographic equipment. He set to work developing improved techniques in spectrography—the separation of light coming from celestial objects into its constituent wavelengths, and the measurement and analysis of those wavelengths (p.113).

Slipher used the 24-in (61-cm) Alvan Clark telescope at the Lowell Observatory to observe the spiral nebulae. Today, people can use the original telescope at the observatory's visitor center.

See also: The Milky Way 88–89 ▪ Examining nebulae 104–05 ▪ Properties of nebulae 114–15 ▪ Measuring the universe 130–37 ▪ The shape of the Milky Way 164–65 ▪ The birth of the universe 168–71 ▪ Beyond the Milky Way 172–77

Measurements of the **blueshifts** and **redshifts** of **spiral-shaped nebulae** show that some are moving toward Earth while others are receding.

→

If the spiral nebulae lie **within the Milky Way Galaxy**, they are moving so fast relative to the rest of the galaxy that they **cannot remain within it for long**.

→

Spiral nebulae may be **separate galaxies** outside the Milky Way.

Studying nebulae

Slipher's initial work and research were directed at the planets, but from 1912, at Lowell's request, he began to study the mysterious spiral nebulae. Lowell had a theory that they were spirals of gas that were coalescing into new solar systems. He asked Slipher to record the spectra of the light from the outer edges of the nebulae, to determine if their chemical makeup resembled that of the solar system's gas giant planets.

The spectra of galaxies moving toward Earth exhibit "blueshifts" and those receding from Earth exhibit "redshifts" because the light waves are squashed or stretched when viewed from Earth. These are called Doppler shifts after the Austrian physicist Christian Doppler, who first explained such phenomena.

WAVELENGTH (nm)

This galaxy is not moving relative to Earth. Light waves coming from it are detected on Earth at their normal, unaffected frequency.

Emission lines in the spectra of stationary galaxies are consistent with the wavelengths of the component gases in the galaxy.

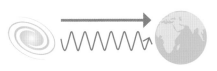

WAVELENGTH (nm)

This galaxy is moving toward Earth. Light waves coming from it are detected on Earth as slightly shortened or of a higher frequency.

Emission lines in the spectra of approaching galaxies are shifted toward the shorter, blue wavelengths: this is a "blueshift."

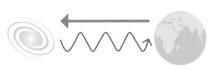

WAVELENGTH (nm)

This galaxy is receding from Earth. Light waves coming from it are detected on Earth as slightly lengthened or of a lower frequency.

Emission lines in the spectra of receding galaxies are shifted toward the longer, red wavelengths: this is a "redshift."

Making small adjustments to its mechanism, Slipher managed to increase the sensitivity of Lowell's spectrograph, a complex 450-lb (200-kg) instrument attached to the eyepiece of the Observatory's 24-inch (61-cm) refractor telescope. During the fall and winter of 1912, he obtained a series of spectrograms from the largest of the spiral nebulae, which was located in the constellation of Andromeda and known at the time as the Andromeda nebula.

The pattern of spectral lines in the nebula's spectrum (like a fingerprint of its composition) indicated a "blueshift"—they were unexpectedly displaced toward the short-wavelength/ high-frequency blue end of the spectrum by what is known as a Doppler shift (see diagram, left). That could only mean that the light waves coming from the Andromeda nebula were being shortened, or compressed, and their frequency raised, because the nebula was rushing toward Earth at a considerable speed. Slipher's calculations revealed that the nebula was approaching at 200 miles per second (300 km per second). Doppler shifts had been measured for astronomical bodies before, but shifts of this size were unprecedented. Slipher »

The galaxy NGC 4565, which Slipher established to be receding at 700 miles/s (1,100 km/s), is also known as the Needle galaxy because of its thin shape when viewed from Earth.

asserted that "we have at the present no other interpretation for it. We may conclude that the Andromeda nebula is approaching the solar system."

Discovering Doppler shifts

Over the next few years, Slipher studied 14 more spiral nebulae and found that nearly all were traveling at incredible speeds relative to Earth. Most remarkably, whereas some were moving toward Earth, most were showing redshifted spectra, where the wavelengths had stretched, meaning they were moving away from Earth. The nebula known as M104 (also called NGC 4594), for example, was flying away at an astonishing speed of nearly 600 miles per second (1,000 km per second). Another called M77, or NGC 1068, was receding at 680 miles per second (1,100 km per second). Altogether, out of the 15 galaxies observed, 11 were significantly redshifted. In 1914, Slipher presented his results to the American Astronomical Society and received a standing ovation.

By the time Slipher had presented his next paper on spiral nebulae in 1917, the ratio of redshifted to blueshifted nebulae had risen to 21:4. In this paper, Slipher noted that the average velocity at which they were approaching or receding— scientifically known as "radial velocity"—was 450 miles per second (700 km per second). This was much faster than any star had ever been measured moving relative to Earth. Slipher found it almost inconceivable that the spiral nebulae could be passing through the Milky Way at such speeds, and he began to suspect that they were not moving through the Milky Way at all, stating: "It has for a long time been suggested that the spiral nebulae are stellar systems seen at great distances ... This theory,

Vesto Slipher

Vesto Slipher was born on a farm in Mulberry, Indiana, in 1875. Soon after graduation, he started working at the Lowell Observatory in Arizona, where he would remain for more than half a century. Most of Slipher's major discoveries occurred in the earlier part of his career. He began by investigating the rotational periods of planets, finding evidence, for example, that Venus's rotation is very slow. Between 1912 and 1914, he made his most significant discovery— that some spiral nebulae are moving at high speed. In 1914, Slipher discovered the rotation of spiral galaxies, measuring spin rates of hundreds of miles per second. He also demonstrated that gas and dust exist in interstellar space. Slipher was director of the Lowell Observatory from 1926 to 1952. During this time, he supervised a search for trans-Neptunian planets, which ledin 1930 to Clyde Tombaugh's discovery of Pluto.

Key work

1915 *Spectrographic Observations of Nebulae*

it seems to me, gains favor in the present observations." Slipher was echoing Kant's suggestion that some nebulae, in particular the spiral ones, could be separate galaxies from the Milky Way.

In 1920, partly prompted by Slipher's findings, a formal debate took place in Washington, D.C., to discuss whether the spiral nebulae were separate galaxies outside the Milky Way. Now referred to as the "Great Debate," two eminent American astronomers advocated opposing positions—Harlow Shapley that the spiral nebulae were part of the Milky Way; and Heber D. Curtis that they were far beyond it. Neither astronomer changed his position as a result of the debate, but many perceptive figures were concluding by this time that the spirals had to be outside the Milky Way.

Slipher's legacy

Despite an enthusiastic response from many in the astronomical community, some still questioned Slipher's findings. For more than a decade, until others began to believe Slipher's ideas and understand the implications arising from them, he was virtually the only person investigating the Doppler shifts of spiral nebulae.

In 1924, a new paper by American astronomer Edwin Hubble put a decisive end to the debate about the nature of spiral nebulae. Hubble had observed a class of stars called Cepheid variables in some nebulae, including the Andromeda nebula. As a result of his observations, Hubble was able to announce that the Andromeda "nebula" and others

Some 4 billion years into the future, the night sky will look like this, as the Andromeda galaxy collides with the Milky Way.

In the great majority of cases the nebula is receding; the largest velocities are positive. The striking preponderance of [these positive velocities] indicates a general fleeing from us or the Milky Way.
Vesto Slipher

like it were far too distant to be part of the Milky Way and so must be galaxies outside it. Slipher's suspicions dating back to 1917 had been proved right. By the time of Hubble's paper, Slipher had measured the radial velocities of

39 spiral nebulae, the majority of which showed high velocities of recession—as much as 775 miles per second (1,125 km per second). Hubble used Slipher's measurements of redshifts in galaxies that he had proved were outside the Milky Way to find a relationship between galaxy redshifts and distances.

By the late 1920s, Hubble had used this result to confirm that the universe is expanding. Thus, Slipher's work in the years 1912–25 played a crucial role in what today is often considered the greatest astronomical discovery of the 20th century, paving the way for further investigations into the motions of galaxies and cosmological theories based on an expanding universe. As for the Andromeda galaxy, it is expected to collide with the Milky Way in about 4 billion years, and together the two are likely to form a new elliptical galaxy. ∎

STARS ARE DOMINATED BY HYDROGEN AND HELIUM
STELLAR COMPOSITION

IN CONTEXT

KEY ASTRONOMER
Cecilia Payne-Gaposchkin
(1900–1979)

BEFORE
1850s Gustav Kirchhoff shows that dark lines in the sun's spectrum are due to light absorption by elements.

1901 Annie Jump Cannon classifies stars by the strength of the dark lines in their spectra.

1920 Indian physicist Meghnad Saha demonstrates how temperature, pressure, and ionization are linked in a star.

AFTER
1928–29 Albrecht Unsöld and William McCrea independently find that hydrogen is a million times more abundant in the sun's atmosphere than any other element.

1933 Danish astrophysicist Bengt Strömgren shows that stars are mainly hydrogen all the way through, not just in their atmospheres.

In 1923, the consensus among astronomers was that the sun and other stars had a similar chemical composition to Earth. This belief was based on the analysis of dark lines (Fraunhofer lines) in star spectra, which are caused by the absorption of light by chemical elements in star atmospheres. The spectra contain strong lines for elements that are common on Earth, such as oxygen and hydrogen, and metals such as magnesium, sodium, and iron, and therefore it was assumed that Earth and stars were made of the same chemical elements, in more or less the same proportions. This established view would be overturned with the arrival that year of British graduate student Cecilia Payne at the Harvard College Observatory (HCO) in Massachusetts.

Star spectra
Payne set to work analyzing the HCO's photographic collection of star spectra. She wanted to clarify the relationship between star spectra and temperatures. Also, because the pattern of absorption lines seemed to vary between the spectra of different classes of star, she wanted to see what differences in chemical composition might exist between these classes.

Since 1901, astronomers at the HCO had classified stars into a sequence of seven main spectral types, and believed that the sequence was related to the stars' surface temperatures. In her doctoral thesis, however, Payne applied an equation formulated by Indian physicist Meghnad Saha in 1920. The equation related a star's spectrum to the ionization (electrical charge separation) of chemical elements in its atmosphere and the ionization of its surface temperature. Payne demonstrated a link between

The reward of the old scientist is the sense of having seen a vague sketch grow into a masterly landscape.
Cecilia Payne-Gaposchkin

See also: The sun's spectrum 112 ▪ The characteristics of stars 122–27 ▪ Nuclear fusion within stars 166–67 ▪ Energy generation 182–83

the spectral classes of stars and their surface temperatures. She also showed that the variation in absorption lines between star spectra was due to varying amounts of ionization at different temperatures, and not to varying abundances of chemical elements.

Payne knew that the intensity of absorption lines in star spectra could give only crude estimates of chemical elements, so other factors needed to be taken into account, such as the ionization states of the atoms of different elements. Using her knowledge of atomic physics, she determined the abundances of 18 elements found in the spectra of many different stars. She found that helium and hydrogen were vastly more abundant than on Earth, making up nearly all the matter in stars.

Astronomers' reaction

In 1925, Payne's thesis was sent to the astronomer Henry Russell for review. Russell declared that Payne's results were "clearly impossible," and pressured her to include a

If you are sure of your facts, you should defend your position.
Cecilia Payne-Gaposchkin

statement saying that the levels of hydrogen and helium she had found were "almost certainly not real." Four years later, however, Russell conceded that Payne was right.

Payne's discoveries were revolutionary. First, she established that most stars are chemically similar. Second, she demonstrated how to determine the temperature of any star from its spectrum. Third, she showed that hydrogen and helium are dominant elements in the universe—a key step toward the Big Bang theory. ▪

Cecilia Payne-Gaposchkin

Cecilia Payne was born in Wendover, England, in 1900. At age 19, she won a scholarship to Newnham College, Cambridge, where she studied botany, physics, and chemistry. After attending a lecture by Arthur Eddington, she switched to astronomy. In 1923, she left for the US to join a new graduate course in astronomy at Harvard College Observatory. Within two years, she had produced her revolutionary doctoral thesis, *Stellar Atmospheres.* Much of her research focused on variable stars and novae (exploding white dwarfs). This work helped explain the Milky Way's structure and the paths of stellar evolution. In 1931, she became a US citizen and in 1934 married Russian astronomer Sergey Gaposchkin. In 1956, she was made professor of astronomy at Harvard University, the first female professor at Harvard. She died in 1979.

Key works

1925 *Stellar Atmospheres*
1938 *Variable Stars*
1957 *Galactic Novae*

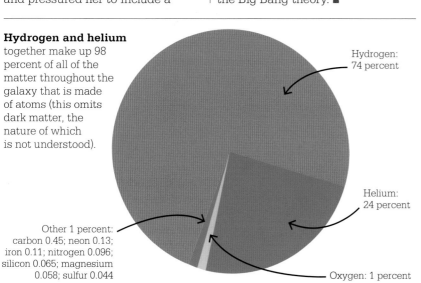

Hydrogen and helium together make up 98 percent of all of the matter throughout the galaxy that is made of atoms (this omits dark matter, the nature of which is not understood).

Hydrogen: 74 percent

Helium: 24 percent

Oxygen: 1 percent

Other 1 percent: carbon 0.45; neon 0.13; iron 0.11; nitrogen 0.096; silicon 0.065; magnesium 0.058; sulfur 0.044

OUR GALAXY IS ROTATING
THE SHAPE OF THE MILKY WAY

IN CONTEXT

KEY ASTRONOMER
Bertil Lindblad (1895–1965)

BEFORE
1904 Jacobus C. Kapteyn
shows how stars can be
divided into two streams
moving in opposite directions.

1917 Vesto Slipher shows that
spiral nebulae are moving
faster than any star.

1920 Harlow Shapley predicts
that the center of the galaxy is
in Sagittarius, estimating it at
50,000 light-years away (now
known to be 26,100 light-years).

AFTER
1927 Jan Oort confirms that
the galaxy is rotating and
proposes that a large mass of
stars forms a bulge at its core.

1929 Edwin Hubble shows
that other galaxies lie far
beyond the Milky Way.

1979 Vera Rubin uses galactic
rotation to show that galaxies
contain invisible dark matter.

I n the 1920s, there were two
opposing views of the universe.
Some astronomers thought
the Milky Way was itself the entire
universe. Others argued that the
spiral nebulae observed were not
cloudy masses on the edge of the
Milky Way, but galaxies in their
own right a vast distance away.

In 1926, a Swedish astronomer
named Bertil Lindblad considered
the likely shape of the Milky Way,
concluding that it took the form
of a rotating spiral. Lindblad was
building on the work of two other

The solar system is orbiting the
center of the Milky Way at a speed
of 140 miles/s (230 km/s). Stars closer
to the center orbit at a faster speed.

astronomers. The first was the
American Harlow Shapley, who
believed that the Milky Way made
up the whole universe. Shapley
suggested that the edge of the
galaxy could be plotted using the
many globular clusters of stars that
had been observed, and that the
center lay in Sagittarius. The second
was the Dutchman Jacobus C.

See also: Spiral galaxies 156–61 ▪ Beyond the Milky Way 172–77 ▪ The Oort cloud 206 ▪ Dark matter 268–71

Kapteyn, who had described a phenomenon he called star streaming. Stars did not move in random directions, Kapteyn said, but appeared to move in groups, going either one way or in the opposite direction. Lindblad himself was a leading expert in measuring the absolute magnitude of stars from their spectra, and was able to calculate their distances from Earth. He combined this data with his observations of the motion of globular clusters and made an interesting discovery.

Spinning in subsystems

Lindblad saw that stars move in subsystems, and each subsystem moves at a different speed. From this, he deduced that Kapteyn's star streaming was in fact evidence of the galaxy rotating, which meant that the Milky Way's stars were all moving in the same direction around a central point. Stars that streamed ahead of the solar system were nearer to the center, and stars that were farther from it appeared to stream in the opposite way because

> Stars in the **same subsystem** appear to move in the **same direction** and at the **same speed**.

> If stars in other subsystems move in the **opposite direction**, it is because they are lagging behind, but they are all moving in the **same direction**.

> The galaxy is shaped like **a spiraling disk** with outer regions moving more slowly than inner ones.

they were lagging behind. As Shapley had predicted, Lindblad placed the galactic center in Sagittarius. He supposed that subsystems farther from the galactic center orbited more slowly than the ones closer in. This was confirmed in 1927 by the observations of Jan Oort, one of Kapteyn's students.

The Milky Way was revealed to be a swirling disk that spun, albeit very slowly, taking 225 million years to complete one orbit. Although

Lindblad had offered no evidence of bodies lying outside the Milky Way, his disk-shaped galactic model with a bulging core gave credence to the idea that similar-looking objects were also galaxies. However, Oort's observations would also reveal a new puzzle. The galaxy appeared to be rotating faster than could be accounted for by the mass of its visible matter. Here was the first hint of a mystery that endures today: dark matter. ▪

Bertil Lindblad

Bertil Lindblad grew up in Örebro, Sweden. He did his undergraduate degree at Uppsala University, north of Stockholm, and became an assistant at the observatory there. It was while working at Uppsala that Lindblad made his observations of the motion of globular clusters that led to his theory of galactic rotation, which was published in 1926. The following year, still barely into his 30s, Lindblad was offered the directorship of the Stockholm Observatory and became the chief astronomer for the Royal Swedish Academy of Sciences. He held

that position until his death, overseeing many improvements. In later years, he was a leading organizer for the European Southern Observatory, which has been located in the high desert of Chile since 1962, and president of the International Astronomical Union.

Key works

1925 *Star-Streaming and the Structure of the Stellar System*
1930 *The Velocity Ellipsoid, Galactic Rotation, and the Dimensions of the Stellar System*

A SLOW PROCESS OF ANNIHILATION OF MATTER
NUCLEAR FUSION WITHIN STARS

IN CONTEXT

KEY ASTRONOMER
Arthur Eddington
(1882–1944)

BEFORE
1890s Briton Lord Kelvin and German Hermann von Helmholz suggest that the sun gets its energy by shrinking.

1896 Physicist Henri Becquerel discovers radioactivity.

1906 Karl Schwarzschild shows that energy can travel through a star by radiation.

AFTER
1931 Robert Atkinson sets out the process by which protons can combine to release energy and build new elements.

1938 German physicist Carl von Weizsäcker discovers that protons can combine into helium in stars by the carbon-nitrogen-oxygen (CNO) cycle.

1939 Hans Bethe details how the proton-proton chain and CNO cycle processes work.

The sun is composed mostly of **hydrogen gas**.

↓

At its **center**, the sun is **hot and dense**.

↓

Conditions are right for **nuclear fusion**, slowly turning **mass into energy** according to the equation $E = mc^2$.

↓

Stars are fueled by a slow process of annihilation of matter.

In the 1920s, British astronomer Arthur Eddington was the first person to explain the processes at work inside stars. He championed the idea that their source of energy is nuclear fusion.

A stable sun
When looking at the sun from Earth, what can actually be seen is the gaseous surface layer in the top 300 miles (500 km) or so, which has a temperature of about 9,900°F (5,500°C). The sun appears to be in equilibrium, meaning that, over the centuries in which astronomers have observed it (a tiny fraction of time in the lifespan of the sun), it has always appeared to be the same size and show the same luminosity. Eddington realized that the gravitational force pulling inward would be balanced not only by the gas's tendency to expand outward but also by the pressure produced by the radiation pouring out of the star.

Eddington was able to show convincingly that all stars are giant balls of hot gas. He calculated how luminous stars of different masses would appear if the gas in their centers, where the temperature and density are very high, followed the same physical laws as cooler, less

See also: The theory of relativity 146–53 ▪ Stellar composition 162–63 ▪ Energy generation 182–83 ▪ Nucleosynthesis 198–99

dense gas. The answers he obtained were a good match for observations of both giant and dwarf stars.

Gas laws and relativity

The physical laws governing the relationships between the pressure, volume, and temperature of a gas were well understood. Since they all have widely spaced molecules, gases behave in similar ways—for instance, Boyle's law (formulated by Irish chemist Robert Boyle) states that, at a constant temperature, the product of the pressure and volume of a given mass of gas is constant.

Since all gases obey the same laws, assuming that the sun is not only gaseous at the surface but throughout allows the calculation of the temperature and pressure at the center.

Eddington used these laws to calculate that the temperature at the center of the sun is about 29,000,000°F (16,000,000°C), with a density 150 times that of water.

To understand what was going on in the sun's center, Eddington now needed Einstein's equation $E = mc^2$ (pp.149–50). This equation states that energy is equal to mass multiplied by the square of the velocity of light. It was the key to unlocking the mystery of the source of the sun's energy, as it showed how mass could be turned into energy. The conditions at the solar center were sufficiently hot and dense to allow nuclear reactions to take place and mass to be destroyed, thus producing the energy that Einstein's equation predicted.

At first, physicists suggested that individual electrons or hydrogen atoms might be the mass that went into Einstein's equation. In 1931, Welsh astrophysicist Robert Atkinson showed that a process in which four hydrogen atoms were fused into one slightly less massive helium atom fit the data from the sun. This process is very slow and produces energy to fuel the sun for billions of years. Here also was evidence of the transmutation of the elements, showing how the composition of the universe changes with time. ▪

It is sound judgment to hope that, in the not too distant future, we shall be competent to understand so simple a thing as a star.
Arthur Eddington

Arthur Eddington

Arthur Eddington was born into a Quaker family, and was educated in mathematics and physics at the universities of Manchester and Cambridge. In 1905, he joined the Royal Observatory, Greenwich, but a few years later he returned to Trinity College, Cambridge, becoming Plumian Professor in 1913 and director of the Cambridge University Observatory in 1914. He lived there for the rest of his life.

In 1919, Eddington sailed to Príncipe Island, West Africa, to observe a total solar eclipse and test Einstein's prediction about the bending of starlight by the sun. He was a brilliant astronomer and mathematician, and able to communicate the most difficult physical idea in a simple and elegant language. This made his books extremely popular, especially his explanations of relativity and quantum mechanics.

Key works

1923 *The Mathematical Theory of Relativity*
1926 *The Internal Constitution of the Stars*

A DAY WITHOUT YESTERDAY

THE BIRTH OF THE UNIVERSE

IN CONTEXT

KEY ASTRONOMER
Georges Lemaître
(1894–1966)

BEFORE
1915 Albert Einstein
publishes his general theory
of relativity, which includes
equations that define various
possible universes.

1922 Alexander Friedmann
finds solutions to Einstein's
equations, indicating that the
universe could be expanding,
contracting, or static.

AFTER
1929 Edwin Hubble observes
that distant galaxies are
moving away from Earth
at a rate proportional to
their distance.

1949 Fred Hoyle coins
the term "Big Bang" for
Lemaître's theory.

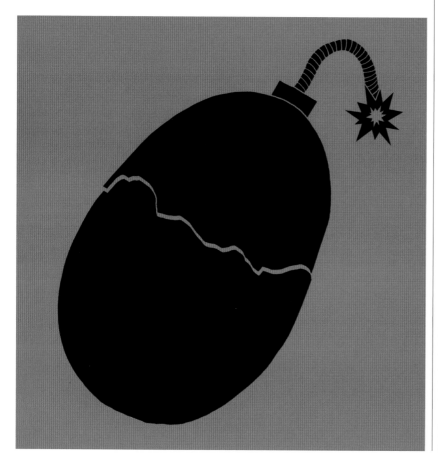

The idea that the universe originated from a tiny object in the form of an egg appears in *The Rigveda*, a collection of Hindu hymns from the 12th century BCE. However, there were few scientific clues to the universe's true origins until Albert Einstein provided a new way of conceiving time and space with his general theory of relativity in 1915. Einstein's insight led many to revisit the idea that the universe started small, among them the Belgian priest Georges Lemaître, whose 1931 proposal would carry echoes of *The Rigveda*.

In the 17th century, Johannes Kepler, observing that the night sky is dark, argued that the universe

See also: Gravitational theory 66–73 ▪ The theory of relativity 146–53 ▪
Spiral galaxies 156–61 ▪ Beyond the Milky Way 172–77 ▪ The primeval atom 196–97

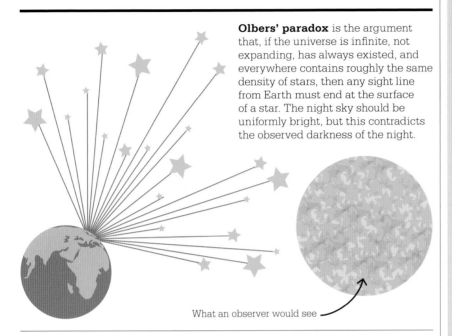

Olbers' paradox is the argument that, if the universe is infinite, not expanding, has always existed, and everywhere contains roughly the same density of stars, then any sight line from Earth must end at the surface of a star. The night sky should be uniformly bright, but this contradicts the observed darkness of the night.

What an observer would see

Georges Lemaître

Georges Lemaître was born in 1894 in Charleroi, Belgium. Following distinguished service in World War I, in 1920 he was awarded a doctoral degree in engineering. He subsequently entered a seminary, where, in his leisure time, he studied mathematics and science.

After his ordination in 1923, Lemaître studied mathematics and solar physics at Cambridge University, studying under Arthur Eddington. In 1927, he was appointed professor of astrophysics at the University of Leuven, Belgium, and published his first major paper on the expanding universe. In 1931, Lemaître put forward his theory of the primeval atom in a report in the journal *Nature*, and his fame soon spread. He died in 1966, shortly after learning of the discovery of cosmic microwave background radiation, which provided evidence for the Big Bang.

Key works

1931 *The Beginning of the World from the Point of View of Quantum Theory*
1946 *The Primeval Atom Hypothesis*

cannot be infinite in both time and space, as otherwise the stars shining from every direction would make the whole sky bright. His argument was restated in 1823 by German astronomer Wilhelm Olbers and became known as Olbers' paradox. Despite this problem, Isaac Newton stated that the universe was static (not getting any bigger or smaller) and infinite in time and space, with its matter distributed more or less uniformly over a large scale. At the end of the 19th century, this was still the prevailing view, and one that Einstein himself initially held.

An unchanging universe?
Einstein's general theory of relativity explains how gravity works at the largest scales. He realized that it could be used to test whether the Newtonian model of the universe could exist long-term without becoming unstable, and to explore which other types of universe might be feasible. The

exact relationship between mass, space, and time was explained in a series of 10 complex equations. These were called Einstein's field equations. Einstein found an initial solution to his equations that suggested the universe is contracting. Since he could not believe this, he introduced a "fix"— an expansion-inducing factor called the cosmological constant—to balance the inward pull of gravity. This allowed for a static universe.

In 1922, Russian mathematician Alexander Friedmann attempted to find solutions to Einstein's field equations. Starting with the assumption that the universe is homogenous (made of more or less the same material everywhere) and spread out evenly in every direction, he found several solutions. These allowed for models in which the universe could be expanding, contracting, or static. Friedmann was probably the first person to use the expression "expanding »

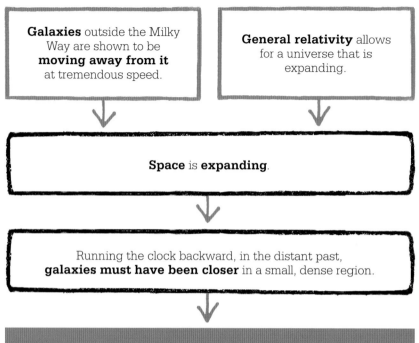

Galaxies outside the Milky Way are shown to be **moving away from it** at tremendous speed.

General relativity allows for a universe that is expanding.

Space is **expanding**.

Running the clock backward, in the distant past, **galaxies must have been closer** in a small, dense region.

The universe began from an explosion of matter on "a day without yesterday."

but your grasp of physics is abominable." However, the British astronomer Arthur Eddington later published a long commentary on Lemaître's 1927 paper, describing it as a "brilliant solution."

In 1929, Hubble released findings showing that there was indeed a relationship between the remoteness of a galaxy and how fast it was receding, confirming for many astronomers that the universe was expanding, and that Lemaître's paper had been correct. For many years the credit for the discovery of the expansion of the universe was given to Hubble, but today most agree it should be shared with Lemaître and possibly also with Alexander Friedmann.

The primeval atom
Lemaître reasoned that, if the universe is expanding and the clock is run backward, then far back in time, all the matter in the universe must have been much closer. In 1931, he suggested that the universe was initially a single, extremely dense particle containing all its matter and energy—a "primeval atom" as he called it, about 30 times the size of the sun. This disintegrated in an explosion,

The radius of space began at zero, and the first stages of the expansion consisted of a rapid expansion determined by the mass of the initial atom.
Georges Lemaître

universe." Einstein first called his work "suspicious," but six months later acknowledged that his results were correct. However, this was Friedmann's final contribution as he died two years later. In 1924, Edwin Hubble showed that many nebulae were galaxies outside the Milky Way. The universe had suddenly become a lot bigger.

The expanding universe
Later in the 1920s, Lemaître entered the debate about the large-scale organization of the universe. He had worked at institutions in the United States, becoming aware of Vesto Slipher's work on receding galaxies and Hubble's measurements of galaxy distances. A competent mathematician, he had also studied Einstein's field equations and found a possible solution to the equations that allowed for an

expanding universe. Putting these various threads together, in 1927, Lemaître published a paper that proposed that the whole universe is expanding and carrying galaxies away from each other and from Earth. He also predicted that galaxies that are more distant from us would be found to be receding at a faster rate than closer ones.

Lemaître's paper was published in an obscure Belgian journal, and as a result, his hypothesis failed to attract much attention at the time. He did, however, communicate his findings to Einstein, telling him of the solution he had found to the field equations allowing for a universe that expands. Einstein introduced Lemaître to Friedmann's work, but remained ambivalent about Lemaître's idea. Famously, Einstein is said to have said: "Your calculations are correct, giving rise to space and time on "a day without yesterday." Lemaître described the beginning of the universe as a burst of fireworks, comparing galaxies to the burning embers spreading out from the center of the blast.

The proposal initially met with scepticism. Einstein found it suspect but was not altogether dismissive. In January 1933, however, Lemaître and Einstein traveled together to California for a series of seminars. By this time, Einstein (who had removed the cosmological constant from his general theory of relativity because it was no longer needed) was in full agreement with Lemaître's theory, calling it "the most beautiful and satisfactory explanation of creation to which I have ever listened."

Lemaître's model also provided a solution to the long-standing problem of Olbers' paradox. In his model, the universe has a finite age, and because the speed of light is also finite, that means that only a finite number of stars can be observed within the given volume of space visible from Earth. The density of stars within this volume is low enough that any line of sight from Earth is unlikely to reach a star.

Refining the idea
Compressed into a tiny point, the universe would be extremely hot. During the 1940s, Russian-American physicist George Gamow and colleagues worked out details of what might have happened during the exceedingly hot first few moments of a Lemaître-style universe. The work showed that a hot early universe, evolving into what is observed today, was theoretically feasible. In a 1949 radio interview, the British astronomer Fred Hoyle coined the

A parallel exists between the Big Bang and the Christian notion of creation from nothing.
George Smoot

term "Big Bang" for the model of the universe Lemaître and Gamow had been developing. Lemaître's hypothesis now had a name.

Lemaître's idea about the original size of the universe is now considered incorrect. Today, cosmologists believe it started from an infinitesimally small point of infinite density called a singularity. ∎

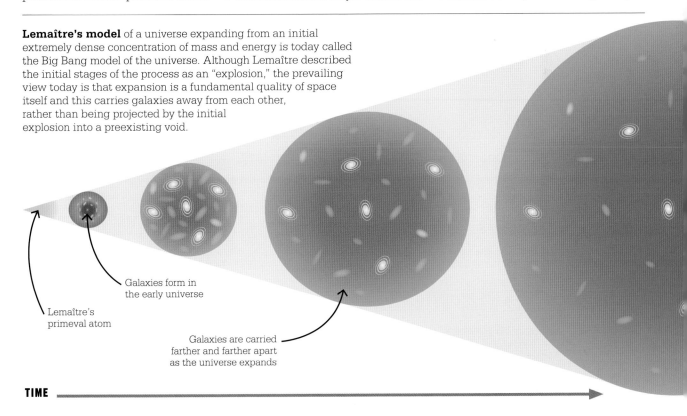

Lemaître's model of a universe expanding from an initial extremely dense concentration of mass and energy is today called the Big Bang model of the universe. Although Lemaître described the initial stages of the process as an "explosion," the prevailing view today is that expansion is a fundamental quality of space itself and this carries galaxies away from each other, rather than being projected by the initial explosion into a preexisting void.

Galaxies form in the early universe

Lemaître's primeval atom

Galaxies are carried farther and farther apart as the universe expands

TIME

THE UNIVERSE IS EXPANDING IN ALL DIRECTIONS

BEYOND THE MILKY WAY

IN CONTEXT

KEY ASTRONOMER
Edwin Hubble (1889–1953)

BEFORE
1907 Henrietta Swan Leavitt shows the link between period and luminosity of Cepheid stars.

1917 Vesto Slipher publishes a table of 25 galactic redshifts.

1924 Hubble shows that the Andromeda galaxy lies well outside the Milky Way.

1927 Georges Lemaître proposes that the universe may be expanding.

AFTER
1998 The Supernova Cosmology Project and High-Z Supernova Search prove that cosmos expansion is accelerating.

2001 Hubble Space Project measures the Hubble Constant (H_0) to within 10 percent.

2015 The Planck Space Observatory puts the age of the universe at 13.799 billion years.

In the early 1920s, American astronomer Edwin Hubble provided proof of the true size of the universe. Working at the Mount Wilson Observatory near Pasadena, California, Hubble used the newly constructed 100-in (2.5-m) Hooker Telescope, at that time the largest in the world, to settle the greatest argument then raging in astronomy. His observations would lead to the startling revelation that the universe is not only far, far larger than previously thought, but is also expanding.

Settling the Great Debate

At the time, the question of whether spiral nebulae were galaxies beyond the Milky Way or a special kind of nebula was the subject of a "Great Debate." In 1920, a meeting was held at the Smithsonian Museum in Washington, D.C., in an attempt to settle the question. Speaking for the "small universe" was Princeton astronomer Harlow Shapley, who contended that the Milky Way comprised the entire universe. Shapley cited as evidence reports that the spiral nebulae were

> Observations indicate that the universe is expanding at an ever-increasing rate. It will expand forever, getting emptier and darker.
> **Stephen Hawking**

rotating, reasoning that this must make them relatively small because otherwise the outer regions would be spinning at speeds faster than the speed of light (these reports were later shown to be wrong). Opposed to Shapley was Heber D. Curtis, who supported the idea that each nebula was far beyond the Milky Way. Curtis cited as evidence the discovery by Vesto Slipher that light from most "spiral nebula" galaxies was shifted to the red part of the electromagnetic spectrum, indicating that they were moving away from Earth

Edwin Hubble

Born in 1889 in Missouri, Edwin Hubble was a gifted athlete as a youth, leading the University of Chicago's basketball team. After graduating with a science degree, he studied law at Oxford University. Returning from England dressed in a cape and behaving like an aristocrat, he was described by Harlow Shapley as "absurdly vain and pompous."

Despite his flair for self-publicity, Hubble was a cautious scientist. He described himself as an observer, and reserved judgment until he had sufficient evidence. He reacted furiously if anyone encroached on his field of research. It is to Hubble's discredit, therefore, that he did not acknowledge that 41 of the 46 redshifts used to formulate his famous law were measured not by him but by Vesto Slipher. Hubble spent his last years campaigning for a Nobel Prize to be awarded in astronomy. He died in 1953.

Key work

1929 *A relationship between distance and radial velocity among extra-galactic nebulae*

See also: Measuring the universe 130–37 ▪ The theory of relativity 146–53 ▪ Spiral galaxies 156–61 ▪ The birth of the universe 168–71 ▪ Space telescopes 188–95 ▪ Curtis (Directory) 337 ▪ Arp (Directory) 339

Hubble is seen here looking through the lense of the Hooker Telescope at Mount Wilson. It was here that he measured galaxy distances and a value for the expansion of the universe.

at enormous speeds—speeds far too high for them to be contained within the Milky Way.

Hubble set out to see whether there was a relationship between the distances of spiral nebulae and their velocity. His strategy was to search for Cepheid variable stars (p.138)—stars whose luminosity changes predictably—within nebulae and to measure their distances from Earth. This provided Hubble with his first big discovery in the winter of 1923.

Beginning with photographic plates of the closest and clearest nebulae, Hubble spotted a Cepheid variable on one of the first plates he reviewed. The distances he calculated for even relatively nearby nebulae were so vast that it effectively killed the Great Debate immediately: NGC 6822 was 700,000 light-years away, while M33 and M31 were 850,000 light-years away. It was instantly clear that the universe extended beyond

the Milky Way. Just as Curtis had maintained, the spiral nebulae were "island universes," or "extra-galactic nebulae" as Hubble termed them. Over time, the term "spiral nebulae" fell into disuse and they are now simply called galaxies.

In the realm of the nebulae

Hubble pressed ahead with his program of measuring the distances to galaxies beyond the Milky Way. Farther out, however, it became impossible to pick out individual Cepheid variables in such faint and fuzzy distant galaxies. He was compelled to fall back on indirect methods, such as the so-called "standard ruler" assumption: reasoning that all galaxies of a similar type are the same size allowed him to estimate the distance to a galaxy by measuring its apparent size and comparing it to the expected

"true" size. Thanks to Slipher's measurements, Hubble already knew that light from most spiral nebulae was redshifted. In addition, the fainter spirals had higher values of redshift, showing that they were moving faster through space. Hubble realized that, if there were indeed a relationship between a galaxy's distance from Earth and its recession velocity, these redshifts would serve as a cosmic yardstick, enabling the distances of the very farthest and faintest galaxies to be calculated, and a ballpark figure to be put on the size of the universe as a whole. Meanwhile, Milton Humason, the assistant astronomer at Mount Wilson, checked Slipher's redshifts and collected new spectra from distant galaxies. It was hard, punishing work, and he and Hubble spent many bitterly cold nights in the observer's cage at the top of the tube telescope on Wilson Mountain in California.

Hubble's landmark paper, "A relationship between distance and radial velocity among extra-galactic nebulae," was »

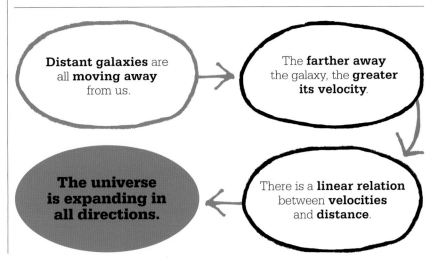

Distant galaxies are all **moving away** from us.

The **farther away** the galaxy, the **greater its velocity**.

There is a **linear relation** between **velocities** and **distance**.

The universe is expanding in all directions.

published in a journal called *Proceedings of the National Academy of Science* in 1929. It contained a straight-line graph that plotted 46 galaxies from near to far against their redshifts. Although there was a considerable scatter, Hubble managed to fit a straight line through the majority. The graph shows that, with the exception of the nearest galaxies, Andromeda and Triangulum, which are encroaching on the Milky Way, all other galaxies are receding. What is more, the farther away they are, the faster their movement.

Toward an interpretation

If, from Earth's perspective within the universe, all galaxies are seen flying away, then the potential explanations are that (a) Earth lies at the center of the universe; or (b) the universe itself originated from a single point and is expanding as a whole.

Objectivity—a kind of foundational law in science—requires that there is no reason

Equipped with his five senses, man explores the universe around him and calls the adventure Science.
Edwin Hubble

to assume that Earth occupies a unique position. Instead, the light from distant nebulae showed that the universe was not static. Many astronomers quickly reached the conclusion that this was due to the expansion of the universe, although Hubble never stated this explicitly.

In reality, Vesto Slipher had indicated the trend in 1919, four years before Hubble made his observations, and Georges Lemaître

had proposed the expansion of the universe from a "primeval atom" in 1927. However, Hubble's result provided a simple link between his redshift-measured velocities and distance, and with it, the convincing proof that the scientific community needed. "Hubble's law," stating that the redshift of galaxies is proportional to their distance from Earth, was accepted almost unanimously.

Einstein's blunder

The revelation that the universe might be expanding made news all over the world—not least for the fact that it directly contradicted a theory of Albert Einstein's. Einstein saw that gravity could eventually cause the universe to collapse under its own weight, so he used a value he called the cosmological constant—a kind of negative pressure—to prevent this from occurring in the field equations of general relativity. He abandoned the idea in the wake of Hubble's discovery.

Einstein and others assumed that the observed velocities were Doppler effects caused by the galaxies' speed of recession, but there were some dissenting voices. Swiss astronomer Fritz Zwicky suggested that the redshift might be due to "tired light" reaching Earth—caused by the interaction of photons with the intervening matter. Hubble himself found it hard to believe that the velocities indicated by the redshifts were actually real, and was happy to use them solely as distance indicators. In fact, the velocities of galaxies observed by Hubble are due to the expansion of spacetime itself.

K-factor

Hubble showed how fast spacetime is expanding by plotting a straight-line graph—which he grandly

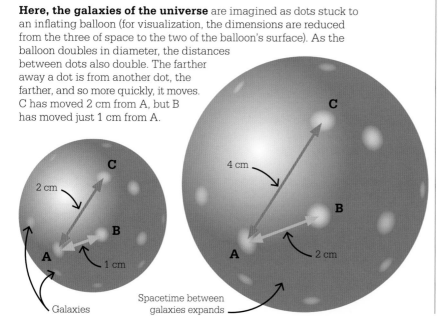

Here, the galaxies of the universe are imagined as dots stuck to an inflating balloon (for visualization, the dimensions are reduced from the three of space to the two of the balloon's surface). As the balloon doubles in diameter, the distances between dots also double. The farther away a dot is from another dot, the farther, and so more quickly, it moves. C has moved 2 cm from A, but B has moved just 1 cm from A.

2 cm
C
1 cm
B
A
Galaxies

4 cm
C
B
2 cm
A
Spacetime between galaxies expands

called the "K-factor." The gradient is described mathematically by a value now known as the Hubble Constant (H_0). This important number determines not only the size of the observable universe but also its age. The Hubble Constant allowed astronomers to work backward and calculate the moment in time of the Big Bang itself, when the radius of the universe was zero.

The initial calculation of H_0 was 300 miles (500 km) per second per megaparsec (one megaparsec is approximately 3.26 million light-years). This presented a problem, since it gave a figure of 2 billion years for the age of the universe, less than half the accepted age of Earth. The discrepancy was found to have been caused by systematic errors in Hubble's

distance measurements. Many were out by a factor of seven due to his method of taking the brightest star in any galaxy—or even the luminosity of the galaxy itself—and assuming it to be a Cepheid variable star. Luckily for Hubble, the inaccuracies were fairly consistent throughout the dataset, allowing him to plot the trend in spite of them.

Hubble Key Project

Calculating the rate of expansion of the cosmos drove the decision to develop the Hubble Space Telescope from its inception in the 1970s to its 1990 launch. NASA made one of the telescope's "Key Projects" determining the Hubble Constant to within 10 percent. As a result, the instrument spent years measuring Cepheid

The ESA's Planck Observatory operated between 2009 and 2013. It produced data that helped to measure many cosmological parameters, including the Hubble Constant.

light curves. The final result, delivered in 2001, gave an age for the universe of 13.7 billion years. This figure was fine-tuned to 13.799 billion years (with an error of 21 million years either way) by data from the Planck Space Observatory in 2015. The most dramatic revision to Hubble's law, however, came in 1998 when astronomers discovered that the universe's expansion is accelerating due to a mysterious and unknown agent known as dark energy, which has led to a renewal of interest in Einstein's so-called blunder, the cosmological constant (pp.298–303). ■

WHITE DWARFS HAVE A MAXIMUM MASS
THE LIFE CYCLES OF STARS

IN CONTEXT

KEY ASTRONOMER
Subrahmanyan Chandrasekhar (1910–1995)

BEFORE
1914 Walter Adams details the spectrum of 40 Eridani B, an unusually faint white star.

1922 Dutch astronomer Willem Luyten coins the term "white dwarf" for low-mass white stellar remnants.

1925 Austrian physicist Wolfgang Pauli formulates the Pauli Exclusion Principle, which states that no two electrons can occupy the same quantum state. This leads to recognition of the phenomenon of electron degeneracy pressure.

AFTER
1937 Fritz Zwicky characterizes a type 1a supernova as the explosion of a white dwarf that exceeded its Chandrasekhar limit.

1972 Astronomers find the first stellar black hole candidate.

In 1930, a young Indian student named Subrahmanyan Chandrasekhar calculated that a star that ends its life with a little more mass than the sun cannot hold itself up against the pull of its own gravity. This was key to understanding the life cycles of stars, and in particular the dim, very hot stars called white dwarfs. This type of star was known to be very dense and made of compact "degenerate" matter composed of atomic nuclei and free electrons. White dwarfs were prevented from collapsing by a phenomenon known as electron degeneracy pressure. This meant that, when electrons were packed very close together, their movement was limited, creating outward pressure.

The Chandrasekhar limit
Chandrasekhar figured out that electron degeneracy pressure can prevent a white dwarf from collapsing only up to an upper limit for the white dwarf's mass, which is about 1.4 times the mass of the sun. Today, it is known that the core of a giant star at the end of its life will collapse into a white dwarf if its mass is below the Chandrasekhar limit, but will collapse to an even denser object—a neutron star or a black hole—if its mass exceeds the limit. This insight was largely ignored by the scientists of the day because neutron stars and black holes were, at this time, still purely theoretical. ∎

The black holes of nature are the most perfect macroscopic objects there are in the universe: the only elements in their construction are our concepts of space and time.
Subrahmanyan Chandrasekhar

See also: Discovering white dwarfs 141 ▪ Nuclear fusion within stars 166–67 ▪ Supernovae 180–81

THE RADIO UNIVERSE
RADIO ASTRONOMY

IN CONTEXT

KEY ASTRONOMER
Karl Jansky (1905–1950)

BEFORE
1887 German physicist Heinrich Hertz demonstrates the existence of radio waves.

1901 Italian inventor Guglielmo Marconi sends a radio signal across the Atlantic, unknowingly bouncing the waves off the ionosphere.

AFTER
1937 American amateur astronomer Grote Reber makes the first radio survey of the sky.

1965 James Peebles proposes that universal background microwave radio waves are the last remnant of the Big Bang.

1967 Antony Hewish and Jocelyn Bell Burnell detect a repeating stellar radio signal, the first pulsar.

1998 Sagittarius A* is shown to be a supermassive black hole at the center of the Milky Way.

While working for Bell Telephone Laboratories in 1930s America, telephone engineer Karl Jansky was given the task of plotting the natural sources of static that might interfere with long-wave radio voice transmissions. To conduct his investigations, Jansky hand-built a directional radio antenna that was 100 ft (30 m) wide and 20 ft (6 m) high. The contraption rotated on four tires salvaged from an old Model T Ford. His colleagues dubbed the device "Jansky's merry-go-round" because the young engineer would systematically rotate the antenna to pinpoint sources of atmospheric radio waves.

Radio astronomy
Jansky matched most sources of radio waves to approaching thunderstorms, but there was a persistent hiss that remained unidentified. The intensity of this static rose and fell once a day, and initially Jansky thought he was detecting radio from the sun. However, the "brightest" spot of radio waves moved through the sky

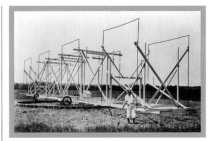

Jansky is pictured here with his hand-built antenna. He published a paper on his work in 1933, but soon after was reassigned by Bell Labs and did no more astronomical work.

following the sidereal day (relative to the stars), not the solar day, and Jansky realized that the radio waves were coming from the constellation Sagittarius, at the heart of the Milky Way: radio waves were "shining" from space just like visible light.

The newspapers reported the discovery of "extraterrestrial radio," and astronomers soon began to copy Jansky's device—in effect the first radio telescope. This opened up the possibility of viewing the universe in a new way—not from its light but from its radio emissions. ∎

See also: Searching for the Big Bang 222–27 ▪ Quasars and pulsars 236–39 ▪ Reber (Directory) 338 ▪ Ryle (Directory) 338–39

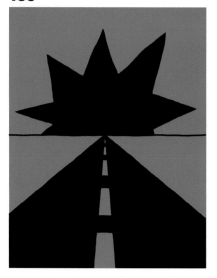

AN EXPLOSIVE TRANSITION TO A NEUTRON STAR
SUPERNOVAE

IN CONTEXT

KEY ASTRONOMERS
Walter Baade (1893–1960)
Fritz Zwicky (1898–1974)

BEFORE
1914 American astronomer Walter Adams first describes white dwarf stars, which are now known to be involved in common novae.

1931 Subrahmanyan Chandrasekhar calculates the greatest mass a white dwarf can have.

AFTER
1967 Antony Hewish and Jocelyn Bell Burnell discover pulsars, which are found to be fast-spinning neutron stars.

1999 A survey of the light from type 1a supernovae shows that the universe's expansion is accelerating due to an unknown quantity known as dark energy.

In 185 CE, Chinese astronomers recorded a phenomenon they called a "guest star." The star had appeared in the direction of Alpha Centauri, the closest star system to Earth, and had shone brightly for eight months before vanishing. This is probably the first recording of a supernova.

Mysterious new stars have appeared several times across the centuries. In 1572, Danish astronomer Tycho Brahe named one a nova, meaning "new." With the development of telescopes, novae became subject to closer scrutiny and were found to be faint stars that lit up with a great intensity for short periods. It was not until the 1930s that two astronomers at Caltech in California, Walter Baade and Fritz Zwicky, calculated that some novae released much more energy than others. For example, they calculated that S Andromedae, a nova seen in 1885, had released the equivalent of 10 million years of the sun's output all at once. Baade and Zwicky dubbed these incredibly energetic events "super-novae."

Core collapse

In 1934, Baade and Zwicky suggested that a supernova was the core of a large star, collapsing under its own gravity after running out of fuel.

Faint stars can become much brighter for short periods, forming **novae**.

Some novae release vastly **more energy than others**.

Some of these **supernovae** are formed by **the collapse of a star** that annihilates its own matter.

The core of the collapsed star is crushed into a **neutron star**, made of material containing only neutrons.

See also: The Tychonic model 44–47 ▪ Quasars and pulsars 236–39 ▪
Dark matter 268–71 ▪ Dark energy 298–303

A supernova in the Large Magellanic Cloud blew out this cloud of shrapnel, captured by the Chandra X-ray Space Observatory. The explosion was caused by the collapse of a massive star.

Fritz Zwicky

Born in Bulgaria to a Swiss father and Czech mother, Fritz Zwicky emigrated to the United States in 1925 to work at Caltech with Robert Millikan, a leading particle physicist. In 1931, he began a collaboration at the Mount Wilson Observatory near Los Angeles with Walter Baade, a German astronomer, who had just arrived from Europe. It was this partnership that led to the discovery of supernovae and neutron stars, but Zwicky's work around this time was also instrumental in another great discovery. Zwicky calculated that the mass of galaxies, as indicated by their gravitational effects, was much greater than the matter that could be measured through observations. He named the missing material *dunkle Materie*, now better known as dark matter. In addition to his theoretical work, Zwicky worked on the development of jet engines and took out more than 50 patents on his inventions.

Key works

1934 *On Supernovae*
(with Walter Baade)
1957 *Morphological Astronomy*

The collapse was so powerful that it annihilated matter, releasing a huge amount of energy in accordance with Einstein's equation $E=mc^2$ (p.149). What was left was a neutron star—a body composed of only neutrons that were packed together like the particles in an atomic nucleus, only on a much larger scale. A neutron star is only about 7 miles (11 km) across, but has huge density and gravitational pull. Neutrons can be packed more closely than atomic nuclei, meaning that a teaspoon of neutron star weighs 10 million tons. The star's escape velocity (the velocity required to escape its gravitational pull) is nearly half the speed of light.

First detection

The concept of a neutron star remained purely hypothetical until 1967, when pulsars were discovered. Pulsars were shown to be rapidly spinning neutron stars. In 1979, a powerful burst of gamma rays was detected. This has since been attributed to a "magnetar"—a kind of neutron star with a magnetic field billions of times greater than Earth's.

Many mysteries remain regarding collapsing stars. Only stars above the Chandrasekhar Limit of 1.4 solar masses (p.178) will become supernovae and form neutron stars. A star above 3 solar masses goes further and becomes a black hole. There may be a halfway stage in which neutron matter degenerates even more into quark particles—the particles from which neutrons and protons are made. Quark stars remain hypothetical, but the search for them is on. ▪

THE SOURCE OF ENERGY IN STARS IS NUCLEAR FUSION
ENERGY GENERATION

IN CONTEXT

KEY ASTRONOMER
Hans Bethe (1906–2005)

BEFORE
1919 Francis Aston discovers that four hydrogen nuclei (protons) have more mass than a helium nucleus.

1929 Welsh astronomer Robert Atkinson and Dutch physicist Fritz Houtermans calculate how the fusion of light nuclei within stars could release energy in accordance with mass–energy equivalence.

AFTER
1946 Ralph Alpher and George Gamow describe how nuclei of helium and some other nuclei could have been synthesized during the Big Bang.

1951 Ernst Öpik describes the triple-alpha process, which converts the nuclei of helium-4 into carbon-12 in the cores of red giant stars.

Until a brilliant young German-born physicist named Hans Bethe figured it out in 1938, no one knew for sure why the sun and other stars emitted so much light, heat, and other radiation, or where their energy came from.

A step toward the correct answer had been made in 1905 with Albert Einstein's special theory of relativity, which proposed that mass and energy have an equivalence. The significance of this discovery was that a small loss of mass could be accompanied by a very large release of energy.

In 1919, British chemist Francis Aston found that an atom of helium (the second lightest element) had a mass slightly less than that of four atoms of hydrogen (the lightest element). Soon afterward, British astrophysicist Arthur Eddington and French physicist Jean Baptiste Perrin independently proposed that stars might obtain their energy by combining four hydrogen nuclei to

Low- to medium-mass stars are fueled by **the proton–proton chain**, turning hydrogen into helium.

High-mass stars are fueled by **the CNO cycle**, which turns hydrogen into helium in the presence of carbon and nitrogen as catalysts.

The **fusion of hydrogen nuclei** to form helium turns **mass into energy**.

The source of energy in stars is nuclear fusion.

See also: The theory of relativity 146–53 ▪ Nuclear fusion within stars 166–67 ▪ The primeval atom 196–97

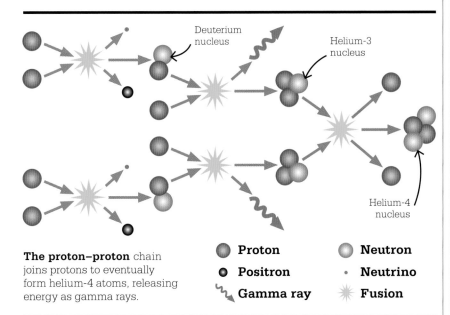

The **proton–proton** chain joins protons to eventually form helium-4 atoms, releasing energy as gamma rays.

● Proton   ○ Neutron

◉ Positron   · Neutrino

〰 Gamma ray   ✳ Fusion

Hans Bethe

Hans Bethe was born in 1906 in Strasbourg, then part of the German Empire. From a very early age, he displayed a high ability in mathematics. By 1928, he had completed a doctorate in physics. With the rise of the Nazi regime, Bethe emigrated first to Britain and then to the US. His work during World War II included three years at the Los Alamos Scientific Laboratory, which was engaged in assembling the first atomic (fission) bomb. After the war, Bethe played an important role in the development of the hydrogen (fusion) bomb. He later campaigned against nuclear testing and the arms race. In addition to his work in astrophysics and nuclear physics, Bethe made major contributions to other fields of physics, including quantum electrodynamics (QED). He continued to work in all these fields until his death in 2005, at age 98.

Key works

1936–37 *Nuclear Physics* (with Robert Bacher and Stanley Livingston)
1939 *Energy Production in Stars*

form a helium nucleus with some loss of mass, which was then converted to energy. Eddington thought this might allow the sun to shine for tens of billions of years. In 1929, Robert Atkinson and Fritz Houtermans calculated how the fusion of light nuclei, rather than atoms, could create energy in stars, but the reactions were unknown.

The proton–proton chain

In 1938, Bethe attended a physics conference in Washington, D.C., to discuss how energy is generated in stars. During the conference, he realized that, given the abundance of hydrogen in stars, the most likely first step in energy generation was one in which two hydrogen nuclei—which are single protons—join to form the nucleus of a deuterium (heavy hydrogen) atom. Bethe knew that this reaction generated energy. He then figured out how two further reaction steps could produce a nucleus of helium-4 (the most common form of helium).

He saw the entire sequence of reactions, called the proton–proton chain, as the main source of energy production in stars up to about the size of the sun.

The CNO cycle

Whereas the core temperature of a star rises slowly as star size increases, the amount of energy it produces rises much more rapidly. The proton–proton chain could not explain this, so Bethe investigated reactions involving heavier atomic nuclei. After hydrogen and helium, the next heaviest element present in appreciable amounts in higher mass stars is carbon, so Bethe looked at possible reactions of carbon nuclei with protons. He found a cycle of reactions, called the CNO (Carbon–Nitrogen–Oxygen) cycle, during which hydrogen nuclei fuse to form helium in the presence of heavier elements, which seemed to work. Bethe's findings were quickly accepted by other physicists. ■

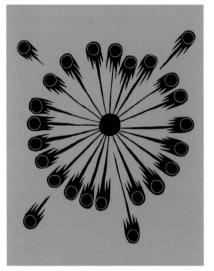

A RESERVOIR OF COMETS EXISTS BEYOND THE PLANETS
THE KUIPER BELT

IN CONTEXT

KEY ASTRONOMER
Kenneth Edgeworth
(1880–1972)

BEFORE
1781 and 1846 The discovery of Uranus and Neptune leads to discussion of where the outer edge of the solar system lies.

1930 Pluto is discovered. Astronomers Frederick C. Leonard and Armin O. Leuschner suggest there may be similar bodies out there.

AFTER
1977 Charles Kowal discovers Chiron, an icy centaur (minor planet) beyond Saturn.

1992 A Trans-Neptunian Object (TNO—an object orbiting at a distance greater than Neptune's) is discovered by David Jewitt and Jane Luu.

2005 The discovery of Eris, a TNO of a similar size to Pluto, and TNOs Haumea and Makemake, lead to Pluto being demoted to a dwarf planet.

In 1943, Irish astronomer Kenneth Edgeworth suggested that beyond Neptune and Pluto, there existed a disk of icy bodies that were formed at the dawn of the solar system, but were too small and widely spaced to accrete into a planet. From time to time, they were nudged into the inner solar system, where they appeared as comets. He published his idea in *Journal of the British Astronomical Association*, a periodical not widely read in the US.

Kuiper belt

In 1951, in the more prestigious *Astrophysical Journal*, a Dutch-American astronomer named Gerard Kuiper suggested that such a disk once existed but had long since been dispersed by the effects of Pluto's gravity. It came to be called the Kuiper Belt, though some astronomers now use "Edgeworth-Kuiper Belt."

Then, in 1980, Uruguayan astronomer Julio Fernández realized that a belt of cometary nuclei beyond Neptune was needed to supply the numbers of short-period comets seen in the inner solar system. Photographs of the region were taken, a few hours apart, and these were then examined to see if any of the objects had moved, indicating that they were much closer than the stars. More than 1,000 objects have now been found in the Kuiper belt. Most are larger than 60 miles (100 km) across, since anything smaller is too faint to detect. ∎

The comets have remained what they were from the beginning—astronomical heaps of gravel without any cohesion.
Kenneth Edgeworth

See also: The Oort cloud 206 ∎ Exploring beyond Neptune 286–87

SOME GALAXIES HAVE ACTIVE REGIONS AT THEIR CENTERS
NUCLEI AND RADIATION

IN CONTEXT

KEY ASTRONOMER
Carl Seyfert (1911–1960)

BEFORE
1908 Edward Fath and Vesto Slipher observe peculiarities in the spectrum of the nebula NGC 1068 (now recognized as a typical Seyfert galaxy).

1936 Edwin Hubble classifies the shapes of galaxies.

AFTER
1951 Cygnus A, one of the strongest sources of radio waves in the sky, is identified as the first radio galaxy.

1963 Dutch astronomer Maarten Schmidt coins the term quasi-stellar radio source (later "quasar") for an object with a starlike appearance that is actually a distant, very bright source of radio waves.

1967 Armenian astrophysicist Benjamin Markarian begins publishing a list of galaxies with strong ultraviolet emissions, many of them Seyfert galaxies.

B
etween 1940 and 1942, American astronomer Carl Seyfert studied a number of spiral galaxies that had compact, particularly bright centers, often bluish in color. His investigations revealed that there were distinctive emission lines in these galaxies' spectra. He published a paper describing galaxies of this type, which later came to be known as Seyfert galaxies. They are usually spiral galaxies with nuclei that produce large amounts of radiation over a broad range of wavelengths, often most strongly in the infrared region, but also often including visible light, radio waves, ultraviolet radiation, X-rays, and gamma rays.

Violent centers

Seyfert galaxies are just one variety of a class of galaxies called active galaxies. These have central regions, termed active galactic nuclei (AGNs), in which an extraordinary amount of violent activity occurs. Quasars are another type of AGN. These are always a vast distance away and

Spiral galaxy NGC 1068 (M 77) is the archetypal Seyfert galaxy. It has an intensely bright active center surrounded by swirls of ionized gas.

produce so much energy that they outshine their host galaxies, which cannot be seen. AGNs are thought to be powered by matter spiraling into massive black holes at their centers. In addition to emitting radiation, many AGNs also send powerful jets of particles into space from the vicinity of their central black holes. Some are associated with vast lobes of material that emit radio waves—active galaxies that feature these "radio lobes" are called radio galaxies. ∎

See also: Spiral galaxies 156–61 ▪ Beyond the Milky Way 172–77 ▪ Quasars and black holes 218–21

THE MATCH OF LUNAR AND EARTH MATERIAL IS TOO PERFECT
THE ORIGIN OF THE MOON

IN CONTEXT

KEY ASTRONOMER
Reginald Daly (1871–1957)

BEFORE
1913 British geologist Arthur Holmes produces the first modern geologic timescale, which proposes that Earth is at least 1.5 billion years old.

AFTER
1969–72 The Apollo missions bring back moon rock to be analyzed on Earth.

1975 Following analysis of moon rock, US astronomer William Hartmann and others return to the giant impact theory to explain the new evidence.

2011 Norwegian–American planetary scientist Erik Asphaug and Swiss astrophysicist Martin Jutzi suggest that the moon formed with a tiny companion moon and the two later collided.

Geologists had broadly put together a story for Earth's billions of years' existence by the early 20th century. But the origin of the moon remained open to speculation. Up until the 1940s, most astronomers subscribed to a theory put forward by George Darwin, son of the naturalist Charles. In 1898, Darwin proposed that the moon formed when a hot and fast-spinning Earth threw out molten rock that coalesced into an orbiting satellite. He suggested that the moon had once been much closer to Earth and was slowly drifting away. This has since been confirmed by measurements showing that the moon is moving away by about 1½ in (3.5 cm) a year.

Astronomers have found evidence that two small planets orbiting the star HD 172555 collided a few thousand years ago. A similar collision involving Earth probably formed the moon.

See also: The discovery of Ceres 94–99 ▪ The composition of comets 207 ▪ Investigating craters 212 ▪ The Space Race 242–49

> **Lunar rocks match the material in Earth's mantle.**

> The moon may have formed in **a giant impact** that knocked Earth's magma into orbit.

> Computer modeling suggests that **a smaller planet hit Earth** 4.3 billion years ago, and that **the collision created the moon**.

At bottom each "exact" science is, and must be, speculative. Its chief tool of research, too rarely used with both courage and judgment, is the regulated imagination.
Reginald Daly

An alternative theory, championed in the 1940s by American chemist Harold Urey, was the "capture" model, in which the moon formed elsewhere in the solar system and fell under the control of Earth's gravity. However, the moon is so big compared to Earth that most thought such an event unlikely.

In 1946, Canadian geologist Reginald Daly presented a third idea. While Daly agreed with Darwin that the moon and Earth were formed from the same material, he hypothesized that the driving force was an impact between Earth and another body, which had thrown material into orbit.

Matching rock
Daly's idea was more or less ignored until the 1970s, when the analysis of moon rocks showed that their mineral content is a very close match to that of Earth's mantle (the layer between the crust and the core). Both are high in silicates but have low levels of metals. If the moon had formed elsewhere, its rocks would be very different from Earth's. If it had formed from the same molten raw materials as Earth, it would be expected to be a mini version of Earth, and have a larger metallic core. However, the rock evidence points to the moon being made from material gouged from Earth's surface after the planet had solidified.

Big Splash
In the last decade, computer modeling of possible impacts has suggested an event now dubbed the Big Splash. In this scenario, a Mars-sized planet (named Theia, after the mother of the moon in Greek mythology) hit Earth 4.3 billion years ago—about 200 million years after Earth's formation. The impact turned both bodies into seething balls of magma. Most of Theia merged with Earth (which explains why the planet has an oversized metallic core) and the "splash" hurled magma into orbit, mostly from the rocky outer region of the planet. This material formed the moon. Although the Big Splash idea is very much a hypothesis at present, it remains the best guess as to the origins of the moon. ▪

Reginald Daly
Geologist Reginald Daly's contributions to the theories of continental drift, plate tectonics, and the rock cycle have proved invaluable in understanding the similarities and differences between Earth and other rocky bodies in the solar system.

Daly's abilities as a geologist became clear when he was surveying the southern border of Canada, from the Pacific Coast, through the Rockies to the Great Plains. The rock samples he collected during this survey led him to become a leading voice on the origins of different rock types. As early as the 1920s, Daly proposed that material ejected from Earth to form the moon was a primary cause for the dynamic character of Earth's crust. The impact theory was a late addition to Daly's work, coming after he had retired as the head of Harvard University's geology department.

Key work

1946 *Origin of the Moon and its Topography*

IMPORTANT
NEW DISCOVERIES
WILL BE MADE WITH FLYING
TELESCOPES

SPACE TELESCOPES

IN CONTEXT

KEY ASTRONOMER
Lyman Spitzer Jr. (1914–1997)

BEFORE
1935 Karl Jansky reveals that celestial objects produce radio waves, offering new ways to view the universe beyond visible light.

1970 NASA launches Uhuru, an orbiting X-ray observatory.

1978 The International Ultraviolet Explorer, the first telescope operated in real time, is launched.

AFTER
1990 The Hubble Space Telescope is launched.

2003 The infrared Spitzer Space Telescope is launched.

2009 The Kepler Telescope is launched to search for extrasolar planets.

2018 Scheduled launch of the infrared James Webb Space Telescope.

I n 1946, a full 11 years before *Sputnik 1*, the first satellite, was launched into Earth's orbit, a 32-year-old astrophysicist named Lyman Strong Spitzer Jr. conceived of a powerful telescope that would one day operate not on Earth's surface but in orbit. High above the opaque atmosphere and the light pollution, this space telescope would have a clear and unprecedented view of the universe. It would be more than four decades before Spitzer's dream was realized, but his patience and tenacity would eventually pay off.

More than light

The discovery of extraterrestrial radio sources by Karl Jansky in 1935 revealed that there were ways of observing the universe other than by visible light. The outbreak of World War II in 1939 interrupted research into this new and exciting field. It was left to an amateur astronomer from Illinois named Grote Reber to take the first steps in radio astronomy. In 1937, Reber had made the first survey of the radio universe using homemade antennae he had built in his backyard. Soon afterward, wartime

> Astronomy may be revolutionized more than any other field of science by observations from above the atmosphere. In a new adventure of discovery, no one can foretell what will be found.
> **Lyman Spitzer Jr.**

researchers found that meteors and sunspots produced radio waves of their own, this time in the microwave band used in radars. If it was possible to discover new objects using radio, then it stood to reason that other forms of electromagnetic radiation, such as infrared, ultraviolet (UV), and X-rays, could be harnessed as tools of observation.

There was a problem, however. Earth's atmosphere, transparent to visible light, is opaque to many

Lyman Spitzer Jr.

Lyman Spitzer Jr. was born in Toledo, Ohio, in 1914. He received a Ph.D. in astrophysics from Princeton under the supervision of Henry Norris Russell. After World War II, he became head of the astrophysics department, and began his 50-year devotion to space telescopes.

As an expert in plasma, Spitzer invented the stellarator in 1950. This device contained hot plasma within a magnetic field and started the search for fusion power that continues today. In 1965, Spitzer joined NASA to develop space observatories, but that year he

triumphed in another field entirely. With his friend Donald Morton, Spitzer became the first to climb Mount Thor, a 5,495-ft (1,675-m) peak in the Canadian Arctic. In 1977, his campaigning for a space telescope paid off, and funding was granted to the Hubble Space Telescope. He lived long enough to see his dream become a reality in 1990.

Key work

1946 *Astronomical Advantages of an Extra-terrestrial Observatory*

See also: Beyond the Milky Way 172–77 ▪ Radio astronomy 179 ▪ Studying distant stars 304–05 ▪
Gravitational waves 328–31

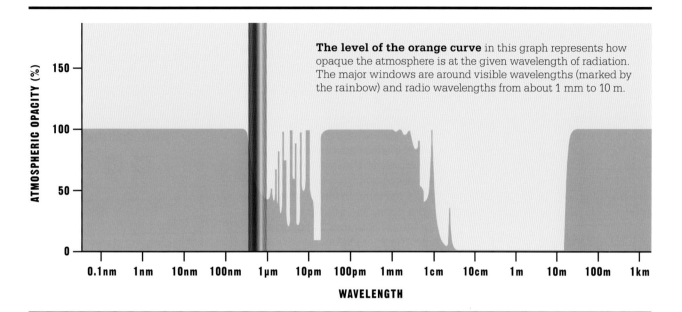

The level of the orange curve in this graph represents how opaque the atmosphere is at the given wavelength of radiation. The major windows are around visible wavelengths (marked by the rainbow) and radio wavelengths from about 1 mm to 10 m.

of these kinds of radiation. The waves are absorbed by the air's molecules, reflected back into space, or scattered in all directions into a meaningless hodgepodge. As a result, it is almost impossible to garner information about most kinds of non-visible radiation from terrestrial observatories.

Spitzer's 1946 paper, entitled *Astronomical Advantages of an Extra-terrestrial Observatory,* highlighted the problem of detecting non-visible radiation. His solution was to put a telescope into space. But Spitzer also highlighted the obstacles to such a proposal: first, the technological challenge of inventing space travel, and second, that of designing an instrument capable of operating in space by remote control from the ground.

Twinkle, twinkle little star

The rest of Spitzer's paper was focused on solving a problem that had frustrated astronomers for centuries—the sky itself. Viewed from Earth, the stars appear to twinkle. This effect is caused by the star's light shifting back and forth, and rising and falling in brightness. This is not a property of starlight, but is caused by Earth's thick atmosphere. The twinkle becomes more marked as magnification increases, making the objects appear shaky and fuzzy in the eyepiece of a telescope or as diffuse smears of light in photographs.

The scientific term for twinkling is scintillation. It is caused when light passes through layer upon layer of turbulent air in the atmosphere. The turbulence itself has no effect on the light, but the density and temperature differences that are making the air churn and swirl do have an effect. As the starlight passes through one pocket of air to another of a different density, it refracts slightly, with some »

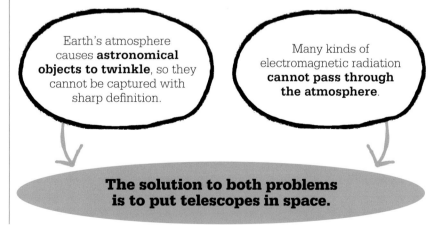

Earth's atmosphere causes **astronomical objects to twinkle,** so they cannot be captured with sharp definition.

Many kinds of electromagnetic radiation **cannot pass through the atmosphere**.

The solution to both problems is to put telescopes in space.

wavelengths bending more than others. As a result, the straight beam of light that traveled to Earth across the cosmos starts to follow an ever-changing and haphazard zigzagging path through the air. A telescope or a naked eye focused on it will see a fluctuation in brightness as some of the light is directed in and out of that line of sight.

The impact twinkling has on capturing sharp astronomical images is called "seeing." When the atmosphere is very still and seeing is good, the image of a distant star in a telescope is a small steady disk. When seeing is poor, the image breaks up into a squirming cluster of dots. An image taken over a period of time

Adaptive optics requires a clear star as a reference point. As these are hard to find, a sodium laser creates a "star" by lighting up dust in the high atmosphere.

is smeared out into a larger disk. The effect is similar to the telescope being out of focus.

Improving the view
Observational conditions change constantly with the atmosphere. Before the 1990s, observers simply waited until distortions were at a minimum. For instance, high winds clear away turbulence, creating near-perfect viewing conditions. In the late 1940s, astronomers started to use movie cameras to film the sky in the hope that, among the thousands of frames filmed over time, there would be the odd "lucky image" that captured the sky in crystal clarity. Another solution was to go higher. Today, the world's most effective terrestrial observatories are invariably built at the top of high, arid mountains, where cloud cover is minimal, and the air above is generally calm.

Our knowledge of stars and interstellar matter must be based primarily on the electromagnetic radiation which reaches us.
Lyman Spitzer Jr.

With the advent of powerful computers in the 1990s, earthbound astronomers began using adaptive optics (AO) to correct the problems of astronomical seeing. AO measures distortions in the arriving light and evens it out, just as a distorted mirror might be used to correct a deformed image to make it look like the original image prior to deformation. AO systems use minutely adjustable mirrors and other optical devices, but they also rely heavily on computers to filter out the atmospheric "noise" from images. Despite the dramatic improvements brought about by AO, however, a large telescope in orbit, which could observe in multiple wavelengths of the spectrum, including visible light, was the ultimate goal for astronomy.

The road to Hubble
As the leading voice in the field, Spitzer had been made head of NASA's task force for developing the Large Space Telescope (LST) program in 1965. In 1968, NASA scored its first space-telescope success with the Orbiting Astronomical Observatory (OAO-2), which took high-quality images in ultraviolent (UV)

A machine polishes the Hubble's mirror. Its 100-in (2.4-m) aperture may seem small today, yet it is the same size as the Hooker Telescope, which was the world's largest telescope until 1948.

light, doing much to raise awareness of the advantages of space-based astronomy.

Spitzer's LST aimed to achieve more dramatic results than the OAO-2, observing near and far objects with the visible light spectrum. His team settled on a 10-ft 5-in (3-m) reflecting telescope and a launch was scheduled for 1979. However, the project became too expensive for its budget. The aperture was reduced to a less costly 100 in (2.4 m), and LST was postponed to 1983. As that year came and went, no launch occurred, but Spitzer persisted and the project continued. In the meantime, LST was renamed the Hubble Space Telescope (HST) after Edwin Hubble, who had first grasped the true scale of the universe (pp.172–77). By now, the telescope's mirrors had been constructed. To help reduce weight, a top layer of low-expansion glass sat on a honeycomb support. The shape of the mirrors was crucial. During construction, they were

held on a support that emulated weightlessness to ensure they did not warp in space. The glass had to be polished into a curve with an accuracy of 10 nanometers. This would make it possible for HST to view everything from UV light to the upper end of the infrared spectrum.

Further delays pushed the launch of HST to 1986, but then tragedy struck with the explosion

of Space Shuttle Challenger on January 28, 1986, with the result that NASA's shuttle fleet was grounded for two years.

Finally, on April 24, 1990, Space Shuttle Discovery hauled the 11-ton HST to its orbit 335 miles (540 km) above Earth. Spitzer had finally realized the dream of his career—a telescope in space unencumbered by the problems of poor seeing and an atmosphere partly opaque to ultraviolet and infrared rays.

Hubble trouble

The problems that had beset the mission on the ground, however, continued in space. The first images sent back by HST were so badly distorted that they were almost worthless. Was HST going to be a worse observational tool than a ground-based telescope? **»**

The Hubble Space Telescope is the realization of Spitzer's vision. It remains one of the finest scientific instruments ever made.

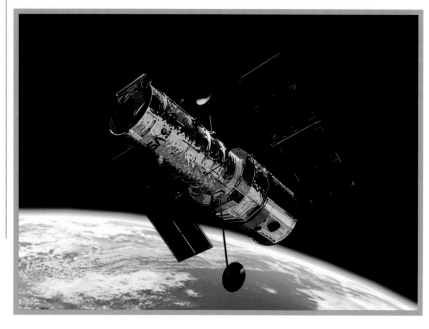

Analysis of the images revealed that the mirror was the wrong shape around the edge. The error was tiny—about 2 millionths of a meter—but enough to send the light captured by the outer part of the primary mirror to the wrong area of the secondary mirror, creating serious aberrations in the images. This was a worrying moment for Spitzer and his team, as it seemed as if HST might be about to prove an embarrassing failure.

Corrective vision

If Hubble was to fulfill its potential, it needed corrective elements added to its optical system. In effect, it was given a pair of eyeglasses. The problem with the primary mirror was precisely calculated by analyzing the telescope's images. The solution was to add carefully designed

US astronaut Andrew Feustel uses a power tool to repair the Hubble Space Telescope during a servicing mission in 2009.

mirrors in front of Hubble's instruments so that the light entering them from the main mirror was correctly focused. Two sets of these mirrors were fitted during a crucial service mission to HST in 1993. They worked perfectly. HST could at last be put to work, and the results were astonishing.

Astronauts serviced HST four more times after 1993 and for the last time in 2009, in one of the final shuttle missions. The shuttles were retired in 2011, after which it would not be possible to service HST again. However, that final service added significant upgrades, which mean that HST may remain in use until 2040.

Ultra deep, ultra clear

Despite its shaky start, HST has surpassed all expectations. The telescope has made 1.2 million observations to date during its 3-billion-mile (5-billion-km) journey around Earth. Despite traveling at 17,000 mph (27,000 km/h), it can pinpoint a position in space to an accuracy of 0.007 arc seconds—which is like hitting a penny coin from 200 miles (300 km) away. It can resolve an object that is 0.05 arc seconds. NASA likened this to standing in Maryland and viewing two fireflies in Tokyo, Japan. Astronomers worldwide began booking HST's time to see objects of interest. The archive of everything it has seen—totaling 100 terabytes and counting—can be viewed on a public website.

Many of HST's observations have looked deep into space— and far back in time. In 1995, the Deep Field image focused on an empty patch of space, one 24-millionth of the total sky. Combining 32 long exposures revealed a number of unknown galaxies that were 12 billion light-years away—light that began its

Nature has thoughtfully provided us with a universe in which radiant energy of almost all wavelengths travels in straight lines over enormous distances with usually rather negligible absorption.
Lyman Spitzer Jr.

Taken in 2004, the Ultra Deep Field reveals thousands of jewel-like galaxies in a variety of shapes, colors, and ages. The red galaxies are the most distant.

journey just 1.5 billion years after the Big Bang. In 2004, the Ultra Deep Field showed objects 13 billion light-years away, and in 2010, HST used infrared radiation to make the eXtreme Deep Field of objects that existed just 480 million years into the history of the universe. To see farther than this will require the infrared James Webb Space Telescope in 2018.

Spitzer in space

HST is the most famous of the four great observatories that are Lyman Spitzer Jr.'s legacy. Between 1991 and 2000, the Compton Gamma Ray Observatory looked at gamma-ray bursts, energetic events that

occur at the edge of the visible universe. The Chandra X-ray Observatory was launched in 1999, and is tasked with finding black holes, infant solar systems, and supernovae. The final member is the Spitzer Space Telescope, which entered space in 2003. One of its tasks was to peer

into nebulae to pick out the hot zones where stars are forming. In 2009, the liquid helium that kept its heat-sensitive detectors cool finally ran out.

Observatories can be placed in orbit around the sun rather than Earth, where it is easier to shield them from the sun's heat and light and they have a wide, unobstructed view of the sky. Today, there are about 30 observatories in orbit, sending back images. NASA's Kepler, which searches for extrasolar planets, and two ESA missions, Herschel and Planck, are examples. All were launched in 2009. Herschel was the largest infrared telescope ever put into space, while Planck studied the cosmic microwave background. In 2015, ESA launched the LISA Pathfinder to test the technology for a space observatory that would detect not electromagnetic waves, but gravity waves. Not even Lyman Spitzer Jr. could have predicted such an advance. ■

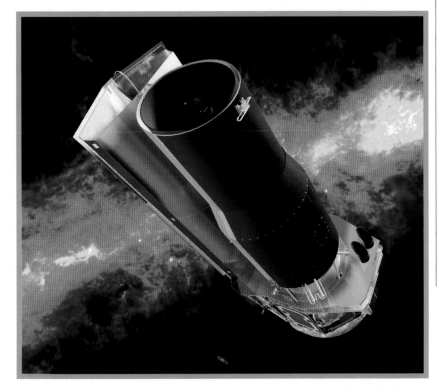

The Spitzer Telescope was named by NASA to honor the vision and contributions of Lyman Spitzer Jr. It was initially called the Space Infrared Telescope Facility.

IT TOOK LESS THAN AN HOUR TO MAKE THE ATOMIC NUCLEI

THE PRIMEVAL ATOM

IN CONTEXT

KEY ASTRONOMERS
George Gamow (1904–1968)
Ralph Alpher (1921–2007)

BEFORE
1939 Hans Bethe describes two ways in which helium can be made from hydrogen in stars.

AFTER
1957 Fred Hoyle and colleagues set out eight processes by which chemical elements can be synthesized from other elements in stars.

1964 German–US physicist Arno Penzias and US astronomer Robert Wilson discover the cosmic microwave background radiation.

1970s The mass of atom-based matter (made of protons and neutrons) as calculated by Big Bang nucleosynthesis is found to be much less than the observed mass of the universe. This puzzle is largely resolved by suggesting the existence of dark matter.

If the Big Bang theory is right, during the universe's **first few moments**, temperatures were **exceedingly high**.

For a **short window of time**, conditions were right for protons and neutrons to combine **to form atomic nuclei**.

It took less than an hour to make the atomic nuclei.

In 1931, Georges Lemaître suggested that the universe originated from the explosion of an initial, extremely dense, "primeval atom" and has been expanding ever since—now known as the Big Bang theory. However, by the mid 1940s, the theory was in need of additional evidence to sustain its credibility.

A Ukrainian physicist named George Gamow began thinking about conditions at the start of the universe as it was proposed by Lemaître. He quickly realized that it would have been unimaginably hot. Matter would have consisted of a frenzy of elementary particles (particles that cannot be broken down into smaller particles)—considered at that time largely to comprise protons, neutrons, and electrons. Temperatures would have been too high for these particles to join up, except very briefly. However, after several seconds of existence, the universe would have expanded and cooled to the point where protons and neutrons might be held together by an interaction

See also: The birth of the universe 168–71 ▪ Energy generation 182–83 ▪ Nucleosynthesis 198–99

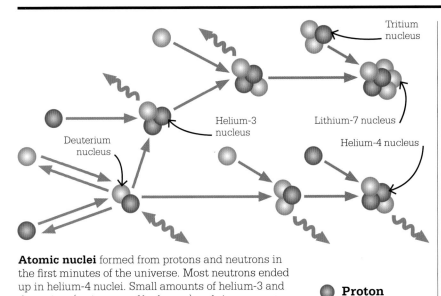

Atomic nuclei formed from protons and neutrons in the first minutes of the universe. Most neutrons ended up in helium-4 nuclei. Small amounts of helium-3 and deuterium (an isotope of hydrogen) and tiny amounts of lithium-7 were also made. Another isotope of hydrogen—tritium—formed, and decayed to helium-3. Energy was released in the form of gamma rays.

● **Proton**
○ **Neutron**
〰 **Gamma ray**

called the strong nuclear force, thus creating an array of atomic nuclei. Gamow believed that, after a few initial "seed" nuclei had assembled from protons and neutrons, others might have been built up by the successive addition of neutrons, with some decaying to protons. At a later time, all the nuclei might have captured electrons to form the atoms of the chemical elements.

Doing the math
Gamow asked an American graduate student, Ralph Alpher, to work out the details of his idea. Alpher and a colleague, Robert Herman, performed extensive mathematical calculations. They found that the right conditions for protons and neutrons to come together existed only for a short window of time of a few minutes. Their calculations showed that most of the universe's neutrons

would have ended up combined with protons in an isotope (one of the possible alternative forms) of helium, helium-4. A small number would have become other small atomic nuclei. In addition, there would have been large

numbers of free protons (hydrogen nuclei) left at the end of the process, plus some unstable nuclei, which would have quickly decayed.

Their calculations showed that the universe would have consisted of about 25 percent helium, with the rest mainly hydrogen. The paper that Alpher and Gamow published also argued that other heavier nuclei might have been created in the Big Bang through successive additions of neutrons.

Correct predictions
It was eventually recognized through the work of scientists such as Fred Hoyle that heavier elements such as carbon were created in stars and supernova explosions. Nevertheless, the Alpher–Gamow theory correctly explained the relative abundances of hydrogen and helium, lending considerable support to the theory that the universe began with a Big Bang. It also correctly predicted the existence of the cosmic background radiation that was discovered in 1964 (pp.222–27). ▪

George Gamow

George Gamow was born in Odessa, Ukraine, in 1904. From 1923, he attended the University of Leningrad, studying under Alexander Friedmann. In 1928, Gamow briefly stayed at the University of Göttingen in Germany, where he developed a theory called quantum tunneling. This theory was used by others to explain how the fusion of light atomic nuclei might create energy inside stars. In 1933, he defected from the Soviet Union while attending

a conference in Brussels. At George Washington University in the US, he turned his attention to the evolution of stars. From 1954, Gamow became interested in genetics and biochemistry. He also wrote a number of popular science books and sci-fi novels.

Key works

1948 *The Origin of Chemical Elements* (also called the Alpher–Bethe–Gamow paper)
1952 *The Creation of the Universe*

STARS ARE FACTORIES FOR THE CHEMICAL ELEMENTS
NUCLEOSYNTHESIS

Heavier elements require **high temperatures** to be created.

Suitable conditions for the creation of many elements occur in the evolution of **giant stars**.

Extreme conditions for other elements occur when giant stars disintegrate in **supernova explosions**.

All but a few elements can be created in stars by **eight distinct processes**.

Stars are factories for the chemical elements.

Until the late 1940s, it was not known where the atoms of most chemical elements in the universe—for example, carbon, oxygen, and iron—had come from, nor how they had been made. It had been established in the 1920s that the two lightest elements, hydrogen and helium, made up most of the universe's matter, and in 1948, George Gamow and Ralph Alpher showed how all of the hydrogen, most of the helium, and tiny amounts of lithium could have been made in the "Big Bang." However, the origin of other elements was a mystery.

Building to iron
The discovery of their origin was made largely thanks to the work of British astronomer Fred Hoyle. Starting from chance conversations with leading astronomers in the US during an academic tour in 1944, he developed an idea that most chemical elements might be created step-by-step by nuclear reactions in stars—a process called nucleosynthesis. Hans Bethe had already shown in 1939 that hydrogen could combine to make helium in star cores, but Bethe offered no suggestions for how

See also: Energy generation 182–83 ▪ The primeval atom 196–97

heavier elements such as iron and carbon could form: the cores of stars were not considered hot enough for such elements to form through nuclear fusion processes. Hoyle, however, thought there could be processes that would sufficiently raise the core temperature of a large enough star.

In 1946, Hoyle showed that in the cores of massive stars where the temperature soars to billions of degrees heavier elements could be made in circumstances called nuclear thermal equilibrium. Such a star would eventually explode as a supernova and eject its heavy elements. In 1954, Hoyle went on to describe how in a massive star that has exhausted its hydrogen fuel its core would contract and heat up before exploding, and helium atoms would start fusing to create carbon. At the end of this phase, carbon atoms would fuse to create heavier, more stable, elements. This could account for the creation of several elements up to iron, which has the most stable of all atomic nuclei.

Building nuclei heavier than iron would be more problematic, since it is an energy-consuming process, whereas creating elements lighter than iron releases energy.

Further developments
There was a flaw in Hoyle's stellar element-building scheme, however. A key step known as the triple alpha process, in which three helium nuclei fuse to make carbon, appeared to be too slow. Hoyle insisted that there must be a mechanism allowing it to happen at a faster speed and, in 1953, a certain property of carbon was discovered that explained it.

Hoyle also explored other processes by which many more elements might be forged in stars. Some of these processes could only occur in the violence of a supernova explosion at the end of a giant star's life (pp.180–81). Hoyle's work thus explained not only where chemical elements came from but also how they came to be dispersed throughout the universe. ▪

Fred Hoyle

Fred Hoyle was born in Yorkshire, England, in 1915. He attended the University of Cambridge from 1933, gaining a degree in mathematics. During World War II, he worked on radar systems for the British Admiralty. In 1957, Hoyle joined the staff of the Hale Observatories in California and the following year became professor of astrophysics at Cambridge University. Aside from his work on the origin of elements in stars, Hoyle is best known as a proponent of the Steady State theory. This claims that as the universe expands, its average density is kept constant as new matter is continuously created. Ironically, Hoyle coined the phrase "Big Bang" for the main rival theory, during a popular radio talk show. From the 1960s, the Steady State theory fell out of favor. In later life, he took particular interest in the presence of organic molecules in comets, which he believed had brought life to Earth.

Key works

1946 *The Synthesis of the Elements from Hydrogen*
1950 *The Nature of the Universe*

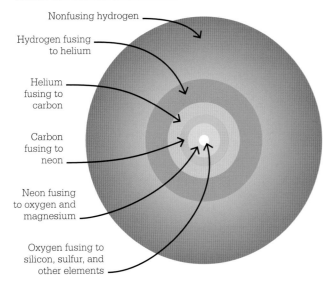

Nonfusing hydrogen

Hydrogen fusing to helium

Helium fusing to carbon

Carbon fusing to neon

Neon fusing to oxygen and magnesium

Oxygen fusing to silicon, sulfur, and other elements

Hoyle showed how in high mass stars several elements ranging in mass from carbon to iron could be created by fusion reactions occurring at the same time in shells around the core of the star. The number of shells would increase as the star aged. This diagram shows shells with element-building reactions in an aging red supergiant.

SITES OF STAR FORMATION

DENSE MOLECULAR CLOUDS

IN CONTEXT

KEY ASTRONOMER
Bart Bok (1906–1983)

BEFORE
1927 American astronomer Edward Barnard catalogs 350 mysterious dark nebulae.

1941 Lyman Spitzer Jr. proposes that stars form from interstellar matter.

AFTER
1980s Herbig-Haro Objects are identified as jets of plasma released by very young stars within star-forming regions.

1993 High-frequency radio astronomers observe protostars in Bok globules.

2010 The Spitzer Space Telescope takes infrared images of the interiors of 32 Bok globules. The images show warm cores to 26 of the globules and evidence that multiple stars are forming within about two-thirds of them, each star at a different stage in its formation.

Bart Bok was an unusual observational astronomer. He made his career studying not that which he could see, but that which he could not. In the 1940s, while observing bright nebulae for evidence of star formation, Bok noticed many small regions that were completely dark. They were surrounded by stars but appeared to be empty holes

The Caterpillar Bok globule in the Carina nebula is shown here in a photograph from the Hubble Space Telescope. Stars are forming behind the dense veils of dust and gas.

in space. In 1947, in collaboration with US astronomer Edith Reilly, Bok proposed that these bodies were dense clouds of gas and dust that were in the process of collapsing under their own gravity, and that a new star was forming inside. The dust, formed from specks of silica, water ice, and frozen gases, was dense enough to block out the light of surrounding stars. As a result, no light came out of the cloud, and the light from any stars behind the cloud (from Earth's perspective) could not pass through. Bok and Reilly likened these clouds to a caterpillar's cocoon, from which a new, brilliant star would one day emerge.

Dark nebula

The dense clouds became known as "Bok globules." In visible light, they appeared only as a silhouette against the backdrop of stars, with a little light passing through their diffuse outer edges, and for many years it was difficult to observe them in any great detail. This meant that Bok and Reilly's proposal remained hypothetical for several decades. By the 1990s, however, a few years after Bok's death, infrared and radio astronomy

See also: Space telescopes 188–95 ▪ Inside giant molecular clouds 276–79 ▪ Ambartsumian (Directory) 338

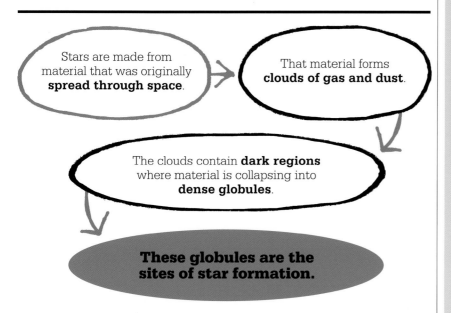

Stars are made from material that was originally **spread through space**.

That material forms **clouds of gas and dust**.

The clouds contain **dark regions** where material is collapsing into **dense globules**.

These globules are the sites of star formation.

Bart Bok

Bartholomeus Jan Bok was born near Amsterdam in 1906. His interest in astronomy began in scout camps, where he was able to observe the stars in clear skies away from the city. Bok began his academic career at two Dutch universities, first at Leiden and then, as a Ph.D. student, at Groningen. In 1929, he opted to transfer his studies to Harvard and work under the supervision of Harlow Shapley. He had fallen in love with Priscilla Fairfield, one of Shapley's researchers, and married her two days after arriving in the US. The pair worked closely together from then on, although Shapley only paid Bok, now going by the Americanized name of Bart.

The Boks worked at Harvard for 30 years before they were invited to set up an observatory in Canberra, Australia, in 1957. In 1966, they returned to the US to run observatories in the southwest. Priscilla died in 1975. Bart continued working until his death in 1983.

Key work

1941 *The Milky Way* (with Priscilla Fairfield Bok)

were able to peer inside the dusty cloud and pick out areas of heat. These areas indicated that Bok's hypothesis was right—stars were indeed forming within.

Bok globules are now seen as a small and dense type of "dark molecular cloud," mostly found in the spiral arms of the Milky Way. They are about one light-year wide and found in H II regions— vast interstellar spaces filled with low-density ionized hydrogen atoms. H II regions form when the ultraviolet emissions of supergiant blue stars ionize the surrounding medium (the matter in interstellar space)—stripping atoms of their electrons to create positively charged ions.

Cold cloud

Bok globules weigh about 50 times the mass of the sun. They mostly comprise molecular hydrogen (H_2), but about 1 percent is dust. The dust comprises particles made of multiple molecules, and is highly

concentrated. The obscuring effect of the dust prevents heat from penetrating the globule, and the temperatures inside are among the coldest measured in the universe—about 10 Kelvin. The outward pressure of the cold gas is weaker than the inward pull of gravity, and a shockwave from a nearby supernova can make the cold clouds collapse. They then become denser and denser until a hot stellar core forms. ▪

For many years I have been a nightwatchman of the Milky Way galaxy.
Bart Bok

NEW WIN
ON THE U
1950–1975

DOWS
NIVERSE

Dutch astronomer **Jan Oort** argues that **a cloud of comets** is to be found orbiting the sun at the edge of the solar system.

In a speech to the US Congress, President **John F. Kennedy** announces the intention to put a **man on the moon** by the end of the decade.

Dutch astronomer **Maarten Schmidt** demonstrates that **quasars**, discovered by radio astronomers in 1960, are distant galaxies.

1950 1961 1963

1959 1962 1964

Giuseppe Cocconi and **Philip Morrison** propose a place on the electromagnetic spectrum to look for **messages from aliens**.

In his book *universe, Life, Intelligence*, Soviet astronomer **Iosif Shkovsky** speculates about **extraterrestrial life**.

US astronomers **Arno Penzias** and **Robert Wilson** discover **cosmic microwave background radiation**, evidence for the Big Bang.

The launch in 1957 by the Soviet Union of the first artificial Earth satellite, Sputnik 1, was a turning point in history, both politically and scientifically. Politically, it provoked the "space race"—a battle for supremacy in space between the Soviet Union and the United States. Scientifically, it opened up new possibilities for astronomy. Telescopes could be put into orbit, giving them a view unhindered by Earth's atmosphere. Robotic explorers could be sent into the solar system to study planets and other bodies from close quarters. NASA's Mariner 2, the first successful mission to another planet, was launched toward Venus in 1962.

Meanwhile, ambitious projects to send people into space carried on. In 1961, Soviet cosmonaut Yuri Gagarin became the first man to orbit Earth. Just eight years later, the Americans succeeded in putting men on the moon. They would bring pieces of our satellite back with them, which would shed new light on the formation of the early solar system.

Seeing from space
Until the mid-20th century, astronomers peered through the narrowest of atmospheric "windows," looking only at visible light. Earth's atmosphere is transparent to only two parts of the electromagnetic spectrum: the narrow waveband we call visible light (with a little ultraviolet and infrared on either side) and the radio band. Astronomers had no means of knowing about the intense output of ultraviolet, X-, and gamma rays from hot, high-energy cosmic sources, which are absorbed by Earth's atmosphere. Cool and concealed constituents of the universe, such as infant stars, were also lurking, waiting for their infrared radiation to be detected.

Radio astronomy
The main kind of "invisible astronomy" open to ground-based observers is radio astronomy. After tentative beginnings in the 1930s, the field developed rapidly in the 1950s. Scientists who had worked in radio science during World War II were instrumental in founding astronomy research groups, such as those at Cambridge and Manchester in the United Kingdom. Also around that time, astronomers at Harvard in the United States identified radio

British mathematician **Roger Penrose** describes spacetime "**singularities**" at the heart of black holes.

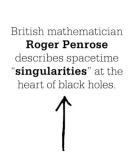

1964

1967

At the University of Cambridge, research student **Jocelyn Bell** detects the radio signal from a **pulsar**, a rapidly spinning neutron star.

OAO-2, the first successful orbiting observatory, is launched by NASA. It is equipped with **ultraviolet telescopes**.

1968

1969

The **Apollo 11** mission completes President Kennedy's project as **Neil Armstrong** sets foot on the moon.

Soviet astrophysicist **Victor Safronov** figures out the mathematics behind the **nebular hypothesis** of the solar system's formation.

1969

1973

NASA launches the **Uhuru** observatory, the first **X-ray telescope** to be placed into orbit.

---

emissions from the hydrogen gas that pervades interstellar space. This discovery allowed the spiral structure of our galaxy tobe mapped for the first time.

In the 1960s, radio astronomers discovered the new phenomena of quasars and pulsars. We now know that "quasi-stellar radio sources"— or quasars for short—are distant galaxies, which have at their heart an immense black hole producing prodigious amounts of energy. Pulsars are neutron stars—bizarre balls of compacted matter— spinning at high speed. Their discovery confirmed theoretical predictions made decades earlier.

All windows wide open

By the early 1970s, the first orbiting observatories were operational, and exploring the ultraviolet, X-, and gamma rays in the skies. Several series of satellites had been launched under programs such as the Small Astronomy Satellites (SAS) and Orbiting Astronomical

It's highly unlikely there are two lots of little green men, on opposite sides of the universe, both deciding to signal to a rather inconspicuous planet Earth.
Jocelyn Bell Burnell

Observatories (OAO). These included the 1970 SAS-1 for X-ray astronomy (named Uhuru, the Swahili word for freedom, in honor of Kenya, from where it was launched) and OAO-3 (named Copernicus, for the 500th anniversary of the astronomer's birth in 1473). Infrared astronomy from orbit took longer to achieve because the telescope must be kept very cold, but the first surveys of the infrared sky were undertaken from the ground.

All parts of the electromagnetic spectrum were now open to investigation and the hunt was even on for the elusive particles known as neutrinos. Other worlds in the solar system had become the targets for future missions. Within three decades, new technology had transformed astronomers' outlook on the universe. ∎

A VAST CLOUD SURROUNDS THE SOLAR SYSTEM
THE OORT CLOUD

IN CONTEXT

KEY ASTRONOMER
Jan Oort (1900–1992)

BEFORE
1705 Edmond Halley correctly predicts the return of a comet.

1932 Estonian astronomer Ernst Öpik proposes that long-period comets originate in an orbiting cloud at the edge of the solar system.

1943 Kenneth Edgeworth suggests that the solar system beyond the orbit of Neptune is occupied by many small bodies, some of which become comets.

1950 Fred Whipple proposes that comets are a conglomerate of ices and rocky material.

AFTER
1992 David Jewitt and Jane Luu discover the first Kuiper belt object other than Pluto.

2014 The Philae lander of the Rosetta spacecraft lands on the comet 67P/Churyumov–Gerasimenko.

In 1950, reviving a theory proposed by Estonian astrophysicist Ernst Öpik, a Dutch astronomer named Jan Oort argued that there is a reservoir of comets at the edge of the solar system. At the time, it was known that two main classes of comets visit the inner solar system— the region comprising the four rocky planets. Short-period comets visit at intervals of less than 200 years, and orbit in the plane in which the planets lie. Long-period comets visit at intervals that are longer than 200 years and have orbits inclined at all directions and angles to the plane of the solar system. The origin of either class was subject to speculation.

Long-period comets
Oort's idea provided a solution to the origin of long-period comets. A comet periodically visiting the inner solar system will eventually collide with the sun or a planet, or will be ejected from the solar system after its orbit is disturbed by passing near a planet. This means that comets cannot simply have been going around in their orbits since the solar system formed. Oort suggested that long-period comets passing into the inner solar system are just a small subset of all the comets orbiting the sun. The comets seen from Earth have been nudged out of the distant comet reservoir, perhaps by a passing star, and have plummeted toward the sun, taking up long, elliptical orbits.

Spherical cloud
Examining the orbits of numerous long-period comets and the farthest distance from the sun they reach, Oort reasoned that the reservoir for the long-period comets is a shell-like, spherical region ranging up to a maximum of about 4.5–19 trillion miles (7.5–30 trillion kilometers) from the sun. This region, envisaged to contain billions or trillions of comets, is now known as the Oort cloud. However, it has since been established that short-period comets probably originate from a disklike region much closer to the sun, the Kuiper belt. ∎

See also: Halley's comet 74–77 ▪ The Kuiper belt 184 ▪ The composition of comets 207 ▪ Exploring beyond Neptune 286–87

COMETS ARE DIRTY SNOWBALLS
THE COMPOSITION OF COMETS

IN CONTEXT

KEY ASTRONOMER
Fred Whipple (1906–2004)

BEFORE
1680 German astronomer Gottfried Kirch is the first to find a comet using a telescope.

1705 Edmond Halley shows that the comet of 1682 is the same object as the comets of 1531 and 1607.

AFTER
2003 A survey by *The Astrophysical Journal* finds that over 50 years Whipple's 1950 and 1951 papers were the most cited papers in astronomy.

2014 Rosetta completes a rendezvous with comet 67P/Churyumov–Gerasimenko and successfully sends the Philae lander onto its surface.

2015 New research suggests comets are like "deep fried ice cream," with an icy crust, a colder and more porous inside, and a topping of organic compounds.

The arrival of a comet can be a spectacular event, with the brightest of them remaining visible even during the day. Yet US astronomer Fred Whipple showed that these dazzling astral visitors are, in fact, extremely dark objects.

In 1950, Whipple proposed that comet nuclei—the "bodies" of these early solar system remnants, as opposed to the bright, gassy tails visible from Earth—are a rough mix of meteoric materials and volatile ices. The ices are mainly frozen water, along with frozen gases such as carbon dioxide, carbon monoxide, methane, and ammonia. The rest is rock and dust. A black crust of tarry organic compounds, similar to crude oil, coats the surface. Comet nuclei are among the darkest objects in the solar system. Only 4 percent of the light that falls on them is reflected. Fresh black asphalt, by comparison, reflects nearly twice as much.

Whipple's "icy conglomerates" concept explained how comets could repeatedly send out trails of vapor as they passed the sun. The idea was accepted, albeit with the catchier name "dirty snowballs" (later modified to "icy dirtball" on the discovery that comets contain more dust than ice). However, Whipple had to wait until 1986 for his ideas to be confirmed. That year, the Giotto spacecraft drew alongside Halley's comet, and took close-range pictures of the dark nucleus usually hidden by the bright coma. ∎

Images from the Giotto spacecraft revealed the nucleus of Halley's comet to be a dark peanut-shaped body, ejecting two bright jets of material.

See also: Halley's comet 74–77 ▪ Asteroids and meteorites 90–91 ▪ Astrophotography 118–19 ▪ The Kuiper belt 184

THE WAY TO THE STARS IS OPEN
THE LAUNCH OF SPUTNIK

IN CONTEXT

KEY ASTRONOMER
Sergei Korolev (1907–1966)

BEFORE
1955 The US announces plans to launch a satellite for the International Geophysical Year.

1955–57 Wernher von Braun, a former Nazi rocket scientist recruited by the US, launches the Jupiter-C rocket, capable of carrying a satellite into orbit.

AFTER
1957 On board Sputnik 2, the dog Laika becomes the first large animal in space.

1958 Juno 1 launches the first US satellite, Explorer 1.

1961 Yuri Gagarin orbits Earth in Vostok 1.

1963 Valentina Tereshkova becomes the first woman and first civilian in space.

1965 Voskhod 2 becomes the first two-man crew in space; Alexei Leonov completes the first spacewalk.

With the launch of Sputnik 1, the world's first artificial satellite in 1957, the Soviet Union won the first round in the superpower Space Race. This momentous accomplishment was achieved primarily through the drive and genius of one man—the tough and pragmatic "Chief Designer" Sergei Korolev, the scientist who masterminded the top-secret space program. Until the collapse of the Soviet Union in 1991, very little was known about Korelev in the West. The Soviets had referred

Vostok 1, designed by Korolev, launched Yuri Gagarin from the Baikanur Cosmodrome on April 12, 1961. During the flight Gagarin said, "I don't see any God up here."

to him only as Chief Designer for fear that the Americans might try to assassinate him.

Korolev had trained as an aircraft designer, but his true talent lay in strategically planning vast and complex projects under extreme political pressure. By 1957, he already had one major "first" under his belt, with the launch

See also: The Space Race 242–49 ▪ Exploring the solar system 260–67 ▪ Exploring Mars 318–25

> The time will come when a spacecraft carrying human beings will leave the Earth and set out on a voyage to distant planets— to remote worlds.
> **Sergei Korolev**

of the world's first intercontinental ballistic missile in 1953. During an extraordinarily successful career, he would go on to have several more, each time catching the US by surprise. (He was aided by the fact that the Soviet space agency could keep its plans secret, while those of its US rivals were announced at press conferences.) In 1957, Korolev launched a dog, Laika, into orbit, which prepared the way for the first man in space in 1961, and the first woman in 1963. Two years later this was followed by the first two-man crew and first spacewalk.

The Space Race

However, it was the launch of Sputnik 1 on October 4, 1957, that was to have the biggest impact on American public opinion. Russia was routinely caricatured in the US media as a backward country, but the launch of Sputnik was undeniable evidence of Soviet technological superiority, and soon stoked Cold War paranoia.

The orbiting "red moon" raised the possibility of nuclear bombs raining down on American cities, and the fears it aroused were seized upon by political opponents of US president Eisenhower.

When the Soviets put the first man in space in 1961, NASA's press officer, woken by a 4:30 a.m. call, said, "We're all asleep down here." The next day's headline read: "Soviets put man in space. Spokesman says US asleep." The perceived technological gap kick-started the US space program and resulted in the Apollo missions.

With Korolev's sudden death in 1966, the Soviets' winning streak ended. Their space program had lost the magnetic personality that held a vast, complex enterprise together, and became embroiled in politics and bureaucracy. It is intriguing to wonder whether the Soviet Union might have put the first man on the moon with Korolev at the helm. Instead, the US gained the initiative and achieved that goal in July 1969. ▪

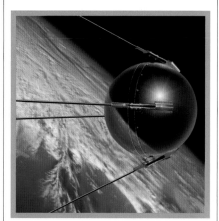

Sputnik 1 was a relatively simple craft, comprising a metal sphere containing radio, batteries, and a thermometer. Its psychological impact on the US was immense.

Sergei Korolev

Born in 1906, Sergei Pavlovich Korolev studied under the aircraft design pioneer Andrei Tupolev, becoming the chief engineer at Russia's Jet Propulsion Research Institute by the mid-1930s. In 1938, however, he was a victim of Stalin's purges. Denounced by his colleagues, Korolev was tortured and sent to the gulag in eastern Siberia, where he worked in a gold mine and contracted scurvy.

Released in 1944, he was appointed head of the secret Scientific Research Insitute No. 88—the Soviet space program. Korolev won political favor with his idea for Sputnik 1—a heavier artificial satellite than the Americans were capable of launching at the time. Larger-than-life, Korolev had a roaring temper and was prone to shouting expletives. But, despite his bearlike frame and limitless energy, he was weaker than he appeared. Korolev had suffered a heart attack in the gulag. He could not turn his neck and his jaw had been broken so badly that it hurt for him to laugh. He died during routine colon surgery in 1966.

THE SEARCH FOR INTERSTELLAR COMMUNICATIONS
RADIO TELESCOPES

IN CONTEXT

KEY ASTRONOMERS
Giuseppe Cocconi (1914–2008)
Philip Morrison (1915–2005)

BEFORE
1924 A "National Radio Silence Day" is instigated to tune in on any potential Martian messages.

1951 US physicists Harold Ewen and E. M. Purcell detect the 21-cm hydrogen line.

AFTER
1961 Frank Drake formulates the Drake Equation to estimate how many intelligent civilizations are likely to lie beyond the solar system.

1977 At Ohio University, Jerry Ehman picks up a sharp signal 30 times the background noise level. This "Wow! signal" has never been detected again.

1999 The SETI@Home network uses the combined power of millions of volunteer desktop computers.

I n September 1959, the scientific journal *Nature* carried a short but hugely influential article: Giuseppe Cocconi and Philip Morrison's "Searching for Interstellar Communications." This introduced an entirely new field of scientific endeavor—speculation on the nature of extraterrestrial life and the possibility of intelligent beings existing outside of Earth. For the first time in scientific history, alien-hunting had been framed as a serious proposition.

The completion, in 1957, of the 250-ft (76-m) Mk 1 radio telescope at Jodrell Bank in England—just in time to track the world's first artificial satellite, Sputnik 1—had brought new possibilities into focus. If equipped with a powerful transmitter, this kind of telescope was capable of communicating across interstellar distances with any civilization that had managed to achieve the corresponding technology. Cocconi and Morrison's paper argued that, on some planet orbiting a distant star, advanced

If there is other **intelligent life** in the universe, it may be **trying to communicate**.

The **21-cm wavelength** emitted by **hydrogen atoms** in the radio band is the same across the universe.

New **radio telescopes** make it possible to **look for messages** in the radio spectrum.

Start the search for interstellar communications at this wavelength.

See also: Life on other planets 228–35 ▪ Exoplanets 288–95

Modulation is a method by which information can be transmitted within a wave signal. The amplitude is kept constant, while the frequency varies.

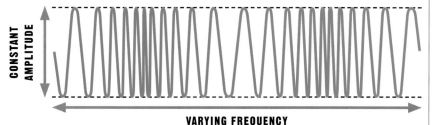

CONSTANT AMPLITUDE

VARYING FREQUENCY

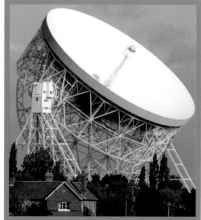

The Lovell (Mk 1) radio telescope at Jodrell Bank, the third-largest radio telescope in the world, was used as part of the Project Phoenix SETI program in the 1990s and 2000s.

societies may already be trying to make contact. The pair suggested looking for signals in the microwave spectrum, identifying likely frequencies and even potential places to start the search for intelligent life.

A place to look

Cocconi and Morrison focused on the "21-cm line," a radiation emission line (characteristic wavelength) of a hydrogen atom. In the high-frequency radio (microwave) band, this 1420 MHz radiation is emitted when protons and electrons within a hydrogen atom change their energy state. Its discovery, in 1951, allowed the distribution of hydrogen across the galaxy to be mapped using radio waves, which, unlike visible light, are not blocked by dust clouds.

Since this line is universal, Cocconi and Morrison argued that it would be known to all intelligent civilizations, and that any search should begin by looking for transmissions around this frequency band. They predicted the most likely form of transmission—a pulse-width-modulated wave, like an FM radio signal, on a loop like a Mayday callout. The modulated wave would carry a constant amplitude but

produce regular pulses of higher frequency. A signal might cycle over long periods, perhaps a number of years.

Future searches

Cocconi and Morrison's ideas dominated the search for extraterrestrial intelligence (SETI) for decades. Acting on the article's recommendations, Frank Drake's pioneering 1960 experiment Project Ozma, at the Green Bank observatory in West Virginia, targeted the close sunlike stars Tau Ceti and Epsilon Eridani, scanning around the 21-cm line. Sadly, the project failed to find

any convincing candidates. Today, many question the wisdom of such limited searches. Instead, SETI researchers look for the chemical or thermal signatures of advanced civilizations, leakage from signals not intended for us, and novel methods of communication using lasers or neutrinos. ▪

Giuseppe Cocconi and Philip Morrison

Giuseppe Cocconi was born in Como, Italy, in 1914. After World War II, he joined Cornell University, New York. Working with his wife, Vanna, Cocconi demonstrated the galactic and extragalactic origins of cosmic rays. Later, he became director of research at CERN (the European Organization of Nuclear Research) in Geneva.

Philip Morrison studied at the University of California, Berkeley, under Robert Oppenheimer. During World War

II, he worked on the Manhattan Project to build the first atomic bomb, and famously shared a car with the core of the Trinity bomb as it was transported to the test site. He later became a vocal antinuclear campaigner. He was a great popularizer of science and narrated the 1977 documentary *Powers of Ten*.

Key work

1959 *Searching for Interstellar Communications*

METEORITES CAN VAPORIZE ON IMPACT
INVESTIGATING CRATERS

IN CONTEXT

KEY ASTRONOMER
Eugene Shoemaker
(1928–1997)

BEFORE
1891 American geologist Grove Gilbert suggests that the moon's craters are the result of meteorite impacts.

1891 Mineralogist Albert E. Foote produces the first geological description of the crater.

AFTER
1980 American physicist Luis Alvarez proposes that a worldwide stratum of shocked (pressurized) quartz between the Cretaceous and Tertiary periods indicates a huge impact, which led to the mass extinction of dinosaurs.

1994 Comet Shoemaker–Levy, codiscovered by Eugene Shoemaker, impacts Jupiter and is observed by the Galileo spacecraft en route to the planet.

So great was the contribution of American geologist Eugene Shoemaker to planetary science that he is the only person whose ashes have been sent to the moon. He was a founding figure of astrogeology, a science that uses the techniques of geology to investigate alien worlds.

Shoemaker's early work hinged on a crater in the Arizona Desert known as Meteor Crater. Early European settlers in nearby Canyon Diablo believed it to be the vent of an

Not going to the moon and banging on it with my own hammer has been the biggest disappointment in life.
Eugene Shoemaker

ancient volcano. Railroad engineers passing through in the 1880s found large iron-rich rocks strewn across the desert, which suggested the crater was caused by the impact of a metallic meteorite. However, this was dismissed because the volume of debris around the rim largely matched the volume of the crater itself. It could not be a meteor crater if there was no meteorite.

In 1903, Daniel Barringer, a mining engineer, searched for the iron meteor under the crater floor, but to no avail. Not until 1960 did Shoemaker find the evidence. The crater contains shocked silica only seen previously at nuclear bomb test sites. These minerals cannot have been made naturally by volcanic forces: only the energy of a meteorite traveling at 40,000 mph (60,000 km/h) could have done it. That energy had vaporized the meteorite, which explained why it was missing. Shoemaker had provided the first proof that large meteorites strike Earth, opening up new possibilities for investigating extraterrestrial objects. ■

See also: Asteroids and meteorites 90–91 ▪ The discovery of Ceres 94–99 ▪ The composition of comets 207

THE SUN RINGS LIKE A BELL
THE SUN'S VIBRATIONS

IN CONTEXT

KEY ASTRONOMER
Robert Leighton (1919–1997)

BEFORE
1954 Canadian astronomer Harry Hemley Plaskett observes the sun's oscillation effect.

AFTER
1970 American physicist Roger Ulrich proposes that the oscillations are from acoustic waves in the solar interior.

1970s Helioseismology opens up a new way of investigating the interior of the sun.

1995 The solar observatory satellite SOHO is launched.

1997 The SOHO team discover "jet streams" of plasma in the convective zone.

1990–2000s Hundreds of thousands of the sun's vibrational modes are identified.

2009 The Kepler satellite measures oscillations in sunlike stars to describe conditions on exoplanets.

I n 1960, American physicist Robert Leighton made observations with a camera that he had devised, which led to the discovery that the sun is "ringing like a bell," as he put it. Working with Robert Noyes and George Simons, Leighton picked up perturbations of the sun's surface using Doppler shift solar cameras. These cameras detected tiny shifts in the frequency of the sun's absorption spectra as its outer layer moved toward or away from Earth.

Five-minute oscillations
The complex patterns of vibrations, with an average period of five minutes (known as "5-minute oscillations"), were at first thought to be a surface phenomenon. Then in 1970, Roger Ulrich explained them as trapped acoustic waves bouncing within the sun from one side to the other, causing the star's surface to wobble as it resonated.

Today, these waves allow scientists to investigate the interior of the sun, in much the same way that acoustic earthquake waves

With his very inquisitive mind, every funny effect that you'd see in nature he'd try to explain.
Gerry Neugebauer
Physicist and colleague of Robert Leighton

reveal the inner composition and structure of Earth. Known as helioseismology, this process is often compared to trying to build a piano by studying the sounds it makes when it falls down a flight of stairs, but it has produced a model of the sun's interior processes. The model places tight constraints on the amount of helium in the star's core, which has important consequences for models of the early universe. ∎

See also: The properties of sunspots 129 ▪ The Homestake experiment 252–53 ▪ Exoplanets 288–95

THE DATA CAN BEST BE EXPLAINED AS X-RAYS FROM SOURCES OUTSIDE THE SOLAR SYSTEM

COSMIC RADIATION

IN CONTEXT

KEY ASTRONOMER
Riccardo Giacconi (1931–)

BEFORE
1895 German physicist Wilhelm Röntgen discovers high-energy radiation, which he names "X-rays."

1949 Solar X-rays are first detected by sounding rockets.

AFTER
1964 Cygnus X-1, the first confirmed black hole binary system, is discovered.

1966 X-rays are detected from galaxy cluster M87, in the Virgo cluster.

1970 Uhuru, the first dedicated X-ray mission, is launched.

1979 X-rays from Jupiter are detected by the Einstein Observatory.

1999 The Chandra X-ray Observatory is launched.

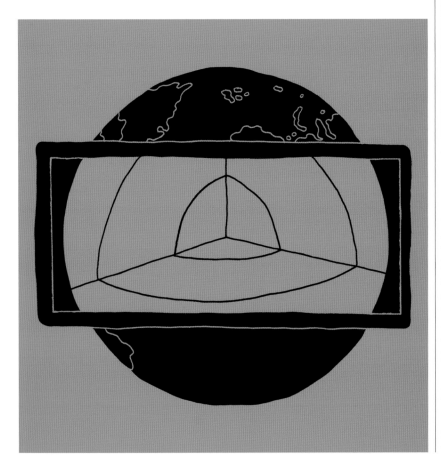

X-rays are a form of high-energy, electromagnetic radiation released by extremely hot objects. In the early 20th century, astronomers realized that space should be flooded with X-rays from the sun. Moreover, the sun's X-ray spectrum would reveal a lot about the processes at work within the star. However, X-ray astronomy was not possible until the advent of rockets and satellites. Despite their energy, X-rays are easily absorbed, which is why they are so good at imaging the body. Water vapor in Earth's atmosphere effectively blocks X-rays from reaching the surface—a good thing for life, because high-energy

See also: The Homestake experiment 252–53 ■ Discovering black holes 254

Nothing is going to happen unless you work with your life's blood.
Riccardo Giacconi

X-rays can cause damage and mutations when they impact on soft, living cells.

The first glimpse of the sun's X-rays came in the late 1940s, during a US Naval Research Laboratory (NRL) program to study Earth's upper atmosphere. A team led by US rocket scientist Herbert Friedman fired German V-2 rockets into space equipped with X-ray detectors—essentially modified Geiger counters. These experiments provided the first incontrovertible evidence of X-rays from the sun. By 1960, researchers were using Aerobee sounding rockets to detect X-rays, and the first X-ray photos of the sun were taken from an Aerobee Hi. Two years later, the first cosmic X-ray source was detected.

Extrasolar X-rays

Riccardo Giacconi, an Italian astrophysicist then working for American Science and Engineering (AS&E), had successfully petitioned NASA to fund his team's X-ray experiment. The team's first rocket misfired in 1960, but by 1961 it had a new, improved experiment ready for launch. This instrument was one hundred times more sensitive than any flown to date. Using a large field of view, the team hoped to observe other X-ray sources in the sky. Success followed a year later: the rocket aimed its camera first at the moon and then away from it. What the camera saw came as a complete surprise to the team. The instrument detected the X-ray "background"—a diffuse signal coming from all directions—and a strong peak of radiation in the direction of the galactic center.

Stars like the sun emit about a million times more photons at visible light frequencies than they do as X-rays. The source of »

Riccardo Giacconi

Born in Genoa, Italy, in 1931, Riccardo Giacconi lived in Milan with his mother, a mathematics and physics high school teacher. She instilled a love of geometry in the young Riccardo. Giacconi's first degree was from the University of Milan. With a Fulbright Scholarship, he moved to Indiana University in the US, and then to Princeton, to study astrophysics.

In 1959, Giacconi joined American Science and Engineering, a small firm in Cambridge, Massachusetts. AS&E built rocket-borne monitoring equipment for measuring electrons and artificial gamma-ray bursts from nuclear weapons. Giacconi was tasked with developing instruments for X-ray astronomy. He was at the heart of most of the breakthroughs in X-ray astronomy, and in 2002, he was awarded a share of the Nobel Prize in Physics for his contributions to astrophysics. In 2016, he was still working in his mid-80s, as principal investigator for the Chandra Deep Field-South project.

Cosmic X-ray radiation is **absorbed** by Earth's atmosphere.

→

Space-based telescopes are needed for X-ray astronomy.

Detectors on balloons and rockets detect X-rays coming from **all over the sky**.

A new view of the universe is revealed in high-energy radiation.

The Chandra X-ray Observatory was launched by NASA in 1999. It was initially planned to operate for five years, but was still in use in 2016.

the X-ray signals, by contrast, radiated a thousand times more X-rays than light. Although a small, barely visible point in the sky, the source was pumping out one thousand times more X-rays than the sun. Furthermore, certain physical processes were taking place within the source and these had never been seen in the laboratory. After weeks of analysis, the team concluded that this must be a new class of stellar object.

Search for the source

There was no candidate in the solar system to account for the intense radiation. The most likely source was named Scorpius X-1 (Sco X-1 for short) after the constellation within which it was located. Herb Friedman at the NRL confirmed the result using a detector with a larger area and better resolution than the AS&E instrument. Sco X-1 is now known to be a double star system and is the brightest, most persistent X-ray source in the skies.

Further launches revealed a sky dotted with X-ray sources, both galactic and extra-galactic. In a short space of time, the team had detected a disparate set of celestial oddities emitting X-rays. These included supernova remnants, binary stars, and black holes. Today, more than 100,000 X-ray sources are known.

Toward Chandra

By the mid-1960s, instruments were becoming ever more sensitive. Detectors were able to record X-rays one thousand times weaker than

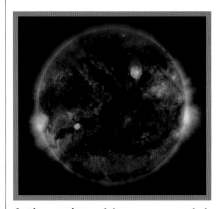

Active regions of the sun are revealed by combining observations from many telescopes. High-energy X-rays are shown in blue; low-energy X-rays green.

Sco X-1 just five years after Giacconi's discovery. Initially proposed by Giacconi in 1963, Uhuru, the first satellite dedicated solely to X-ray astronomy, was launched in 1970. It spent three years mapping X-rays. This all-sky survey located 300 sources, including a bizarre object in the center of the Andromeda galaxy, and it earmarked Cyg X-1 as a potential black hole. Uhuru also found that the gaps in galaxy clusters are strong sources of X-rays. These apparently empty regions are in fact filled by a low-density gas at millions of degrees Kelvin. Although thinly spread, this "intercluster medium" contains more mass than that of all of the cluster's galaxies combined.

In 1977, NASA launched its High Energy Astronomy Observatory (HEAO) program. HEAO-2, renamed the Einstein Observatory, was equipped with highly sensitive detectors and revolutionized X-ray astronomy. With its fused quartz mirrors, the telescope was a million times more sensitive than that of Giacconi's 1961 discovery rocket. Einstein observed X-rays emanating from stars and galaxies, and even from planetary aurorae on Jupiter.

The universe is popping
all over the place.
Riccardo Giacconi

Eager to probe the X-ray background further, Giacconi once again proposed an advanced telescope. In 1999, this became the Chandra X-Ray Observatory, the third of the orbiting Great Observatories. Chandra is the most powerful X-ray telescope ever built, tens of billions of times more sensitive than the early detectors. Its phenomenal performance outstripped all expectations and its mission lifetime was tripled from five to 15 years. As of 2016, however, its mission is ongoing. Chandra's outstanding technical firsts include detecting sound waves coming from a supermassive black hole. The X-ray data, when combined with optical observations from the Hubble Space Telescope and infrared data from the Spitzer Space Telescope, have provided stunning images of the cosmos.

Realm of the X-rays

X-ray astronomy observes the highest-energy objects in space: colliding galaxies, black holes, neutron stars, and supernovae. The energy source behind this

activity is gravity. As matter falls toward a massive concentration of material, particles collide and accumulate. They give up their energy by emitting photons, which at these speeds have X-ray wavelengths (0.01–10 nanometers, or billionths of a meter)—equivalent to temperatures of tens of million of degrees. The same mechanism, is at work in a wide range of dramatic phenomena: active stars more massive than the sun, for example, produce strong solar winds and significant amounts of X-rays. "X-ray binary star" systems, in which mass transfers from one star to its partner, also produce intense radiation.

Seeing black holes

When stars explode at the end of their lives, the blast waves from the supernova compress the interstellar medium, causing the gas to release X-rays. Left within what remains of the supernova, the massive star continues life as a neutron star or a black hole. Turbulence generated by material being torn apart as it is sucked into a black hole will also produce X-rays. The radiation being pumped out causes the outer layers of the supernova remnant to fluoresce in a range of colors.

Certain galaxies have centers that outshine all the billions of stars in the galaxy itself, with emissions that are bright at all wavelengths. The center of such an "active galactic nucleus" is assumed to contain a supermassive black hole. Material falling toward the centers of galaxy clusters—the largest structures in the universe—also shines in X-rays, and is not visible in other light frequencies. Chandra has now taken two "deep field" images of the X-ray background—23- and 11-day exposures of the northern and southern hemispheres of the sky. X-ray instruments of the future may help scientists see how black holes are distributed. ■

Observations in the X-ray spectrum reveal hidden structures. The larger blobs in this patch of sky from an ESA X-ray survey are galaxy clusters; smaller dots are black holes.

BRIGHTER THAN A GALAXY, BUT IT LOOKS LIKE A STAR

QUASARS AND BLACK HOLES

IN CONTEXT

KEY ASTRONOMER
Maarten Schmidt (1929–)

BEFORE
1935 Karl Jansky develops the first radio telescope.

1937 Radio engineer Grote Reber makes the first radio survey of the sky.

1955 The Radio Astronomy Group at Cambridge begins to map the northern hemisphere at 159 MHz.

AFTER
1967 Jocelyn Bell Burnell, of the Radio Astronomy Group, detects the first pulsars.

1972 The first physical, rather than theoretical, candidate for a black hole is identified in the Cygnus X-1 system.

1998 Andrea Ghez detects a black hole four million times as massive as the sun at the center of the Milky Way.

By the end of the 1950s, radio astronomy had given a new way to look to the sky. In addition to imaging celestial objects with light, surveys of the sky could use radio emissions from space, showing up previously unseen features. Radio waves were found to come from the sun, the stars, and the center of the Milky Way, but there were also mysterious invisible radio sources. In 1963, Maarten Schmidt, a Dutch astronomer working with the Hale Telescope at Palomar Observatory, California, managed to catch a glimpse of the light from one of these objects. When he looked at its redshift, he discovered something

The sky has many **powerful radio sources** that seem to be **invisible**.

These are found to be **distant, bright, fast-moving**, starlike objects called **quasars**.

Quasars are **active galactic nuclei**, in which a **black hole** is eating up the galaxy's stars.

It is likely that **all galaxies** have a black hole at the center and **have been quasars** in the past.

startling. The object was 2.5 billion light-years away, which meant that it was unimaginably bright. Its absolute magnitude was −26.7 (the lower the figure, the brighter the object). The object in Schmidt's eyepiece was 4 trillion times brighter than the sun (magnitude +4.83)— brighter than the whole of the Milky Way put together.

Schmidt named the body a quasi-stellar radio source, which was later shortened to quasar. Before Schmidt, the object had been known as 3C 273. The 3C referred to the 3rd Cambridge Catalogue of Radio Sources (produced by the Radio Astronomy Group) and 273 because it was the 273rd object

to be located in that survey. 3C 273 had been spotted in 1959, although the first quasar to be identified (or what would be later termed a quasar) was 3C 48, which had been found shortly before.

Improving radio astronomy

Radio astronomy had started in the 1930s after the accidental discovery of cosmic radio sources by Karl Jansky. Interrupted by World War II, and helped somewhat by the development of radar technology, surveys using radio telescopes did not start in earnest until 1950. Early surveys were hindered by the low frequency of 81.5 MHz (megahertz— or million cycles per second) used by early radio receivers. At that frequency, it was difficult to pinpoint the location of signals with a low flux density. (Flux density is a measure of the strength of a signal, and is measured in watts per square meter per hertz, simplified as the unit jansky [Jy].)

In 2001, the Hubble Space Telescope captured a glimpse of one of the most distant and luminous quasars ever seen (circled). It dates to less than one billion years after the Big Bang.

In 1955, the Radio Astronomy Group at Cambridge University began a survey using a radio interferometer, which picked up signals at 159 MHz. This was better at resolving faint radio sources, and led to the discovery of the first two quasars.

The light from both objects was invisible to the optical telescopes available to the Cambridge researchers at the time. However, their measurements of the flux density told them that these radio sources were very compact. »

Understanding [of quasars] has not developed very much in 50 years. You only see a point source; you don't see its structure. It's a difficult thing to get hold of.
Maarten Schmidt
speaking in 2013

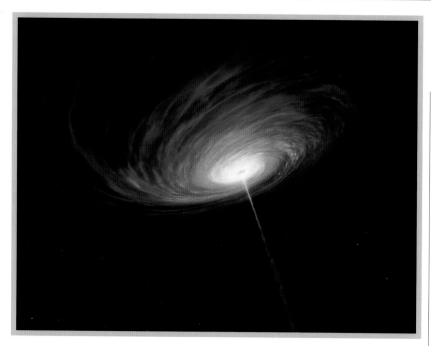

An artist's impression shows the possible structure of quasar 3C 279. A disk of material rotates around a black hole a billion times as massive as the sun.

In 1962, 3C 273 was occulted, or covered, several times by the moon. By watching for the reappearance of the radio source from behind the lunar disk, astronomers were able to get a very precise location of the source. Maarten Schmidt used those measurements to take a look at it through the Hale Telescope, then the largest optical telescope in the world. He found 3C 273 to be the brightest object yet known. He published his findings in *Nature* in March 1963, and in the same issue, two other astronomers, Jesse Greenstein and Thomas Matthews, presented data on the redshift of 3C 48, which showed that the object was moving away at one third of the speed of light, making it the fastest-moving object yet discovered.

By the early 1970s, hundreds of quasars had been identified. Many were even more distant than 3C 48 and 3C 273; today, most of the quasars that have been found are located about 12 billion light-years away. In addition, quasars are mostly brighter than the first observations suggested, with luminosities up to 100 times that of the Milky Way.

White holes?

The debate now began as to what these things actually were. One suggestion was that the enormous redshifts seen in quasars were not the result of the expansion of space, but were the result of the light crawling out of a large gravity well. Such a well would be created by a truly monstrous star, with a gravitational field close to that of a black hole. However, calculations showed that such a star could never be stable.

Another proposal was that a quasar was the opening of a white hole. A white hole is the opposite of a black hole. This idea was proposed in 1964, and white holes remain entirely hypothetical. They are generally ignored as a theory today, but in the 1960s and '70s, black holes were also unobserved phenomena, so the concept of white holes carried more weight. The idea is based on a complex interpretation of the Einstein field equations of general relativity, which proposes that a black hole that exists in the future would link to a white hole that exists in the past. A white hole is, therefore, a region of space where light and matter can leave but cannot enter. This would match the focused streams of radiation and matter that were being observed firing out of quasars. The question remained over where all that energy came from. The answer offered was that it has come through a wormhole, or Einstein–Rosen bridge, a theoretical feature of spacetime that connects the future to the past.

Small bangs

Currently, the only event that is accepted as anything like a white hole is the Big Bang itself, and some theories suggest that the material entering black holes

Twinkle, twinkle, quasi-star
Biggest puzzle from afar
How unlike the other ones
Brighter than a billion suns
Twinkle, twinkle, quasi-star
How I wonder what you are.
George Gamow

may emerge in another universe as "small bang" events. Nevertheless, as the understanding of black holes grew, the white-hole explanation of quasars faded away.

Supermassive black hole

Quasars are too luminous and energetic to be using nuclear fusion, the process that powers stars, to produce their energy. However, theoretical work on black holes showed that a region of material, known as the accretion disk, would form around an event horizon. Since this material was steadily pulled into the black hole, it would heat up to millions of degrees. A supermassive black hole, with a mass billions of times greater than the sun, would produce an accretion disk that matched the output observed in quasars.

The accretion disk theory also matched up with the beams of plasma, known as relativistic jets, that blasted out in opposite directions from some quasars. These are caused by the spin of the black hole, which creates a magnetic field and focuses matter and radiation into two streams.

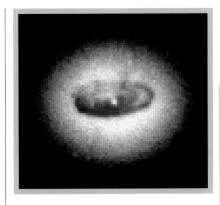

Superheated plasma blasts out at close to the speed of light from each stream.

Today's understanding of quasars began to crystallize in the 1980s. The accepted view is that a quasar is a supermassive black hole—or perhaps two—at the heart of a galaxy, that is eating up the stellar material. A galaxy that behaves like this is said to have an active nucleus, and it appears that quasars are just one manifestation of these so-called active galaxies.

An active galaxy is detected as a quasar when the relativistic jets are angled to Earth's line of sight. Therefore, the object is detected chiefly from its radio emissions. If the jets are perpendicular to Earth's

The Hubble telescope took this image of the active galactic nucleus of the elliptical galaxy NGC 4261. The disk of dust is 800 light-years wide.

line of sight, then they can never really be detected, and instead Earth sees a radio galaxy—a galaxy that is pumping out a loud radio source. If the relativistic jets are directed right at Earth, an excellent view is gained of the active nucleus in an object, known as a blazar.

Most quasars are ancient objects, and Earth sees their activity from when the universe was young. Unlike in other active galaxies, the brilliance of a quasar's nucleus makes it hard to discern much of the galaxy around it. It is thought that young galaxies always have active nuclei, and that once there is no material left for their black hole to swallow they become quieter places, like the Milky Way today. However, galactic collisions, in which one galaxy merges with another, can activate the nucleus again. It is likely that the Milky Way, which is on track to collide with Andromeda in 4 billion years, is destined to become a quasar itself one day. ∎

Maarten Schmidt

Born in Groningen, the Netherlands, Maarten Schmidt went to his home city's university and studied alongside Jan Oort. Schmidt earned his doctorate, before emigrating to the US and taking up a post at Caltech's Palomar Observatory. He became a leading expert on star formation, encapsulated by the Schmidt law, which relates the density of interstellar gas clouds to the rate of star formation inside them. Schmidt also became one of the chief investigators of quasars. After a conference on the subject in 1964, Schmidt and the other

leading figures in the field, including William Fowler and Subrahmanyan Chandrasekhar, were leaving on the same plane, which experienced a dicey takeoff. Fowler is said to have quipped: "If this plane crashes, at least we'll get a new start on this quasar problem." Schmidt went on to occupy several eminent roles in astronomical institutions.

Key work

1963 *3C 273: A Star-Like Object with Large Red-Shift*

AN OCEAN OF WHISPERS

LEFT OVER FROM OUR ERUPTIVE

CREATIONS

SEARCHING FOR THE BIG BANG

IN CONTEXT

KEY ASTRONOMERS
Robert H. Dicke (1916–1997)
James Peebles (1935–)

BEFORE
1927 Georges Lemaître proposes his "hypothesis of the primeval atom."

1948 Ralph Alpher and Robert Herman predict that radiation from the Big Bang would now have a temperature of 5 K.

1957 Soviet astronomer Tigran Shmaonov reports a "radioemission background" of 4 +/- 3 K, but does not connect this finding with the Big Bang.

AFTER
1992 The COBE results confirm black-body curve and anisotropy (tiny variations) of the CMB.

2010 WMAP measures tiny temperature variations of 0.00002 K in the CMB.

2013 The Planck team releases a detailed map of the CMB.

The discovery of the universe's "first light" is one of the most fundamental scientific findings of all time. A full 99.9 percent of all photons (light particles) arriving at Earth are associated with this cosmic microwave background (CMB). It has been traveling for more than 13 billion years, reaching us from a time near the very dawn of the universe. The CMB is the thermal radiation emitted when the universe was at a temperature of about 4,000 K.

The scientist who usually takes the credit for the prediction of the CMB is physicist George Gamow (pp.196–97). An expanding universe implied a point at which it was once squeezed into a tiny volume. Gamow saw that this in turn implied a hot beginning, and realized that such a hot "Big Bang" would have left its signature in the sky. In 1948, his doctoral students Ralph Alpher and Robert Herman worked out the details of this "fireball radiation." Cooled by the expansion of the universe, over 13 billion years, they deduced that it should take the form today of radio-frequency radiation as if emitted by an object at 3 K—just

> There is no point in attempting a half-hearted experiment with an inadequate apparatus.
> **Robert H. Dicke**

above absolute zero. Seemingly unaware of Alpher's and Herman's work, Robert H. Dicke, working at the Princeton University "Rad Lab," independently predicted the CMB in the early 1960s. Dicke asked his team of postgraduates to find it. David Wilkinson and Peter Roll were to build a machine to detect it, while James Peebles was to "think about the theory."

Echoes of the Big Bang
Gamow had assumed that the faint signal of the CMB would be indistinguishable from radio waves flooding in from other astronomical objects, but Alpher and Herman

---

The Big Bang is a **disputed theory**.

One of the predictions of the Big Bang theory is a **cosmic microwave background** radiation at **a temperature of about 3 K**, with a spectrum very nearly that of **a black body**.

The **background radiation** is discovered at about 3 K. Further studies show that it has **a spectrum** that is almost exactly that of **a black body**.

The Big Bang is **no longer a disputed** scientific theory.

A theoretical "black body" absorbs all the radiation that hits it, then emits radiation at different intensities (measured as spectral radiance) across different wavelengths, depending on its temperature, as shown here.

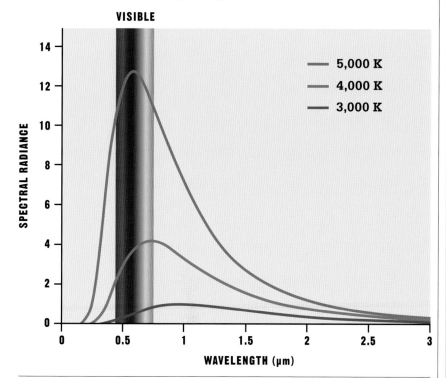

Robert H. Dicke

Bob Dicke was born in St. Louis, Missouri, in 1916, but grew up in Rochester, NY. Fascinated by science from a young age, he began an engineering degree before switching to physics. After graduating from Princeton in 1939, Dicke worked at MIT's Radiation Laboratory during World War II, developing microwave radar. He and his wife, Anne, returned to Princeton following the war and remained there for the rest of their lives. Dicke's research was initially centered around radiation, and he formulated a new quantum theory to explain the emission of coherent radiation produced by a theoretically ideal laser. His interest in radiation led him to team up with James Peebles and predict the existence of the CMB. By the 1960s, Dicke's interests had spread to theories of gravitation. He developed high-precision experiments to test general relativity more robustly and produced an alternative theory of gravitation. An imaginative experimentalist and a prolific inventor, Dicke held more than 50 patents, ranging from lasers to designs for clothes dryers.

showed that it would have two distinguishing features. It would come from every direction in the sky, and the energy curve would have the telltale shape of an object very close to thermal equilibrium, a so-called black body.

Alpher and Herman stopped there, advised that the radio telescopes of the day would not be able to pick up such a quiet hiss. But Dicke thought otherwise. During World War II, while working on radar systems, he had built such a machine: the Dicke radiometer, which collects a microwave signal and measures its power. Dicke added a switch to filter out "noise." The device is still used today in space telescopes and satellites. Choosing

a suitable bandwidth in which to conduct the search for the radiation was the next important stage, since so many things produce radio waves. For example, the sky is filled with microwave wavelengths around the 8¼-in (21-cm) mark, emitted by atoms of hydrogen. It seemed logical to start at a dark part of the spectrum. In the spring of 1964, Wilkinson and Roll began looking at the 1¼-in (3-cm) band, but they were beaten in their quest by a piece of serendipity.

Holmdel Horn
Less than an hour's drive from Princeton University is the Holmdel Horn—a giant radio antenna, built by Bell Laboratories for satellite »

communications. In 1964, it was being used by two radio astronomers, Robert Wilson and Arno Penzias, who were trying to detect a halo of cold gas around the Milky Way. Penzias and Wilson were looking at the 2¾-in (7-cm) range, but they could not get rid of a stubborn low-level hiss that was ruining their measurements.

The pair painstakingly eliminated potential sources of interference, dusting down plugs and checking circuits. At first, they thought that the noise was coming from New York, but pointing their telescope away from the city did not help. Then they thought it might be high-altitude static from a nuclear bomb test or an unknown radio source in the solar system but, over the course of the year, the signal never varied its ceaseless soft hiss. In desperation, they even removed a nesting pair of pigeons and chipped away an accumulation of "white dielectric material" (bird droppings).

Stumped, Penzias contacted a colleague, who directed him to James Peebles at Princeton. Taking the call for Peebles, Dicke knew immediately what the Bell Laboratories scientists had found. He hung up the telephone and said to his colleagues, "Well, boys, we've been scooped."

The only way is up

The discovery of the cosmic background radiation is one of the three experimental pillars of the Big Bang theory—the other two being Hubble's Law and the cosmic abundances of the elements hydrogen and helium (pp.196–97). Big Bang theorists had predicted exactly what had now been found: radiation coming from all directions at 3 K.

The Big Bang had until then been a highly contested idea, and many scientists—including Penzias and Wilson—still favored Fred

Hoyle's Steady State model, in which an expanding universe essentially remains unchanging due to the constant creation of matter. According to this theory, the universe does not change in appearance over time.

Steady State theorists suggested that the microwave background was the result of scattered starlight from distant galaxies. To demonstrate to a sceptical physics community that the signals were indeed the relic fireball radiation predicted by Big Bang theory, it was necessary to confirm Alpher's and Herman's condition that the radiation should match that of a theoretical black body. This required measuring the CMB at different frequencies. The next obvious move was to take an even more accurate determination of the spectrum by launching a space-based receiver.

During the 1970s and 1980s, Herb Gush, a physicist at the University of British Columbia, fired sounding rockets into space in order to observe the CMB without interference from Earth's atmosphere. This indicated that the

Robert Wilson and Arno Penzias pose in front of the Holmdel Horn antenna following the announcement in 1978 of their Nobel Prize for discovering the CMB.

The CMB represents the outer shell of the observable universe. Just beyond it lies the moment of the Big Bang, shown here as a series of flashes.

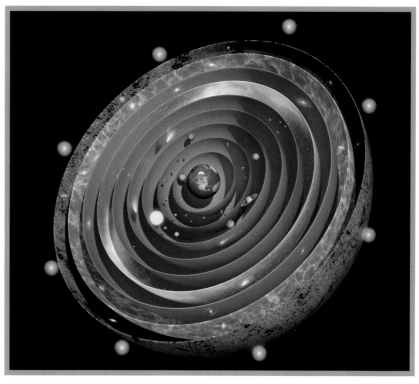

CMB had the black-body signature of a system in thermal equilibrium, with no heat flowing from one part to another. This astounding—and often overlooked—result verified that the signal was thermal in origin. Unfortunately, the rocket's hot exhaust gases often marred Gush's measurements, preventing him from producing a definitive result.

Edged out

By 1989, Gush had finally developed an instrument that could compare the CMB spectrum with that of an onboard radiator that approximated a black body. However, problems with a malfunctioning vibrator delayed the launch until early 1990. The results were immediate and striking, but Gush had missed his opportunity for a scoop by a few weeks. NASA's Cosmic Background Explorer (COBE) satellite, launched in late 1989, already had the shape of the spectrum with a temperature near 2.7 K. In the end, Gush's rocket data confirmed COBE's results and not the other way around. Peebles was later to state that Gush had deserved a Nobel Prize for his work.

COBE's results returned a near-perfect match with the theoretical black-body spectrum and revealed, for the first time, a faint unevenness in the background radiation. Subsequent missions—such as NASA's Wilkinson Microwave Anisotropy Probe (WMAP), launched in 2001, and ESA's Planck spacecraft, launched in 2009—have mapped the CMB's "lumpiness" in greater detail. ∎

James Peebles

Phillip James Edwin Peebles was born in 1935 in Winnipeg, Canada. After graduating from the University of Manitoba, he earned a doctorate at Princeton University, where he found himself "surrounded by all these people who knew so much more than I." Working under Robert H. Dicke, Peebles found he was retreading old paths. He began to focus on the constraints that the CMB put onto the early universe—specifically on the creation of atomic nuclei in the Big Bang and on how small temperature differences affect models of structure formation in the universe. Peebles also made important contributions to theories of dark matter and dark energy. With typical modesty, he says that his cold dark matter model became popular because it was easy to analyze. He is currently Albert Einstein professor of science at Princeton.

Key works

1971 *Physical Cosmology*
1980 *Large Scale Structure of the Universe*
1993 *Principles of Physical Cosmology*

THE SEARCH FOR EXTRATERRESTRIAL INTELLIGENCE IS A SEARCH FOR OURSELVES

LIFE ON OTHER PLANETS

IN CONTEXT

KEY ASTRONOMER
Carl Sagan (1934–1996)

BEFORE
1865 German physician Hermann Eberhard Richter suggests that planets could be seeded by simple life carried by comets.

AFTER
1973 Brandon Carter sets out an early version of the anthropic principle, which states that the universe is necessarily a certain way; otherwise humans would not be here to experience it.

1977 Voyagers 1 and 2 are launched, carrying images and sounds from Earth.

2009–2016 NASA space telescope Kepler discovers 3,443 exoplanets in 2,571 planetary systems.

2015 Kepler discovers the first Earth-sized planet in a habitable zone of another star.

The Copernican principle states the working assumption that Earth is not special—it is no more than an average planet, orbiting a medium-sized star, in an unremarkable part of an ordinary galaxy. If Earth is not unique, then there is little reason to think that other planets cannot also harbor life. Given the number of stars in the universe—in the order of 10^{23}—this might be a statistical certainty. Over the centuries, many thinkers, such as American Carl Sagan, have pondered the possibility.

Is Earth alone?

In the 16th century, the Italian monk Giordano Bruno proposed that the stars were other suns, each of which could have its own solar system. Life could even populate these other earths. Believing that the universe was infinite, Bruno also insisted that it could have no center. Bruno was tried by the Roman Inquisition for these and other heretical beliefs and burned at the stake in 1600.

Throughout history, various astronomers have claimed to see evidence for life on other planets of the solar system. In the 1890s, American astronomer Percival

Do there exist many worlds, or is there but a single world? This is one of the most noble and exalted questions in the study of Nature.
Albertus Magnus
13th-century scholar

Lowell claimed to have mapped artificial "canals" on Mars, while the dense clouds of Venus were imagined, by Swedish chemist Svante Arrhenius in 1918, to hide from view a lush surface blooming with life. It is now known that the clouds are acidic, while the surface of Venus is an inhospitable 864°F (462°C). However, these are just two planets out of potentially billions.

The universe's immensity and the apparent universality of its physical laws make it seem likely that microbial life exists elsewhere.

Carl Sagan

Carl Sagan is one of the most widely known 20th-century scientists. His deep, honeyed tones are the instantly recognizable voice of the documentary series *Cosmos*. Sagan was raised in a working-class Jewish area of New York and as a boy was an avid reader of science fiction. A talented pupil, he went to the University of Chicago in 1951 on a full scholarship. Sagan received his Ph.D. in 1960, showing that the high surface temperatures of Venus are due to a runaway greenhouse effect. Sagan conducted pioneering research in planetary science and exobiology (the biology of extraterrestrial life), which many in mainstream astronomy viewed with suspicion. In 1985, he wrote the sci-fi book *Contact*, which was later turned into a movie. With his visionary, positive, and humanist outlook, the Cornell University professor inspired a new generation of astronomers.

Key works

1966 *Intelligent Life in the Universe* (with Iosif Shklovsky)
1983 *Cosmos*

See also: Radio telescopes 210–11 ▪ The Space Race ▪ Exploring the solar system 260–67 ▪
Exoplanets 288–95 ▪ Shklovsky (Directory) 338 ▪ Carter (Directory) 339 ▪ Tarter (Directory) 339

Indeed, life may have arisen elsewhere and been transported to Earth. Greek philosopher Anaxagoras first suggested this idea of "panspermia" in the 5th century BCE. Naturalist Charles Darwin briefly turned to this idea while working on his theory of evolution by natural selection, troubled that the accepted figure for Earth's age did not give enough time for complex organisms to evolve. Earth is now known to be vastly older than was believed in Darwin's time, so panspermia is not needed to explain the origin of life on the planet.

Recent discoveries show that comets can carry many of the basic chemical components of life, but the exact mechanism by which life on Earth began remains a mystery. Solving that mystery should give a far better idea of how likely life is elsewhere.

Where is everyone?

One day, over lunch in Los Alamos in 1950, Italian scientist Enrico Fermi asked a simple question: "Where are

Computer simulations have shown that it is theoretically possible for simple single-cell life forms to exist inside comets or asteroids and to survive an impact like this with Earth.

they?" He reasoned that, even if only a small proportion of planets play host to intelligent life, given the unimaginable numbers of stars within the galaxy, one might expect a large number of civilizations to exist on other planets. At least some of them may have chosen to send messages or tried to visit Earth themselves. Earth has been producing electromagnetic signals for 90 years or so, since the dawn of radio and television broadcasting. These modulated radio waves—expanding and extending some 90 light-years in all directions—should be a giveaway of a technologically advanced society to any potential spacefaring intelligence.

In 1959, Giuseppe Cocconi and Philip Morrison suggested a bandwidth to search for alien radio messages. A year later, Frank Drake, »

There is an **immense number of stars** in the universe.

Most of these stars have **planetary systems**.

Life may have arisen on **many planets**.

If life **survives for long enough**, it may become **intelligent** enough to start looking for life elsewhere, as humans have done.

The search for extraterrestrial intelligence is a search for ourselves.

at the National Radio Astronomy Observatory in Green Bank, West Virginia, set out to look for them. Drake founded Project Ozma, named for the queen of author L. Frank Baum's imaginary Land of Oz—a place "difficult to reach and populated by exotic beings." After a briefly exciting and noisy encounter with some top-secret military radio-jamming equipment, Drake and his team were met with silence. More than 50 years later, the silence has not yet been broken.

Order of the Dolphin

Drake drew together a diverse group of scientists to lay the foundations and protocols for the search for extraterrestrial intelligence (SETI). The group jokingly called itself the Order of the Dolphin, in reference to the work of neuroscientist John Lilly, who pioneered the science of speaking to dolphins. As one of the few people dealing with interspecies communication, Lilly was an important part of the group, which also included a young astronomer Carl Sagan, who was an expert on planetary atmospheres.

In preparation for the Order's first meeting in 1961, Drake came up with a formula for the number of alien civilizations in the galaxy:

$$N = R_\star \times f_p \times n_e \times f_l \times f_i \times f_c \times L$$

The total (N) was reached by multiplying the factors necessary for intelligent extraterrestrials to evolve and be discovered. It depends on the rate at which stars suitable for intelligent life form ($R_\star$); the fraction of these stars that are orbited by planets (f_p); the number of planets in any given planetary system that can support life (n_e); the fraction of these planets upon which life actually appears (f_l); the proportion of life-bearing planets that go on to produce intelligent life (f_i); the proportion of civilizations that develop technology that betrays

> The search for extraterrestrial life is one of those few circumstances where both a success and a failure would be a success by all standards.
> **Carl Sagan**

detectable signs of their existence (f_c); and, finally, the length of time such civilizations survive (L).

With these terms in place, bounding limits could theoretically be placed on each one. In 1961, however, not a single one was known with any confidence. Delegates at the meeting concluded that N was approximately equal to L, and a potential 1,000 to 100 million civilizations might exist in the galaxy. Although values for some of the variables in the Drake equation have been narrowed down over the intervening years, modern estimates of N still vary wildly. Some scientists argue that the figure may be zero.

Message in a bottle

In 1966, Sagan cowrote *Intelligent Life in the Universe*, perhaps the first comprehensive discussion of planetary science and exobiology. The book was an expanded and revised version of an earlier edition, published in 1962 by the Soviet astronomer and astrophysicist Iosif Shklovsky. Although highly speculative, the book ignited discussion among scientists. It inspired NASA's Project Cyclops report, an influential document now referred to as the "SETI Bible."

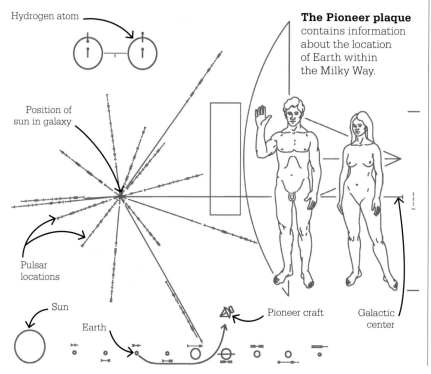

Hydrogen atom

Position of sun in galaxy

Pulsar locations

Sun

Earth

Pioneer craft

Galactic center

The Pioneer plaque contains information about the location of Earth within the Milky Way.

In 1971, Sagan approached NASA with the idea of sending a message on its Pioneer spacecraft. Sagan and Drake worked on a design that would advertise Earth's existence to alien civilizations and help them locate Earth in the cosmos. The graphics on the Pioneer plaque establish a unit of measurement using the 21-cm hydrogen emission line. Units defined by reference to Earth-based phenomena, such as meters and seconds, would be meaningless to extraterrestrial scientists. By choosing units from properties of nature, the hope was that they would be understood universally.

All the images on the plaque were scaled in terms of these units. A map of bright and distinctive pulsars points the direction to Earth, and Pioneer's route is traced on a simple pictogram of the solar system. Images of a man and a woman were drawn by Sagan's artist wife, Linda Salzmann Sagan.

Pioneer 10 and Pioneer 11, launched in 1972 and 1973, were fitted with Sagan's plaque, etched on a 6- by 9-in (152 x 199 mm) gold-anodized aluminum plate. Critics warned that it would attract the

We began as wanderers, and we are wanderers still. We have lingered long enough on the shores of the cosmic ocean. We are ready at last to set sail for the stars.
Carl Sagan

The Arecibo message was broadcast into space a single time in 1974. Coded in binary, its message is arranged in 73 rows of 23 columns.

unwanted attentions of power-hungry (or just hungry) aliens. Feminist groups were unhappy that the man waved in greeting, while the woman's pose angled her body (they thought) submissively toward the male figure. Salzmann responded that women are smaller, on average; that having both figures waving might be interpreted as it being the natural arm position; and that she merely wished to show how the hips moved. Sagan had initially wanted the man and woman to be holding hands, but decided it might make the Earthlings look like a single creature with two heads.

The Arecibo message

While the search continued for beacons set up by intelligent beings and likely star systems, Drake and Sagan decided to send planet Earth's own "we are here" signal. The 3-minute burst of 1,000 kW radio waves was designed to cross distances separating stars. Beamed out from the Arecibo radio dish in Puerto Rico in November 1974, the interstellar message was aimed at the globular cluster M13, a group of about 300,000 stars 25,000 light-years from Earth.

Instead of pictograms, the Arecibo message took the form of densely packed mathematical code, consisting of 1,679 binary digits (chosen because 1,679 is a product of two prime numbers, 73 and 23). The digital message contained the numbers 1 to 10 and information about the identity of the sender—details about DNA, the overall shape and dimensions of a human, and the position of planet Earth.

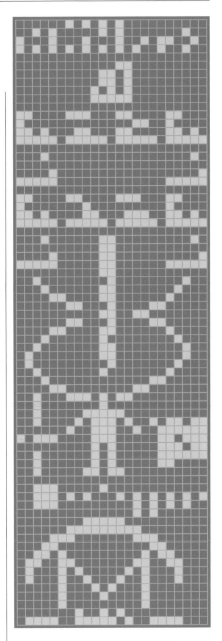

One of the hopes that accompanied the robotic explorers, as they were dispatched across the solar system from the 1960s onward, was that they might uncover some indication of extraterrestrial life within the solar system itself, even if it were only single-celled organisms. The spacecraft that »

Carl Sagan stands next to a model of the Viking 1 lander. The lander sent back signals from the surface of Mars from 1976 to 1982. Its instruments found no sign of life.

touched down on planets, such as NASA's Viking landers on Mars, carried experiments to test for signs of life. To date, no indication of life, either past or present, has been found, although some unexplored corners of the solar system remain candidates for life, such as the deep oceans thought to lie beneath the frozen surface of Jupiter's moon Europa.

Continued silence

No reply has ever been received to any of the 10 interstellar radio messages sent since 1962 and no communication has been detected. However, there have been false alarms. The most famous of these came in 1977, when an inexplicably powerful blast of radio signals was recorded coming from the direction of the Chi Sagittarii star system by Jerry Ehman at Ohio State

The SETI Institute's Allen Telescope Array in California is used daily to search for possible alien communications, as well as for radio astronomy.

University. He circled the signals on the readout and wrote "Wow!" next to them. The "Wow!" signal was never found again, however, and recent research suggests that it may have come from a hydrogen cloud surrounding a comet.

Given the vast distances between stars, it is still early days, however. The Arecibo message will not reach its target stars for another 25,000 years. Neither the Pioneer plaques nor the gold-plated disks carried by Voyagers 1 and 2—the Voyager Golden Record—are headed toward any particular

star system. Unless they are intercepted, they are destined to wander the Milky Way eternally. Sagan, for his part, believed that to find life or to fail was a win-win scenario—either result would show something important about the nature of the universe.

SETI in the modern age

NASA struggled to maintain its SETI funding, and today SETI is privately funded. Since the 1980s, the mantle has been taken up by the SETI Institute, based in Mountain View, California. UCal, Berkeley, through its SETI@home initiative, harnesses a network of volunteer computers to trawl Arecibo Observatory data for patterns that might indicate an unnatural radio source. Meanwhile, in 2016, China announced the completion of the largest ever radio telescope, the Five-hundred-meter Aperture Spherical Telescope (FAST). Among other things, FAST will search for extraterrestrial communications. It will eventually be made available to researchers from around the world.

Voyager 1 sent back this image of Earth from beyond the orbit of Pluto. The "pale blue dot" appears in a band of scattered sunlight.

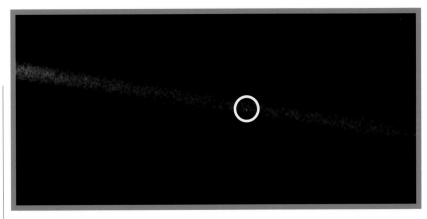

In recent years, the focus of SETI has moved away from merely listening for messages. Efforts have been directed toward picking up biochemical signs of life or indications of advanced technology. Alien life should leave its signature in evolved planetary atmospheres, volatile molecules, or complex organic chemicals that could only be created by life processes. Highly technological societies may have learned how to harvest the energy of their star. A "Dyson sphere" megastructure completely or partially surrounding a star to capture its energy would affect the star's observed output. It may also be possible to observe signs of asteroid mining or directly detect extraterrestrial spacecraft.

A cautious approach

In 2015, the Breakthrough Initiatives program was launched with the backing of Russian billionaire Yuri Milner. In addition to a $1,000,000 prize pool for SETI research and a plan to send a fleet of spacecraft to a nearby star, an open competition was announced to design a digital message to be sent to an extraterrestrial civilization. The Breakthrough Message project aims to accurately and artistically represent humanity and Earth, but pledges not to transmit any message until the risks and reward of contacting advanced civilizations have been debated.

Looking at ourselves

In 1990, Carl Sagan persuaded Voyager 1's controllers to swivel its camera back toward Earth. From 4 billion miles (6 billion kilometers) away, the craft captured the "Pale Blue Dot" image. Sagan wrote: "Everyone you love, everyone you know, everyone you ever heard of, every human being who ever was, lived out their lives on a mote of dust suspended in a sunbeam." Sagan stressed the importance of looking at ourselves: "The Earth is the only world known so far to harbor life. There is nowhere else, at least in the near future, to which our species could migrate. Visit, yes. Settle, not yet. Like it or not, for the moment the Earth is where we make our stand."

SETI represents a series of questions whose answering would tell us about Earth's place in the universe: whether the Copernican principle is correct, and if so, where else life has evolved. The answers could eventually provide humans with a way to transcend their origins and become a galactic species. ∎

We are almost certainly not the first intelligent species to undertake the search ... Their perseverance will be our greatest asset in our beginning listening phase.
Project Cyclops Report, NASA

IT HAS TO BE SOME NEW KIND OF STAR

QUASARS AND PULSARS

IN CONTEXT

KEY ASTRONOMERS
Antony Hewish (1924–)
Jocelyn Bell Burnell (1943–)

BEFORE
1932 English physicist James Chadwick discovers the neutron.

1934 Walter Baade and Fritz Zwicky propose that stars that explode as supernovae leave behind collapsed remnants made of closely packed neutrons, which they name neutron stars.

AFTER
1974 American astrophysicists Joseph Taylor and Russell Hulse discover two neutron stars, one of them a pulsar, orbiting each other.

1982 American astrophysicist Donald Backer and colleagues discover the first millisecond pulsar, which spins 642 times per second.

I n the late 1950s, astronomers across the world started to find mysterious, compact sources of radio signals in the sky without any corresponding visible objects. Eventually a source of these radio waves was identified—a faint point of light, which became known as a quasar. In 1963, Dutch astronomer Maarten Schmidt discovered a quasar that was hugely distant (2.5 billion light-years away). The fact that it was so easily detected meant it must be pouring out energy.

Searching for quasars
By the mid 1960s, many radio astronomers were searching for new quasars. One such figure

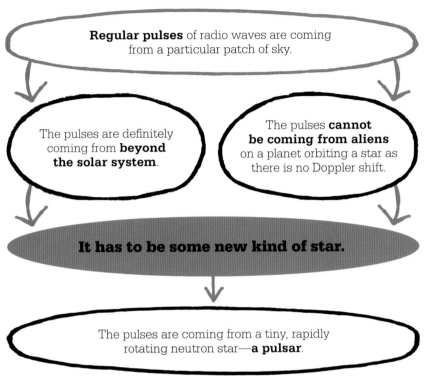

Regular pulses of radio waves are coming from a particular patch of sky.

The pulses are definitely coming from **beyond the solar system**.

The pulses **cannot be coming from aliens** on a planet orbiting a star as there is no Doppler shift.

It has to be some new kind of star.

The pulses are coming from a tiny, rapidly rotating neutron star—**a pulsar**.

Members of the Cambridge radio astronomy group built the new telescope themselves. Among them was a Ph.D. student named Jocelyn Bell. When the telescope started operating in July 1967, Bell was made responsible for operating it and analyzing the data, under the supervision of Hewish. Part of her job was to monitor output data from the telescope, made by pen recorders on chart-recorder printouts. Examining about 100 ft (30 meters) of chart paper every day, Bell quickly learned to recognize scintillating sources.

Little Green Man 1

About two months into the project, Bell noticed an unusual pattern of signals, which she described as "scruff." It looked far too regular and had too high a frequency to be coming from a quasar. Checking back through her records, she found it had appeared in the data before and always came from the same patch of sky. Intrigued, Bell started making more regular chart recordings of the same area of sky. At the end of November 1967, she found the signal again. It was a series of pulses, equally spaced and always 1.33 seconds apart. »

was Antony Hewish, part of a radio astronomy research group at Cambridge University. Hewish had been working on a new technique in radio astronomy based on a phenomenon called interplanetary scintillation (IPS), which is a "twinkling," or fluctuation, in the intensity of radio emissions from compact radio sources. The twinkling of sources of visible light, such as stars, is caused by disturbances in Earth's atmosphere that the light has to pass through (p.189). The twinkling of radio sources, however, is caused by streams of charged particles emanating from the sun. As radio waves pass through this "solar wind," they are diffracted, meaning that the waves spread out, making the radio source appear to twinkle.

Hewish hoped that IPS could be used to find quasars. Radio waves coming from a compact source, such as a quasar, twinkle more than radiation from a less compact source, such as a galaxy, and so quasars should twinkle more than other radio sources. Hewish and his team built a large radio telescope designed specifically to detect IPS. It covered an area of nearly 4.5 acres (2 hectares), took two years to construct, and required more than 120 miles (190 km) of cable to carry all the signals.

This image of a pulsar in the Crab nebula, a well-known supernova remnant, was taken in space by the Chandra X-ray Observatory. The white dot at the center is the neutron star.

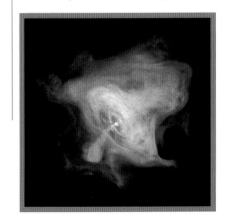

Bell showed the signal, dubbed "Little Green Man 1" (LGM-1), to Hewish. His initial reaction was that a pulse occurring every 1.33 seconds was far too fast for something as large as a star, and the signal must be due to human activity. Together, Bell and Hewish ruled out various human-related sources, including radar reflected from the moon, Earth-based radio transmissions, and artificial satellites in peculiar orbits. A second telescope was also found to pick up the pulses, which proved that they could not be due to an equipment fault, and calculations showed that they were coming from well outside the solar system.

Hewish had to revise his opinion that the signals had a human origin. The possibility that they were being sent by extra-terrestrials could not be ruled out. The team measured the duration of each pulse and found it was only 16 milliseconds. This short duration suggested that the source could be no larger than a small planet. But a planet—or an alien civilization living on a planet—was unlikely,

My eureka moment was in the dead of night, the early hours of the morning. But when the result poured out of the charts ... you realize instantly how significant this is—what it is you've really landed on—and it's great!
Jocelyn Bell Burnell

A pulsar is a spinning neutron star with an intensely strong magnetic field. It emits beams of radiation from its north and south poles.

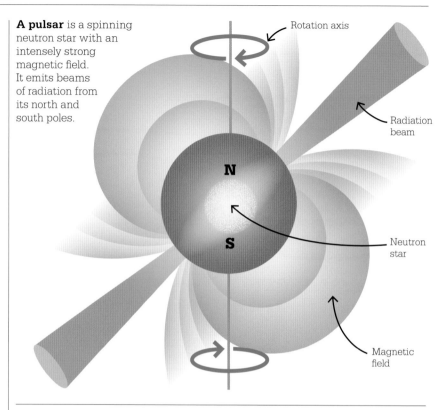

Rotation axis

Radiation beam

Neutron star

Magnetic field

since the signal would show slight changes in frequency, called Doppler shifts (p.159), as a planet orbited its star.

Publishing dilemma

Hewish, Bell, and their colleagues were unsure how to publish their findings. While it seemed unlikely that the signals were being sent by an alien civilization, no one had any other explanation. Bell returned to her chart analysis, and soon found another "scruff" in a different part of the sky. She discovered it was due to another pulsating signal, this time slightly faster, with pulses every 1.2 seconds. Now she was reassured that the pulses must have some natural explanation—two sets of aliens in different places would surely not be sending signals to Earth at the same time and at nearly the same frequency.

By January 1968, Hewish and Bell had found four pulsing sources in total, which they decided to call "pulsars." They wrote a paper describing the first source, suggesting that it might be due to pulsed emissions from a theoretical type of superdense collapsed star called a neutron star. Objects of this type had been predicted as long ago as 1934, but up to that time had never been detected.

Explaining the pulses

Three months later, Thomas Gold, an Austrian–American astronomer at Cornell University in the US, published a fuller explanation for the pulsed signals. He agreed that each set of radio signals was coming from a neutron star, but proposed that the star was rapidly spinning. A star like this would not need to be emitting pulsed radiation to account for the pattern of signals observed. Instead, it could be emitting a steady radio signal in a beam that it swept around in circles, just like a beam of light from a lighthouse. When the pulsar's beam (or perhaps one of its two beams) was pointing at Earth, a signal would be detected, which would show up as the sort of short pulse that Bell had noticed on printouts. When the beam had passed by Earth, the signal would stop until the beam came around again. Challenged about the pulsation rates, which implied extremely rapid spinning, Gold explained that neutron stars could be expected to behave in this way because of the way in which they form—from the collapse of stellar cores in supernova explosions.

Confirming the hypothesis

Initially, Gold's explanations were not well received by the astronomy community. However, they became widely accepted after the discovery of a pulsar in the Crab nebula, a well-known supernova remnant. Over subsequent years, many more

pulsars were found. They are now known to be rapidly rotating neutron stars with intense electromagnetic fields, which emit beams of electromagnetic radiation from their north and south poles. These beams are often, but not always, radio waves and sometimes other forms of radiation, including in some cases visible light. One reason for the excitement regarding the discovery of pulsars was that it increased the likelihood that another theoretical phenomenon— black holes—might also be detected and proven. Like neutron stars, black holes are objects that could result from the gravitational collapse of a stellar core following a supernova explosion.

In 1974, Hewish and Martin Ryle shared a Nobel Prize: "Ryle for his observations and inventions … and Hewish for his decisive role in the discovery of pulsars." However, Jocelyn Bell Burnell was told that she would not share the award with them because she had still been a student at the time of her work. She graciously accepted that decision. ∎

Jocelyn Bell Burnell

Jocelyn Bell was born in 1943 in Belfast, Northern Ireland. After earning a physics degree from Glasgow University in 1965, she moved to Cambridge University, where she studied for a Ph.D. There, she joined the team that built a radio telescope to detect quasars. In 1968, Bell became a research fellow at the University of Southampton and changed her last name to Bell Burnell when she married. She has held astronomy and physics-related positions in London, Edinburgh, and at the Open University, where, from 1991 to 2001, she was professor of physics. From 2008 to 2010, she was President of the Institute of Physics. Bell Burnell has received numerous awards for her professional contributions, including the Herschel Medal of the Royal Astronomical Society in 1989. In 2016, she was visiting professor of astrophysics at Oxford University.

Key work

1968 *Observation of a Rapidly Pulsating Radio Source* (with Antony Hewish and others)

A spinning neutron star that is emitting radiation beams can be detected as a pulsar on Earth if, as it spins, one or possibly both of its radiation beams recurrently point in Earth's direction as they are swept around through space. The pulsar will then be detected as a very regular series of signal "blips."

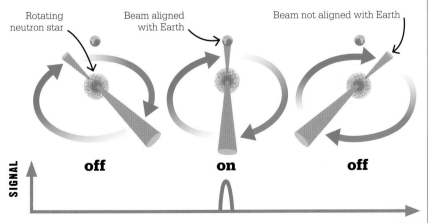

Rotating neutron star

Beam aligned with Earth

Beam not aligned with Earth

off **on** **off**

SIGNAL

TIME

GALAXIES CHANGE OVER TIME

UNDERSTANDING STELLAR EVOLUTION

IN CONTEXT

KEY ASTRONOMER
Beatrice Tinsley (1941–1981)

BEFORE
1926 Edwin Hubble produces a classification for galaxies based on their shapes.

Early 1960s American Allan Sandage proposes that disk galaxies form through the collapse of large gas clouds. He estimates distances to remote galaxies based on the idea that the brightest galaxies have similar luminosity.

AFTER
1977 Canadian Brent Tully and American Richard Fisher find a link between the luminosity and rotation of spiral galaxies. This is useful in estimating distances to spiral galaxies.

1979 Vera Rubin uncovers a discrepancy between actual and predicted rotational velocities in spiral galaxies, giving evidence of unseen "dark matter" in such galaxies.

U ntil a young New Zealand astronomer named Beatrice Tinsley published a highly original thesis in 1966, the methods used by cosmologists to calculate the distances to remote galaxies had been flawed. Accurate data for these distances was important because it would help to answer some of cosmology's most fundamental questions, namely the average density of the universe, its age, and its rate of expansion.

One method used in the 1960s was based on the idea that galaxies of the same type (giant ellipticals, for example) should all have roughly the same intrinsic brightness. On this basis, it was thought that distances to faraway galaxies should be obtainable simply by measuring their light output and comparing it to those of nearby galaxies of the same type whose distance was known.

Tinsley's argument

Tinsley challenged this approach, saying it was crude and unreliable. In calculating galaxy distances, she argued, more consideration had to be given to the fact that galaxies

Beatrice Tinsley

Beatrice Hill was born in Chester, UK, in 1941, and moved with her family to New Zealand when she was four years old. In 1961, she received a degree in physics from Canterbury University, and in the same year married classmate Brian Tinsley. In 1963, they moved to Dallas, Texas, where her husband had been offered a university job. Beatrice wasn't allowed to work at the same university, so she took a teaching job at the University of Texas at Austin.

In 1966, she completed her Ph.D. with a thesis on the evolution of galaxies. Tinsley soon became an influential figure in the field of cosmology. In 1974, she took a position as assistant professor at Yale University, becoming the university's first female professor of astronomy in 1978. She died of cancer in 1981, at just 40. Mount Tinsley in New Zealand is named in her honor.

Key work

1966 *Evolution of Galaxies and its Significance for Cosmology*

See also: Measuring the universe 130–37 ▪ The birth of the universe 168–71 ▪ Beyond the Milky Way 172–77 ▪ Nuclei and radiation 185 ▪ The primeval atom 196–97 ▪ Dark matter 268–71

The **apparent brightness** of galaxies from Earth depends on both their distance and their age.

Distant galaxies are seen through telescopes as they were millions or billions of years ago.

Remote galaxies differ from closer galaxies in how they appear from Earth in part because they are being seen at **an earlier stage of evolution**.

When **measuring the distance** of galaxies, it is necessary to take into account **how old** they are.

evolve over time. The light from remote galaxies can take millions or billions of years to reach Earth, where the galaxies are seen as they were in the distant past. The farther away they are, the earlier they appear in their stage of evolution. In other words, a distant galaxy that appears elliptical may be quite unlike a closer, known elliptical galaxy. In calculating distances to remote galaxies, she argued, corrections needed to be applied, based on the factors that change as galaxies evolve—specifically the abundance of different chemical elements and the rate of star birth.

Tinsley outlined ways in which galaxies evolve in terms of their brightness, shape, and color. The stars and the non-stellar material (gas and dust) in galaxies change over long periods. For example, some stars eventually turn into giant stars and get brighter as they age, star formation rates change as gas and dust are used up, and the interstellar medium (the matter between stars) is enriched with elements heavier than helium and hydrogen as old stars die.

Galaxy models

Tinsley's thesis was described as "extraordinary and profound" by her peers at the University of Texas. For the rest of her short career, she continued studying the ways in which different populations (groups) of stars age and affect the observable qualities of galaxies. From this, she developed models of the ways in which galaxies evolve, which combined an understanding of stellar evolution with knowledge of the motions of stars and nuclear physics. Today, these models form the basis for studies of galactic evolution. They also provide information about what protogalaxies (galaxies in their infancy) might look like. Tinsley's work also contributed to research on whether the universe is open (will expand forever) or closed (will eventually stop expanding and contract). One of the most insightful theoreticians in astronomy over the past century, Tinsley has been described as "opening doors to the future study of the evolutions of stars, galaxies, and even the universe itself." ▪

This artwork depicts the night sky from a hypothetical planet within the Milky Way, when the galaxy was just 3 billion years old. The sky is ablaze with the hydrogen clouds of new star birth.

WE CHOOSE TO GO TO THE MOON

THE SPACE RACE

IN CONTEXT

KEY ORGANIZATION
NASA—Apollo (1961–1972)

BEFORE
1957 The Soviet Union surprises the US by launching the first satellite, Sputnik 1.

1961 Cosmonaut Yuri Gagarin is the first person to travel into space and to orbit Earth.

AFTER
1975 The first joint US–USSR space project officially marks the end of the Space Race.

1994–1998 US and Russian space agencies share skills and expertise during the Shuttle–Mir program.

2008 Indian lunar mission Chandrayaan-1 finds evidence of widespread water ice on the moon's surface.

2015 Chinese rover Yutu discovers distinct layers of rock, including a new type of basalt, on the moon.

In the early 1960s, the US lagged behind the Soviet Union in the "Space Race." The Soviets had launched the first satellite in 1957, and on April 16, 1961, Yuri Gagarin became the first human in space. In response, in 1961 US President John F. Kennedy publicly committed to landing a man on the moon before the end of the decade. The project was carefully chosen—landing on the moon was so far beyond the capabilities of either protagonist that the Soviets' early lead might not seem so significant.

Despite the reservations of many at the time regarding a moon landing's scientific value, especially given the dangers and technical complexities involved, human spaceflight was now the focus of the US space program. NASA managers felt that with enough funding they could put a man on the moon by 1967. NASA administrator James E. Webb suggested another two years be added as a contingency.

In those six years from 1961 to 1967, NASA tripled its workforce, even though most of the planning, designing, and building of the

> I believe that this nation should commit itself to achieving the goal, before this decade is out, of landing a man on the moon and returning him safely to the Earth.
> **John F. Kennedy**

hardware was undertaken by private industry, research institutes, and universities. NASA claimed that only the construction of the Panama Canal and the Manhattan Project to develop the nuclear bomb rivaled the effort and expense of the Apollo program.

Which way to the moon?

At the time of Kennedy's historic announcement, the US boasted a grand total of 15 minutes of human spaceflight. To move from here to a

Gene Kranz

Perhaps the embodiment of the NASA spirit is not the heroic astronauts but the legendary Apollo flight director Gene Kranz. Born in 1933, Kranz was fascinated by space from an early age. He served as a pilot with the US Air Force before leaving to pursue rocket research with the McDonnell Aircraft Corporation and then NASA.

Prominent and colorful, with a brutally close-cut flattop hairstyle, Kranz was unmistakable in Mission Control, dressed in his dapper white "mission" vests made by his wife.

Although he never actually spoke the words "Failure is not an option"—they were written for his character in the movie *Apollo 13*—they sum up his attitude. Kranz's address to his Flight Control staff after the Apollo 1 disaster has gone down in history as a masterpiece of motivational speaking. In it, he stated the Kranz Dictum— "tough and competent"—that would guide Mission Control. Kranz was awarded the Presidential Medal of Freedom in 1970 for successfully returning Apollo 13 to Earth.

See also: The launch of Sputnik 208–09 ▪ Understanding comets 306–11 ▪ Exploring Mars 318–25

Project Mercury astronaut John Glenn enters the Friendship 7 on February 20, 1962. His mission, lasting just under five hours, was the US's first manned orbital spaceflight.

with it the very real risk of leaving a crew stranded in space should anything go wrong. After much debate and lobbying, influential figures, such as Wernher von Braun, director of NASA's Marshall Space Flight Center, threw their backing behind LOR, and in 1962, LOR was chosen. This was the first of many leaps of faith for Apollo.

Technological hurdles

On February 20, 1962, John Glenn became the first American to orbit Earth, looping three times around the planet in Friendship 7, as part of the US's first spaceflight program, Project Mercury, which ran from 1958 to 1963. Three more successful Mercury flights followed, but there was a big difference between operations in low Earth orbit and landing on the moon. An entire new fleet of launch vehicles was required. Unlike Mercury spacecraft, which carried a single »

moon landing, many technological hurdles needed to be overcome. One of the first was the method of getting to the moon. Three options, known as mission architectures, were on the table. The direct ascent (DA) profile, or "all-the-way," required an enormous multistage rocket with enough fuel on board to transport the crew back to Earth. This was initially the favored approach. However, it was also the most expensive, and doubts were raised over the feasibility of building such a monster rocket before the 1969 deadline.

In the Earth-orbit rendezvous (EOR) profile, a moon-bound rocket ship would be assembled in space and dock with modules that had already been placed in orbit. Lifting things into space is the most energy-consuming part of any off-Earth mission, but multiple rocket launches would sidestep the need for a single spaceship. This was the safest option, but it would be slow.

The real weight-savings came with the lunar-orbit rendezvous (LOR) profile. Here, a smaller rocket would put a three-part spaceship on course to the moon. At the moon, a command module would remain in orbit with the fuel for the journey home, while a lightweight two-stage lunar lander would be sent to the surface. This quick and comparatively cheap option carried

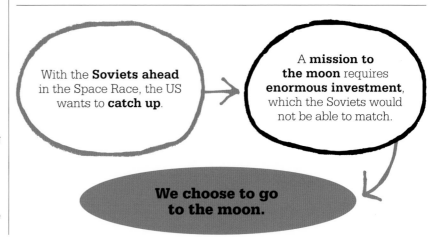

With the **Soviets ahead** in the Space Race, the US wants to **catch up**.

A **mission to the moon** requires **enormous investment**, which the Soviets would not be able to match.

We choose to go to the moon.

> From this day forward, Flight Control will be known by two words: "Tough and Competent."
> **Gene Kranz**

astronaut, Apollo missions would need a crew of three. In addition, a more reliable power source was needed and much more experience in space. The world's first fuel cells were built to provide the power.

Project Gemini, NASA's second human spaceflight program, provided the skills, with endurance spaceflights, orbital maneuvers, and space walks. Scientists also needed to know more about the moon's surface. A deep layer of dust could swallow up a spacecraft and prevent it from leaving, clog up the thrusters, or cause the electronics to malfunction.

Unmanned fact-finding missions were mounted in parallel with Apollo, but the first wave of

The Saturn V rocket was developed for the Apollo program. Many private corporations were involved in its production, including Boeing, Chrysler, Lockheed, and Douglas.

robotic explorers dispatched to the moon was an unmitigated failure. Six Ranger landers failed on launch, missed the moon, or crashed on impact, causing the program to be nicknamed "shoot and hope." Luckily, the final three Rangers were more successful.

Between 1966 and 1967, five Lunar Orbiter satellites were placed in orbit around the moon. They mapped 99 percent of the surface and helped to identify potential Apollo landing sites. NASA's seven Surveyor spacecraft also demonstrated the feasibility of a soft landing on the lunar soil.

A gamble and a disaster

At 363 ft (110.5 m), Saturn V—the heavy-lift booster that carried the Apollo astronauts out of Earth's atmosphere—is still the tallest, heaviest, and most powerful rocket ever built. "Man-rating" the rocket (certifying it to carry a human crew) proved particularly troublesome. The mammoth engines generated vibrations that threatened to break

the rocket apart. Knowing that the project was behind schedule, NASA's associate administrator for manned spaceflight, George Mueller, pioneered a daring "all-up" testing regime. Rather than the cautious stage-by-stage approach favored by von Braun, Mueller had the entire Apollo–Saturn systems tested together.

While striving for perfection, the NASA engineers developed a new engineering concept: that of redundancy. Key or critical components were duplicated in order to increase overall reliability. The Mercury and Gemini projects had taught engineers to expect unforeseen risks. A fully assembled Apollo vehicle had 5.6 million parts, and 1.5 million systems, subsystems, and assemblies.

The Lunar Orbiter satellites took images of potential landing sites. In 1966, Lunar Orbiter 2 sent back this image of Copernicus Crater, one of the first-ever close-up views of the moon.

Even with 99.9 percent reliability, the engineers could anticipate 5,600 defects. Nevertheless, over its 17 unmanned and 15 manned flights, the Saturn boosters had shown 100 percent reliability. With two partially successful test flights under its belt, Mueller declared that the next launch would carry astronauts.

Until 1967, progress had been smooth, despite the breakneck pace. Then disaster struck. An electrical short-circuit during a launch rehearsal started a fire that incinerated the Apollo 1 crew inside the Command Module. The toxic smoke and intensity of the fire in a pressurized, pure-oxygen atmosphere killed Virgil "Gus" Grissom, Ed White, and Roger Chaffee in less than five minutes. In the wake of this tragedy, the next five Apollo missions were unmanned tests. Modifications were made, resulting in a safer spacecraft with a new gas-operated hatch,

Apollo riding his chariot across the sun was appropriate to the grand scale of the proposed program.
Abe Silverstein

a 60–40 oxygen–nitrogen mix in the cockpit, and fireproof wiring throughout.

Earth's place in space

Apollo 8 was the first manned spacecraft to leave Earth's orbit. On Christmas Eve 1968, Frank Borman, James Lovell, and Bill Anders looped around the far side of the moon and witnessed the astounding sight of Earth rising from behind the moon's surface. For the first time, humans could see their home from space—a startlingly blue world lost in the immensity of the void. As Anders put it: "We came all this way to explore the moon, and the most important thing is that we discovered the Earth."

The crew was also the first to pass through the Van Allen radiation belts. This zone of charged particles extends up to 15,000 miles (24,000 km) from Earth, and was initially thought to be a serious barrier to human space travel. As it turned out, it resulted in a dosage of radiation only equivalent to a chest X-ray.

Finally, the program was ready for the last step—to take real steps on the moon itself. On July 21, 1969, an estimated »

In 1968, Apollo 8 broadcast live from moon orbit. Images taken from the spacecraft by astronaut Bill Anders included the iconic Earthrise.

Neil Armstrong took this famous photograph of Buzz Aldrin on the surface of the moon. Armstrong's reflection, standing next to the lunar module, can be seen in Aldrin's visor.

not continually repaved, and so the youngest moon rocks are the same age as Earth's oldest. The moon is not entirely geologically inactive, however, and occasionally has moonquakes that last for hours.

One Apollo 11 experiment remains active and has been returning data since 1969. Reflectors planted on the lunar surface bounce back laser beams fired from Earth, enabling scientists to calculate the distance to the moon to within an accuracy of a couple of millimeters. This gives precise measurements of the moon's orbit, and the rate at which it is drifting away from Earth (about 1½ in [3.8 cm] per year).

Apollo's legacy

On December 19, 1972, the sonic boom over the South Pacific, as the Apollo 17 capsule thumped into Earth's atmosphere, sounded the end of the Apollo program. In total, 12 men had walked on the moon. At the time, it was widely assumed that routine flights to Mars would soon be a reality, but in the intervening 40 years,

global audience of 500 million tuned in to watch Neil Armstrong land the Lunar Module and step out onto the surface of the moon, closely followed by Buzz Aldrin. It was the culmination of nearly a decade of collaborative effort and effectively ended the Space Race.

There were six more missions to the moon following Apollo 11, including the near-disaster of Apollo 13, whose lunar landing in 1970 was aborted after an oxygen tank exploded on board. The crew was returned safely to Earth on the crippled spacecraft in a real-life drama that played out in front of a worldwide television audience.

Learning about the moon

Before Apollo, much of what was known about the physical nature of Earth's only natural satellite was speculation but, with the political goals achieved, here was an opportunity to find out about an alien world firsthand. Each of the six landing missions carried a kit of scientific tools—the Apollo Lunar Surface Experiments Package (ASLEP). Apollo's instruments tested the internal structure of the moon, detecting seismic vibrations that would indicate a "moonquake." Other experiments measured the moon's gravitational and magnetic fields, heat flow from its surface, and the composition and pressure of the lunar atmosphere.

Thanks to Apollo, scientists have compelling evidence from analysis of moon rock that the moon was once a part of Earth (pp.186–87). Like Earth, the moon also has internal layers and was most likely molten at some point in its early history. Unlike Earth, however, the moon has no liquid water. Since it has no moving geological plates, its surface is

Houston. Tranquility Base here. The Eagle has landed.
Neil Armstrong

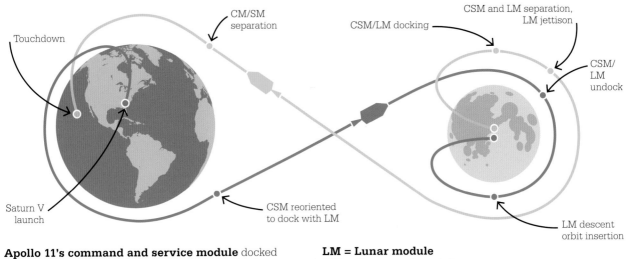

Apollo 11's command and service module docked with the lunar module in orbit before heading for the moon. Before touchdown, the service module was jettisoned, and only the command module returned to Earth.

LM = Lunar module
CM = Command module
SM = Service module
CSM = Command service module

—— Outbound
—— Inbound

scientific priorities changed, politicians worried about costs, and human space travel has not ventured farther than Earth's orbit.

For many, the decision to end manned moon missions was a wasted opportunity, caused by a lack of imagination and leadership. However, the end of the acute Cold War competition that gave rise to the Apollo program heralded a new era of international cooperation for NASA, with Skylab, Mir, and the International Space Station.

Gene Cernan, the last man on the moon, predicted that it could be another 100 years before humankind appreciates the true significance of the Apollo missions. One result could be that it may have made the US smarter—the intake for doctoral degrees at American universities tripled during the 1960s, particularly in the field of physics.

On the final three Apollo missions, astronauts explored the surface of the moon on lunar rovers. The rovers were abandoned and can still be seen where they were left behind.

Apollo contracts also nurtured nascent industries, such as computing and semiconductors. Several employees of the California-based Fairchild Semiconductors went on to found new companies, including Intel, a technology giant. The Santa Clara area where these firms were based has become

today's Silicon Valley. But perhaps Apollo's real legacy is the idea of Earth as a fragile oasis of life in space. Photos taken from orbit, such as the "Blue Marble" and "Earthrise" (p.247), fed into a growing awareness of planet Earth as a single entity, and the need for careful stewardship. ∎

THE PLANETS FORMED FROM A DISK OF GAS AND DUST
THE NEBULAR HYPOTHESIS

IN CONTEXT

KEY ASTRONOMER
Viktor Safronov (1917–1999)

BEFORE
1755 German philosopher Immanuel Kant argues that the solar system formed out of a large gas cloud that collapsed.

1796 Pierre-Simon Laplace develops a model of solar system formation that is similar to Kant's.

1905 The American geologist Thomas Chrowder Chamberlin first proposes that planets develop from particles that he calls "planetesimals."

AFTER
1980s Several apparently young stars, such as Beta Pictoris, are found to be surrounded by disks of cool dust.

1983 The Infrared Astronomical Satellite is launched. It observes that many stars have an excess of infrared radiation that could be explained if they were orbited by disks of cooler material.

For centuries, astronomers have proposed various models to explain how the sun and planets formed. During the 18th and 19th centuries, the nebular hypothesis came to prominence. This proposed that the solar system emerged from a giant cloud of gas and dust that collapsed and started rotating. Most of the material collected in the center, forming the sun, while the rest flattened into a spinning disk of material from which the planets and smaller objects condensed.

A version of this hypothesis was put forward by Frenchman Pierre-Simon Laplace in 1796.

In the late 1960s, Viktor Safronov was working in Moscow on how planets could form in a nebula. He wrote an important paper in 1969, which was unknown outside the Soviet Union until an English version was published in 1972. Safronov's theory, which today is known as the solar nebular disk model (SNDM), was essentially a modified, mathematically more fully formed version of the nebular hypothesis.

In the **disk of material** orbiting the early sun, **particles occasionally collided**.

→

In these collisions, some slower-moving **particles stuck together**, forming larger particles.

Over time, **larger planetesimals** formed. These aggregated into a few large bodies, leading to the **emergence of the planets**.

The planets formed from a disk of gas and dust.

See also: Gravitational disturbances 92–93 ▪ The Kuiper belt 184 ▪ The Oort cloud 206 ▪ Inside giant molecular clouds 276–79

Viktor Safronov

Viktor Sergeevich Safronov was born in Velikiye Luki near Moscow in 1917, and graduated from Moscow State University Department of Mechanics and Mathematics in 1941. In 1948, he obtained a doctorate degree. Safronov spent a considerable part of his career working at the Schmidt Institute of the Physics of the Earth, part of the Academy of Sciences in Moscow. There, he met his wife, Eugenia Ruskol, who for a time collaborated with him in his research. From the 1950s to the 1990s, he worked on modeling the idea that the planets formed within a disk of gas and dust. Today, Safronov's planetesimal hypothesis of planet formation is widely accepted, although alternative theories exist. After the fall of the Soviet Union in 1991, he had the opportunity to explain his ideas in the West.

Key work

1972 *Evolution of the Protoplanetary Cloud and Formation of the Earth*

Up to the 1940s, astronomers considered that the nebular hypothesis contained a major flaw known as the "angular momentum problem." They calculated that if the solar system had formed out of a contracting, rotating cloud the sun should be spinning much faster than it actually is. During the first half of the 20th century, a number of alternative hypotheses competed with the nebular hypothesis. One suggested that planets might have formed from material pulled out of the sun by a passing star; another that the sun passed through a dense interstellar cloud and emerged enveloped in the gas and dust from which planets coalesced. Eventually, solid reasons emerged for rejecting these alternatives.

Safronov's theory develops

Undeterred, Safronov studied in detail how planets might have formed in the disk of material proposed by Laplace. This disk would have consisted of a collection of dust grains, ice particles, and gas molecules, all orbiting the early sun. Safronov's breakthrough came when he calculated the effect on such a system of some particles colliding. He figured out the speeds at which they would collide. Particles traveling at fast speeds would simply bounce off each other. But slower-moving particles would stick together, resulting in larger particles. As they grew bigger, the gravity of each particle would cause them to coalesce, forming larger objects called planetesimals.

The larger objects would attract more mass, and the largest planetesimals would grow larger and larger, until they had gathered everything that lay within their gravitational reach. After a few million years, only a few planet-sized bodies would remain.

By the 1980s, there was wide agreement over the SNDM. One researcher suggested that the angular momentum problem could be solved by dust grains in the original disk slowing down rotation in the center. Others incorporated Safronov's ideas into computer models, which suggested that systems of particles orbiting the early sun could indeed have formed into a handful of planets. Recent observation of disks of cool dust surrounding apparently young stars lend further support to the SNDM. ▪

In Safronov's model, the planets formed out of dust and ice particles, which stuck together within a disk of material spinning around the newly formed sun.

1 A large cloud of gas and dust starts contracting and slowly rotating.

2 The cloud flattens out into a spinning disk with a denser, hotter center, which forms the sun.

3 Solar radiation heats up the inner solar system.

4 Planetesimals rich in iron and silicate dust begin to form.

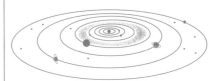

5 The solar system forms.

SOLAR NEUTRINOS CAN ONLY BE SEEN WITH A VERY LARGE DETECTOR

THE HOMESTAKE EXPERIMENT

IN CONTEXT

KEY ASTRONOMER
Ray Davis (1914–2006)

BEFORE
1930 Austrian physicist
Wolfgang Pauli proposes
the existence of neutrinos.

1939 Hans Bethe outlines
two main processes by which
stars obtain their energy.

1956 American physicists
Clyde Cowan and Frederick
Reines confirm the existence
of the antineutrino, the
antiparticle of the neutrino.

AFTER
1989 The Kamiokande II
experiment in Japan,
organized by Masatoshi
Koshiba, demonstrates
conclusively that the sun
is a source of neutrinos and
confirms Davis's abnormally
low detection rate.

If the sun obtains energy from **nuclear fusion**, fast-moving low-mass particles called **neutrinos should be produced**.

Neutrinos **barely interact** with other particles, but they may interact in a form of **radioactive decay**.

The **detection rate** in the interaction is likely to be **very low**.

A very large detector is required.

During the first half of the 20th century, scientists figured out a process by which the sun produces energy by fusing hydrogen into helium. In the sun's core, four hydrogen nuclei, which are single protons, are changed into a helium nucleus, two positrons (also called anti-electrons), and two tiny ghostlike particles called neutrinos, with the release of energy. The neutrinos produced were envisaged to escape easily from the sun.

Although this theory was accepted by the 1950s, it was not proven. In 1955, an American chemist named Ray Davis set out to show that the sun produces energetic neutrinos by detecting just a few of them. He faced a huge problem in achieving this goal, however. Apart from the fact that their existence was uncertain, scientists thought that neutrinos had zero electric charge and a tiny mass (if any at all), and very rarely interacted with other particles. If the sun fuses hydrogen,

See also: Cosmic rays 140 ▪ Energy generation 182–83 ▪ Gravitational waves 328–31

> Neutrino physics is largely an art of learning a great deal by observing nothing.
> **Haim Harari**
> *Israeli physicist*

scientists thought, billions of neutrinos should be passing through every square centimeter of Earth's surface every second, but perhaps only one in one hundred billion might interact with atomic matter.

Davis thought that neutrinos might be detectable through their involvement in a type of radioactive decay called beta decay. In theory, an energetic neutrino should be able to convert a neutron in an atomic nucleus into a proton. In his experiments, Davis found that, on very rare occasions, a neutrino passing through a tank of a chlorine-containing substance would interact with the nucleus of a stable chlorine atom to produce a nucleus of an unstable isotope of argon called argon-87.

Homestake experiment

In 1964, in what was called the Homestake experiment, Davis began a trial using a large tank of a chlorine-containing chemical as a detector. An acquaintance of Davis, astrophysicist John Bahcall calculated the theoretical number of neutrinos of different energies that the sun should be producing

and, from this, the rate at which argon-37 should be produced in the tank. Davis began counting the actual numbers of argon-37 atoms produced.

Although Davis's experiment showed conclusively that the sun does produce neutrinos, only about one-third of the number of argon-37 atoms predicted by Bahcall were detected. The discrepancy between the number of neutrino interactions predicted and those detected became known as "the solar neutrino problem."

Building on his 1989 work, in 1999, Masatoshi Koshiba discovered the cause of the discrepancy at Japan's huge Super-Kamiokande neutrino detector. Neutrinos were found to oscillate between three different types— the electron neutrino, the muon neutrino, and the tau neutrino— while traveling through space. Davis's experiment had detected only the electron neutrinos. ∎

Davis's neutrino detector was placed deep underground to protect it from cosmic rays (another possible source of neutrinos).

Ray Davis

Raymond Davis was born in Washington, D.C., in 1914. He earned a Ph.D. in physical chemistry from Yale University in 1943. Davis spent the later years of World War II in Utah, observing the results of chemical weapons tests. From 1946, he worked at a laboratory in Ohio, carrying out research on radioactive chemical elements. In 1948, he joined Brookhaven National Laboratory, on Long Island, which was dedicated to finding peaceful uses for nuclear power. He spent the rest of his career studying neutrinos. Davis retired from Brookhaven in 1984, but continued his involvement in the Homestake experiment until it ended in the late 1990s.

Davis met his wife, Anna, at Brookhaven Laboratory and together they had five children. He shared the Nobel Prize in Physics in 2002 with Masatoshi Koshiba for pioneering contributions to astrophysics. He died in Blue Point, New York, age 91 in 2006.

Key work

1964 *Solar Neutrinos II, Experimental*

A STAR THAT WE COULDN'T SEE

DISCOVERING BLACK HOLES

IN CONTEXT

KEY ASTRONOMERS
Louise Webster (1941–1990)
Paul Murdin (1942–)
Tom Bolton (1943–)

BEFORE
1783 English clergyman John Michell suggests the existence of a star whose gravity is so strong that not even light can escape it.

1964 Cosmic X-rays are detected by Geiger counters in sounding rockets.

1970 Uhuru, the first X-ray observatory satellite, is launched.

AFTER
1975 Stephen Hawking makes a bet with theoretical physicist Kip Thorne that Cygnus X-1 is not a black hole.

1990 Hawking concedes the bet and buys Thorne a subscription to *Penthouse* magazine.

2011 Further observations give Cygnus X-1 an expected mass of 14.8 suns (14.8 solar masses).

B lack holes are invisible. They allow no matter to escape, and, with the exception of low-level Hawking radiation at the event horizon, even swallow up electromagnetic light energy. Due to the difficulty of detecting an invisible object, black holes remained purely theoretical concepts up to the mid-20th century. However, such a concentrated mass ought still to create observable effects. As it is dragged into a black hole, matter will be heated to millions of degrees as it is ripped apart by gravitational forces, pouring out X-rays into space in the process.

In the 1960s, astronomers looked for cosmic X-ray sources with a series of balloon- and rocket-launched detectors. Many of the hundreds of sources that they found were assumed to be "X-ray binaries"—star systems in which a superdense stellar remnant, such as a neutron star, tears material away from its visible companion. Among the first of these X-ray binaries to be discovered, in 1964, was a strong source close to active star-forming regions of the Milky Way, in the constellation of Cygnus. In 1973, Australian Louise Webster, Briton Paul Murdin, and American Tom Bolton independently took measurements of the blue supergiant star HDE 226868. They revealed that it orbits an object far too massive to be a neutron star. The only candidate for the invisible partner, Cygnus X-1, was a black hole. Black holes were now more than mere theoretical entities. ■

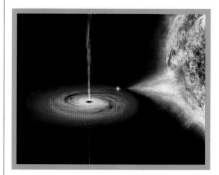

An artist's impression shows matter flowing from the blue supergiant star HDE 226868 into its black hole partner, Cygnus X-1. The star is losing one solar mass of material every 400,000 years.

See also: Supernovae 180–81 ▪ Cosmic radiation 214–17 ▪ Hawking radiation 255

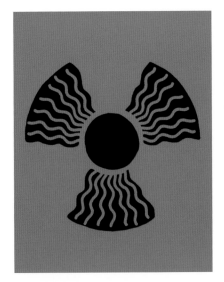

BLACK HOLES EMIT RADIATION

HAWKING RADIATION

IN CONTEXT

KEY ASTRONOMER
Stephen Hawking (1942–)

BEFORE
1916 Karl Schwarzschild
provides a solution to the field
equations of general relativity,
allowing him to describe the
gravitational field around a
black holelike object.

1963 New Zealand
mathematician Roy Kerr
describes the properties
of a rotating singularity.

1965 British mathematician
Roger Penrose shows that the
gravitational collapse of a giant
star could result in a singularity.

1967 American physicist John
Wheeler coins the term "black
hole" for the types of objects
described by Schwarzschild,
Kerr, Penrose, and others.

AFTER
2004 Stephen Hawking retracts
an earlier claim that any object
entering a black hole is
completely lost to the universe.

The mathematical theory of black holes was pioneered in the 1910s by the German physicist Karl Schwarzschild. The object described by Schwarzschild was a non-rotating mass concentrated at a point of infinite density, called a singularity. At a distance from this, known as the Schwarzschild radius, was an imaginary spherical surface called the event horizon. The gravity on the singularity side of this surface was so great that nothing—not even light—could escape. In subsequent decades, black hole theory evolved in various ways, but black holes continued to be regarded as entirely black, emitting no light.

Virtual particles
In 1974, a big change occurred in black hole theory. British physicist Stephen Hawking proposed that black holes emit particles, known today as Hawking radiation. Hawking maintained that black holes are not completely black, since they emit radiation of some sort, if not necessarily light.

Quantum theory predicts that, throughout space, pairs of "virtual" particles and their antiparticles should continually appear out of nothing, then annihilate (cancel out back to nothing). One of each pair has positive energy, the other negative energy.

Some of these particle–antiparticle pairs will appear just outside the event horizon of a black hole. It is possible that one member of the pair could escape—observed as an emission of (positive) radiation energy—while the other falls into the black hole. In order to preserve the same total energy in the system, the particle that fell into the black hole must have had negative energy. This causes the black hole to slowly lose mass-energy—a process called black hole evaporation.

Hawking radiation remains a theoretical prediction. If it proves correct, it means that black holes do not last forever, which has implications regarding the fate of the universe, since it had been thought that black holes would be among the last objects in existence. ∎

See also: Curves in spacetime 154–55 ▪ The life cycles of stars 178 ▪ The heart of the Milky Way 297 ▪ Michell (Directory) 335

THE TR
OF TEC
1975–PRESENT

UMPH
NOLOGY

NASA launches the two **Voyager spacecraft** on a mission to tour the outer planets.

1977

US astronomer **Vera Rubin** publishes data showing that the rotational speeds of galaxies indicate the presence of invisible **"dark matter."**

1980

The **Hubble Space Telescope** enters orbit. It provides the best-ever images in the **visible and near-visible** spectrum.

1990

1979

US cosmologist **Alan Guth** develops the idea that the early universe experienced a period of **rapid inflation**.

1986

American **Frank Shu** and colleagues present a new model for **star formation**.

1995

The first **brown dwarfs** are detected, confirming a theoretical prediction made in 1962 by **Shiv S. Kumar**.

Most major discoveries in astronomy have been made possible by advances in technology. Recent developments have provided powerful tools to collect radiation from space and to process vast amounts of data, and the pace of discovery has accelerated at a breathtaking rate. Microelectronics and computing capability, in particular, have opened up new possibilities over the last 40 years.

Telescopes and detectors

The New Technology Telescope (NTT) opened by the European Southern Observatory (ESO) in the Chilean Andes in 1989 is an example of a telescope with revolutionary innovations that have subsequently become standard equipment. Its main and secondary mirrors are flexible, but kept precisely in shape by a network of computer-controlled supports called actuators.

ESO's choice of Chile was part of the trend for astronomers to scour the world, testing for the best sites where the air is clear, still, and dry, and the sky free from light pollution. Another major center for astronomy was established at the summit of the volcano Mauna Kea on the Big Island of Hawaii in 1967. This prime site is now home to 13 telescopes.

Until the early 1970s, all astronomical imaging was carried out by means of conventional photography. Then, in the mid-1970s, a completely new way of recording an image electronically became a practical reality. This was the charge-coupled device (CCD). CCDs are electronic circuits with light-sensitive pixels that generate electrical charges when light photons land on them. They are far superior to photography for sensing faint light and recording an

We're going to need a definitive quantum theory of gravity, which is part of a grand unified theory—it's the main missing piece.
Kip Thorne

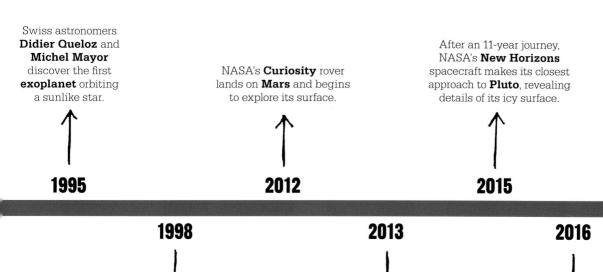

Swiss astronomers **Didier Queloz** and **Michel Mayor** discover the first **exoplanet** orbiting a sunlike star.

1995

1998

Cosmic expansion is found to be accelerating, suggesting the existence of a mysterious **"dark energy."**

NASA's **Curiosity** rover lands on **Mars** and begins to explore its surface.

2012

2013

The **European Southern Observatory** opens its **Atacama Large Millimeter Array**, a giant radio telescope in Chile.

After an 11-year journey, NASA's **New Horizons** spacecraft makes its closest approach to **Pluto**, revealing details of its icy surface.

2015

2016

The **Ligo Scientific Collaboration** announces the detection of **gravitational waves**, confirming Einstein's general theory of relativity.

---

object's brightness accurately, and they made visible objects that had previously been too faint to detect, such as the small, icy worlds of the Kuiper belt beyond Neptune.

Computing power

Fast, reliable computers and an immense capacity for storing data have been key not only to the way that telescopes and their instruments are constructed, but also to making sense of the astronomical data they collect. One major project alone, the Sloan Digital Sky Survey, has collected information about 500 million celestial objects since it began in 2000. This database has been used to create a three-dimensional map showing how galaxies are distributed across the universe, revealing its largest structures.

Computers are indispensable to theorists, too. Huge computing power makes it possible to gain insight into what observations are telling us about the way the universe works by creating simulations based on the laws of physics. For example, computers allow scientists to model ways in which the solar system may have formed and evolved.

Space exploration has now pushed right out to the edge of the solar system, and no region of the planetary system remains unexplored on some level. The New Horizons mission passed Pluto in 2015 and is moving through the Kuiper belt, while the Voyager spacecraft, launched back in 1977, are now sending back data from interstellar space. With the advent of the internet, missions can also

be followed in real time, as the latest images from the Hubble Space Telescope or the Curiosity rover on Mars are made instantly available.

Landmark discoveries

Of the many discoveries in recent decades that have had an impact on our understanding, three stand out. The surprise finding in 1998 that the universe's expansion is accelerating showed that there is a gap in fundamental theory. By contrast, the detection of gravitational waves in 2016 confirmed Einstein's 100-year-old theoretical prediction. Meanwhile, the discovery of the first extrasolar planet in 1995, and thousands more since, has energized the search for alien life. It is impossible even to speculate where the next 20 years may lead. ∎

A GRAND TOUR OF THE GIANT PLANETS

EXPLORING THE SOLAR SYSTEM

IN CONTEXT

KEY ORGANIZATION
NASA—Voyager mission
(1977–)

BEFORE
1962 Mariner 2 passes Venus in the first planetary flyby.

1965 Mariner 4 is the first craft to visit Mars.

1970 Venera 7 makes the first soft landing on Venus.

1973 Pioneer 10 is the first spacecraft to cross the asteroid belt en route to Jupiter.

1976 Viking 1 sends pictures from the surface of Mars.

AFTER
1995 The Galileo spacecraft goes into orbit around Jupiter.

1997 Sojourner is the first rover to land on Mars.

2005 The Cassini orbiter releases the Huygens probe, which touches down on Titan.

2015 New Horizons makes the first flyby of Pluto and Charon.

O
n August 20, 1977, the Voyager 2 spacecraft was launched from Cape Canaveral in Florida. Two weeks later, its sister craft Voyager 1 was launched. Thus began the most ambitious exploration of the solar system ever. The launch was the culmination of more than a decade's work. The core mission would run for 12 years, but an interstellar mission continues.

Going interplanetary

By the early 1960s, both the Soviet and US space agencies were sending missions to other planets. There were more failures than successes, but over the decade robotic spacecraft began sending back close-up images of Venus and Mars. The NASA craft were part of the Mariner program, run largely from the Jet Propulsion Laboratory (JPL) in California. The mathematicians at JPL perfected the art of the "flyby"—sending a spacecraft on a trajectory that had it fly past a planet close enough to photograph and observe it, albeit too quickly to enter its orbit. In 1965, a graduate student named Gary Flandro, who was working at JPL for the summer, was given the task of figuring out routes to the

A Titan 3E rocket lifts off carrying the Voyager 1 spacecraft as its payload. The Titan 3E was the most powerful launch vehicle of its time.

outer planets and discovered that, in 1978, all the outer planets would be on the same side of the sun. His calculations revealed that this had not happened since 1801, and would not occur again until 2153.

Flandro saw the opportunity for a Grand Tour of the outer solar system, but the distances involved were far beyond the capabilities of the spacecraft of the day. In 1965, Mars's alignment made it the

See also: Life on other planets 228–35 ▪ The nebular hypothesis 250–51 ▪ Exoplanets 288–95 ▪
Understanding comets 306–11 ▪ Studying Pluto 314–17

closest planet to Earth at that time at 35 million miles (56 million km), but Neptune was 2.5 billion miles (4 billion km) away, and a journey to it would take several years.

Planetary slingshots

A Grand Tour spacecraft would have to change course several times in order to fly past all its planets. Flandro's plan had to use gravity assists to fling the craft from planet to planet. Also known as a gravitational slingshot or swing-by, a gravity assist had first been used by the Soviet Luna 3, which had swung around the far side of the moon in 1959, photographing as it went. It had never been used to guide spacecraft as far from Earth as the outer planets. The planned slingshot required the craft to approach the planet head-on, traveling in the opposite direction

An artist's impression shows Voyager 1 in space. This craft and its twin, Voyager 2, communicate with Earth via radio waves transmitted and received by a 12-ft (3.7-m) dish.

The best way to learn about planets is to send **robotic spacecraft** to them.

All the **outer planets** will be **close together** for a short period.

A **Grand Tour** could send probes to study them throughout this period.

The Voyager program makes a Grand Tour of the giant planets.

from the planet's orbital motion. The planet's gravity would speed up the craft as it made a loop around the planet. It would then slow down again as it headed off into space, having done an about turn. If the motion of the planet were ignored, the craft's escape speed would be more or less equal to its approach speed. However, taking the motion of the planet into account, the craft would leave the

planet having added approximately twice the speed of the planet to its own velocity. A slingshot would not only redirect the craft, but also accelerate it on to its next target.

Taking a Grand Tour

In 1968, NASA set up the Outer Planets Working Group. It proposed the Planetary Grand Tour mission, which would send one spacecraft to visit Jupiter, Saturn, and Pluto, »

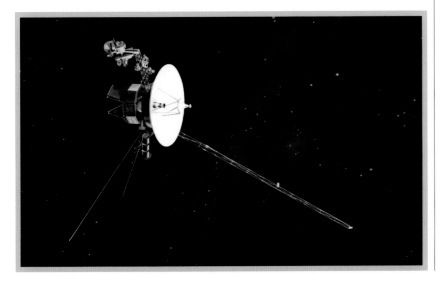

It was a chance that came around once in every 176 years and we prepared for it. Out of that emerged the greatest mission of planetary exploration to this day.
Charles Kohlhase

Jupiter's moon Europa was photographed by Voyager 2. It is covered in a thick crust of ice, which has been fractured and filled in with material from the moon's interior.

and a second one toward Uranus and Neptune. The plan required a new long-range spacecraft and costs grew steadily. Then in 1971, NASA canceled the Grand Tour as it needed cash to fund the Space Shuttle program.

The exploration of the outer planets was handed back to the Mariner program. The mission was named Mariner Jupiter–Saturn, or MJS77 for short—77 referred to the launch year. To reduce costs, Pluto was removed from the tour itinerary. Instead, one craft was to visit Jupiter, Saturn, and finally Saturn's huge moon Titan. Titan was considered more intriguing than distant Pluto. It was larger than Mercury, and thought at the time to be the largest moon in the solar system. It was also the only

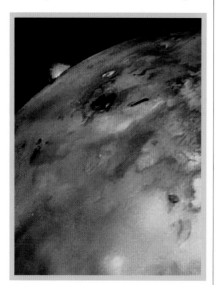

Voyager 1 captured a 100-mile- (150-km-) high eruption on Jupiter's moon Io. Strongly affected by Jupiter's gravity, Io is the most volcanically active place in the solar system.

moon known to have its own atmosphere. This change meant that the mission would be budgeted as an exploration of the two gas giants, not a Grand Tour. However, the spacecraft, code-named JST, was to have a back-up, JSX. Its mission would also include Jupiter and Saturn if JST failed. The X represented an unknown quantity. If required, JSX would go to Titan, but if JST achieved its mission, then JSX would be sent to Uranus and Neptune.

Mission profile

In 1974, mission design manager Charles Kohlhase began to make a master plan for the MJS77 mission. He had to consider every aspect, from the spacecrafts' design, size, and launch system to the many variables they would encounter along their routes—the radiation levels, light conditions, and contingencies for altering the missions. It took Kohlhase and his team eight months to eventually settle on two trajectories that met

all the criteria and would take the spacecraft as close to as many points of interest as possible.

Neither Kohlhase nor anyone else working on MSJ77 liked the name. As the launch date approached, a competition for a new name was organized. Nomad and Pilgrim made the shortlist, but by the time the two identical spacecraft were ready, they were known as Voyager 1 and Voyager 2. At 1,590 lb (720 kg), the two Voyager spacecrafts were nearly 50 percent heavier than any previous flyby craft. About 220 lb (100 kg) of that was scientific equipment, comprising two cameras, magnetic field sensors, spectrometers that would analyze light and other radiation to show which chemicals were present in atmospheres, and particle detectors for investigating cosmic rays. In addition, the radio system could be used for a variety of experiments, such as probing atmospheres and Saturn's rings. The spacecrafts' trajectories were to be controlled by 16 hydrazine thrusters. However, it would be too dark beyond the asteroid belt for solar panels to generate enough electricity for the spacecraft, and batteries would not last nearly long enough. The answer was nuclear power in the form of radioisotope thermoelectric generators (RTG), held out on a boom to isolate them from sensitive equipment. Each RTG contained 24 balls of plutonium that gave out heat, which was converted into an electric current by thermocouples. The power supply was built to last for nearly 50 years.

Jupiter and its moons

By December 1977, Voyager 1 had overtaken Voyager 2, which was taking a more circular trajectory. It reached the Jupiter system in January 1978. Most of Voyager 1's important discoveries were made in a frenetic 48-hour period around March 5, when it made its closest approach, coming within 217,000 miles (349,000 km) of the planet's cloud tops. In addition to sending back images, Voyager 1 analyzed

The latter half of the next decade abounds in interesting multiple planet opportunities. Of particular interest is the 1978 "grand tour" which would make possible close-up observation of all planets of the outer solar system.
Gary Flandro

the compositions of the clouds and measured the planet's immense magnetic field. It also showed that Jupiter had a faint ring system. Its most memorable discoveries came from the flybys of the Galilean moons. These were not sparse, cratered balls but active worlds. Photographs of Io showed the largest volcanic eruptions ever seen, spurting ash clouds into orbit. Fresh measurements of Ganymede revealed that it superseded even

Titan in size, while images of Europa's eerily smooth yellowish disk had astronomers puzzled. Voyager 2 arrived at Jupiter more than a year later, and did not approach as close as Voyager 1, but it took some of the mission's most »

Voyager 2's images of Saturn's rings revealed a complex structure made of small satellites, none of which were larger than 3–6 miles (5–9 km) across.

Voyager 2 sent back this image of Neptune's ice moon Triton. During the craft's flyby, only the southern ice cap was visible. Highly reflective, it is made of frozen nitrogen and methane.

The Titan trajectory then sent the craft over Saturn's pole and away to the edge of the solar system.

Voyager 2 arrived at Saturn in August 1981, and was able to study the planet's rings and atmosphere in more detail, but its camera failed during much of the flyby. Fortunately, it was restored, and the order was given to continue to the ice giants.

Uranus and Neptune

Voyager 2 is the only craft to have visited the ice giants Uranus and Neptune. It took 4.5 years to travel from Saturn to Uranus, where the craft passed 50,500 miles (81,500 km) above the pale blue atmosphere. It looked at the planet's thin rings and discovered 11 new moons, all of which are now named after Shakespearian characters, as is the rule for Uranus. The most curious thing to be examined

iconic images of Io transiting Jupiter. Voyager 2 also got a closer look at Europa, showing that it was covered in a crust of water ice riven by cracks. Later analysis revealed that these cracks were caused by upwellings in a liquid ocean under the crust, an ocean that is estimated to hold at least twice as much water as Earth and which is thought by scientists to be a prime candidate for the presence of alien life.

Titan and Saturn

By November 12, 1980, Voyager 1 was skimming 77,000 miles (124,000 km) above the atmosphere of Saturn. On the approach, and despite some instrument failures, it revealed details of the rings, which were made of billions of chunks of water ice and were as thin as 30 ft (10 m) in places. Kohlhase had sent Voyager 1 to visit Titan before approaching

Saturn to prevent any damage caused by Saturn's atmosphere and rings from endangering this crucial phase. The spacecraft swung behind Titan so the sun's light shone through the atmosphere, allowing measurements of its thickness and composition.

Charles Kohlhase

Charles "Charley" Kohlhase was born in Knoxville, Tennessee, and graduated with a degree in physics. He briefly served in the US Navy before joining JPL in 1960, where he turned his life-long fascination with exploration into work on the Mariner and Viking projects, before joining the Voyager team. In 1997, Kohlhase left Voyager to design the Cassini–Huygens mission to Saturn, which succeeded in dropping a lander onto the surface of Titan in

2005. In the late 1970s, he worked with computer artists to create accurate animations of space missions for advancing the public understanding of NASA's work. Now retired, Kohlhase remains involved in several projects that blend art and space science, aiming to educate and inspire the next generation of rocket scientists and interplanetary explorers.

Key work

1989 *The Voyager Neptune travel guide*

The golden records carried by the Voyager spacecraft included a selection of music, greetings in 55 different languages, and images of humans, animals, and plants.

on this otherwise relatively quiet planet is the tilt of its axis, which is roughly 90°. As a result, Uranus does not spin as it orbits, but "rolls" around the sun.

The final port of call was Neptune, reached in August 1989. This deep-blue planet was found to have the strongest winds in the solar system, up to 1,500 mph (2,400 km/h)—nine times stronger than anything experienced on Earth. The Voyagers' mission controllers were able to abandon caution as the planetary mission drew to an end. Without regard for the safety of its final trajectory, Voyager 2 was redirected to fly past Neptune's moon Triton. The images of the huge ice moon showed geysers blasting fountains of slush from the surface.

Continuing mission

The Voyager program continues and the two craft are still in touch with NASA. As of 2016, Voyager 1 was 12.5 billion miles (20 billion km) and Voyager 2 was 10 billion miles (16 billion km) away. Six times a year, the craft spin around to measure the cosmic rays around them. This data shows that the craft are approaching the edge of the heliosphere, the region of space that is influenced by the sun. Soon they will enter interstellar space and measure the cosmic wind from ancient stellar explosions.

In 2025, the two spacecraft will power down and go quiet forever, but their mission may still not yet be complete. A committee chaired by Carl Sagan selected content for a gold-plated phonograph record (its analog groove would be easier to read than a digital format). They included greetings from the world, the sounds and sights of Earth, and even human brain waves. The record is a calling card from humankind to an alien civilization. The Voyagers are not heading for any star systems; the closest they will get is when Voyager 1 passes 1.6 light-years from a star in 40,000 years' time. In all likelihood, they

The spacecraft will be encountered and the record played only if there are advanced spacefaring civilizations in interstellar space. But the launching of this "bottle" into the cosmic "ocean" says something very hopeful about life on this planet.
Carl Sagan

will never be found by intelligent life, but the golden records are a symbol of the hope with which the two interplanetary spacecraft were sent on their way. ■

By 2005, the Voyagers had reached the termination shock, where the solar wind slows and becomes turbulent as it mixes with the interstellar medium (matter in the space between star systems), entering the heliosheath region. By 2016, they were nearing the heliopause, where the solar wind is stopped by the interstellar medium.

Heliosheath

Voyager 1

Solar wind

Earth

Voyager 2

Termination shock

Heliopause

MOST OF THE UNIVERSE IS MISSING

DARK MATTER

IN CONTEXT

KEY ASTRONOMER
Vera Rubin (1928–)

BEFORE
1925 Bertil Lindblad
calculates the likely
shape of the Milky Way.

1932 Jan Oort finds that
the rotational speeds of the
Milky Way galaxy do not
match the presumed mass.

1933 Fritz Zwicky suggests
that a majority of the
universe is made up of
dark, invisible matter.

AFTER
1999 It is discovered that
dark energy is accelerating
the expansion of the universe.

2016 The LIGO experiment
detects gravitational waves,
which offer a new method to
map the distribution of dark
matter across the universe.

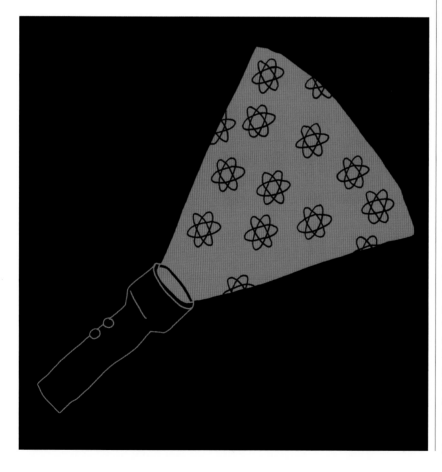

Isaac Newton's universal law of gravitation is adequate for making calculations to launch a satellite into orbit, land a crew on the moon, and send a spacecraft on a grand tour of the planets. Newton's clear mathematics works well enough for most things on the solar system scale, but not on grander scales of the universe, where Einstein's relativistic theory of gravity is needed (pp.146–53). Nevertheless, Newton's law of gravitation was all that was required to reveal one of the biggest—and as yet unsolved—mysteries of astronomy: dark matter. In 1980, American astronomer Vera Rubin presented clear evidence that dark

See also: Gravitational theory 66–73 ▪ Gravitational disturbances 92–93 ▪ The shape of the Milky Way 164–65 ▪
Supernovae 180–81 ▪ The Oort cloud 206 ▪ Dark energy 298–303

We became astronomers
thinking we were studying the
universe, and now we learn
that we are just studying the
5 percent that is luminous.
Vera Rubin

matter exists. Thanks to Rubin, the
general public learned that most of
the universe appears to be missing.

Throughout the 1960s and
1970s, the science of astronomy
was dominated by projects on a
grand scale, as researchers used
massive instruments, often in
remote parts of the world, to search
for exotic objects, such as black
holes, pulsars, or quasars. By
contrast, Rubin was looking for
a research area that would allow

her to stay in her home city of
Washington, D.C., and raise her four
children. She chose to study the
rotation of galaxies, specifically
looking at the anomalous behavior
of the outer regions of galaxies.

Spinning spirals

The problem Rubin tackled was
the fact that huge disks of stars
in nearby galaxies did not move
in a way that was consistent with
Newton's law of gravity: their
outer regions moved too quickly.
This curiosity was not new, but it
had previously been largely ignored.

Since the 1920s, when Bertil
Lindblad and others showed that
the Milky Way—and by extension
many other galaxies—were disks
of stars moving around a central
point, it was assumed that galaxies
were orbital systems just like any
other. In the solar system, near
objects orbit at a faster speed than
distant ones, so Mercury is moving
much more rapidly than Neptune.
This is because, following Newton,
gravity decreases with a square of
the distance. When the velocities »

The **outer regions** of
galaxies move much **more
quickly** than expected.

To stop spinning
galaxies from disintegrating,
they must contain **a lot more
mass** than can be seen.

This mass comes from
invisible dark matter—
there is **six times more** dark
matter in the universe than
ordinary matter.

**Most of
the universe
is missing.**

Vera Rubin

Born Vera Cooper in Philadelphia,
Rubin earned her first degree from
Vassar College in upstate New
York, and then applied to go to
Princeton. Her application was
ignored because women were
barred from joining the university's
graduate astronomy program until
1975. Instead Rubin pursued her
studies at Cornell University,
where she studied under greats
such as Richard Feynman and
Hans Bethe. She subsequently
earned a Ph.D. from Georgetown
University in Washington, D.C.,
supervised by George Gamow.
Her thesis, published in 1954,

concluded that galaxies would
clump together, a fact that was
not fully explored until the work
of John Huchra in the late 1970s.
After teaching at a college in
Maryland, Rubin returned to
Georgetown, and then moved
to the Carnegie Institution of
Washington in 1965. It was here
that she conducted her work on
galactic rotation, and she has
remained there ever since.

Key work

1997 *Bright Galaxies,
Dark Matters*

of the planets are plotted against their distance from the sun, the data forms a smooth downward "rotation curve." It followed that plotting the orbital speeds of stars at different distances from the galactic center should produce a similar curve.

In 1932, Dutch astronomer Jan Oort was the first person to provide observational proof that the galaxy was a single orbital system made up of a swirling spiral of stars, in which the sun completed a 225-million-year orbit. However, in the course of his calculations, Oort found that the motion of the galaxy suggested that it was twice as massive as the total mass of visible stars. He concluded that there must be some hidden source of mass. A year later, the Swiss–American Fritz Zwicky was studying the relative motion of galaxies in the Coma cluster. He found, again, that their motion suggested the mass of what could be seen was not the only stuff there. He named the missing material *dunkle Materie* or "dark matter."

Oort's early measurements were inaccurate, while Zwicky's initial assessment was that "dark matter" was 400 times more abundant than the matter of visible material—a huge overestimate. This meant that their findings were dismissed as measurement errors. In 1939, American Horace Babcock again found anomalies in the rotation of Andromeda and suggested that there was some mechanism by which light from the missing matter was being absorbed in the galactic core.

Galactic rotation curve

More than 20 years later, Rubin returned to the problem of galactic rotation. Like Babcock, she chose to focus on the rotation of the Andromeda galaxy, the Milky Way's nearest galactic neighbor. She worked with her colleague Kent Ford at the Carnegie Institution of Washington to measure the velocities of objects in the outer region of the galaxy. They did this using a sensitive spectrograph, which allowed them to detect the

No observational problem will not be solved by more data.
Vera Rubin

redshift and blueshift of objects, and calculate their relative speeds away from and toward Earth.

After several years of slow but careful work, Rubin had enough data to plot a rotation curve for the galaxy. Instead of swooping downward like the curve of the solar system, the speed data of the galactic curve stayed relatively level with distance. This meant that the outer regions of Andromeda were moving at the same speed as the areas nearer to the center. If the galaxy's mass was limited to what could be observed using telescopes, the outer regions of Andromeda would be moving faster than escape velocity, and they should simply fly off into space. However, they were clearly being held in place by the galaxy's overall mass. Rubin calculated that the total galactic mass required to hold the outer regions in orbit was about seven times greater than the visible mass. The ratio of matter to dark matter is today thought to be around 1:6.

What is dark matter?

Rubin's galactic rotation curve, widely disseminated in 1980, was the visual proof that dark matter existed. As further evidence mounted, the mystery as to what it might be remained. Dark matter

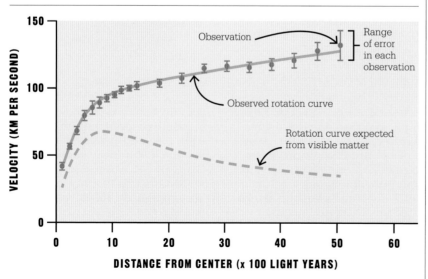

In the absence of dark matter, the velocities of objects in the outer regions of galaxies would be slower than the observed values. Here, the observed rotation curve is plotted against the expected curve from visible matter alone.

Dark matter could be evidence that the universe is one of many that exist next to one another, in separate spatial dimensions, in a bubblelike multiverse.

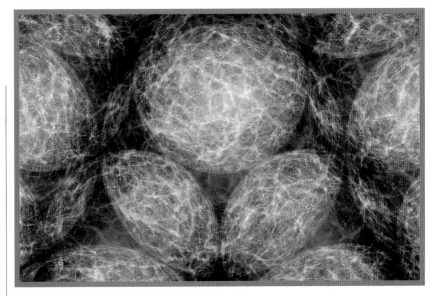

cannot directly be observed; only its effects are detectable, and the only effects that can be detected are from its gravity. It does not interact with the electromagnetic force, meaning that it does not absorb heat, light, or other radiation, nor does it emit any. Dark matter may be completely invisible.

Possible sources

The simplest solution to the dark matter problem is the most literal. It comprises ultra-dense bodies of ordinary matter that are too dark to observe. Astronomers have dubbed these MACHOs, which stands for Massive Compact Halo Objects. MACHOs include objects like black holes, neutron stars, and white and brown dwarfs. They occupy the galactic halo, a dark and diffuse region that surrounds the main, shining disk of a galaxy—and this is why it is difficult to see them. MACHOs are clearly out there, but by current estimates they would only account for a tiny proportion of dark matter. An alternative idea

is the WIMP—Weakly Interacting Massive Particle. This concept is based largely on an idea in particle physics called supersymmetry. It proposes a new explanation of energy and ordinary matter. Energy and matter form two distinct groups of subatomic particles, and supersymmetry proposes that these groups interact thanks to the actions of "super particles," or sparticles. Dark matter WIMPs might be sparticles that escaped from their partners in the early period of the universe, or they may be objects that are there all along.

Finally, dark matter might be the observable effect of another universe, or perhaps several, that exist in a spatial dimension different from this universe. Their matter could be very close, a few centimeters away, but because the radiation from each universe is trapped inside its own spacetime, one universe can never see another.

A huge ring of dark matter, which formed long ago in the collision between two massive galaxy clusters, is shown around the edge of this Hubble Space Telescope image in lighter blue.

However, the gravitational effects of the matter in the hidden universes leaks through into this one through the warping of spacetime.

Providing an explanation for dark matter remains one of the biggest prizes in astronomy. However, in 1999, a possibly even more puzzling phenomenon was uncovered. It was discovered that 68 percent of the universe was neither matter nor dark matter, but so-called dark energy. Dark matter makes up 27 percent; visible matter comprises a mere 5 percent. ∎

For the moment we might very well call them DUNNOS (for Dark Unknown Nonreflective Nondetectable Objects Somewhere).
Bill Bryson

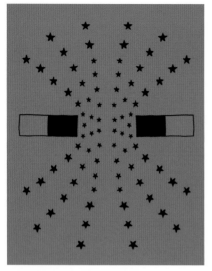

NEGATIVE PRESSURES PRODUCE REPULSIVE GRAVITY
COSMIC INFLATION

IN CONTEXT

KEY ASTRONOMER
Alan Guth (1947–)

BEFORE
1927 Georges Lemaître proposes that the universe arose from a single primordial atom. This is later named the Big Bang theory.

1947 George Gamow and Ralph Alpher describe how the elements hydrogen and helium were formed in the early universe.

1964 The cosmic microwave background is discovered to be a remnant of the Big Bang.

AFTER
1999 Dark energy is found to be accelerating the expansion of the universe.

2014 BICEP2 withdraws claims of finding evidence of inflation.

2016 LIGO detects gravitational waves, offering a new way to observe the structure of spacetime.

By the 1970s, cosmologists were grappling with a variety of puzzles thrown up by the Big Bang theory. In an attempt to solve them, Alan Guth proposed a stage of rapid inflation in the early universe, caused by the effects predicted by quantum theory.

The puzzles
One of the problems with the Big Bang theory came from the Grand Unified Theory (GUT), which describes how the forces of the universe (aside from gravity) arose a fraction of a second after the Big Bang. The GUT predicted that high temperatures at this time would create bizarre features, such as so-called magnetic monopoles (particles with only one magnetic pole). There are, however, none to be found, which suggests the universe cooled faster than expected.

A second problem arose from the way space is amazingly "flat," meaning that it expands according to "normal" Euclidean geometry (see diagram opposite). A flat universe would only have arisen if the density of the early universe matched a certain critical figure. Varying it slightly one way or the other would have resulted in curved universes.

The final issue was the horizon problem. Light from the edge of the observable universe has taken the entire life of the universe to reach Earth. As light's speed is constant, scientists know that it has not

---

The Big Bang theory **predicts features** that **are not seen** in the current universe.

→ The **first stage** of the universe after the Big Bang may have been a period of rapid expansion called **inflation**.

Inflation explains many of the features of the universe, but there is no evidence that it is correct.

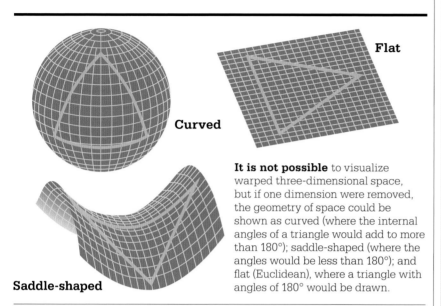

Flat

Curved

Saddle-shaped

It **is not possible** to visualize warped three-dimensional space, but if one dimension were removed, the geometry of space could be shown as curved (where the internal angles of a triangle would add to more than 180°); saddle-shaped (where the angles would be less than 180°); and flat (Euclidean), where a triangle with angles of 180° would be drawn.

had time to shine to the opposite edge of the universe. So if no light, energy, or matter has ever passed between the edges of the universe, this leaves a puzzle as to why space appears so similar in every direction.

The solution

Guth's theoretical solution to these problems was to inflate the early universe using a quantum effect called a false vacuum, where positive matter energy was created as space expanded, equally balanced by an increase in gravity (a form of negative energy). In the first 10^{-35} seconds after the Big Bang, space doubled in size 100 times over, going from a billionth of the size of a subatomic particle to the size of a marble. This means that, at the very beginning, the edges were close enough to mix and become uniform, thus solving the horizon problem. During inflation, space expanded faster than the speed of light. (The speed of light is only a speed limit through space.) The inflation

cooled the universe rapidly, thus solving the GUT problem, and locked in the uniformity seen today. Finally, the inflation ended as the density of the universe evened out at a value needed for a flat universe. In 2014, BICEP2, an experiment at the South Pole, reported ripples in space consistent with cosmic inflation. However, the claim was soon withdrawn. Cosmic inflation remains unproven, but it is the best current theory for the Big Bang. ∎

The recent developments in cosmology strongly suggest that the universe may be the ultimate free lunch.
Alan Guth

Alan Guth

Born in New Jersey, Alan Guth received his doctorate in 1972 and specialized in particle physics, pursuing research into quarks (elementary particles). By the late 1970s, he had worked at MIT, Princeton, Columbia, Cornell, and Stanford, as he searched the country for a long-term academic position. While at Columbia, Guth became interested in the Grand Unified Theory (GUT), which had been proposed in 1974. He began developing his inflation theory in 1978 while at Cornell, after hearing about the problem of the flatness of the universe, and shortly afterward, the problems associated with GUT. While at Stanford, he then came across the horizon problem, and went on to publish his famous theory in 1981. He is now a professor at MIT, where he helps with the search for evidence of cosmic inflation.

Key works

1997 *The Inflationary Universe: The Quest for a New Theory of Cosmic Origins*
2002 *Inflation and the New Era of High-Precision Cosmology*

GALAXIES APPEAR TO BE ON THE SURFACES OF BUBBLELIKE STRUCTURES
REDSHIFT SURVEYS

IN CONTEXT

KEY ASTRONOMERS
Margaret Geller (1947–)
John Huchra (1948–2010)

BEFORE
1842 Christian Doppler describes how wavelengths can change due to relative motion.

1912 Vesto Slipher discovers that galaxies are redshifted by the Doppler effect.

1929 Edwin Hubble uses redshift to show that distant galaxies are moving away faster than nearer ones.

1980 Alan Guth proposes that a rapid expansion, called cosmic inflation, shaped the universe.

AFTER
1998 The Sloan Digital Sky Survey finds walls, galaxy sheets, and filaments many hundreds of light-years long.

1999 A redshift survey of supernovae reveals that the universe's expansion is speeding up.

Since the 1920s, the study of the redshift of distant galaxies has revealed the scale of space and the way in which the universe is expanding in all directions. Redshift occurs when a light source is moving away from the observer (p.159). In the 1980s, redshift surveys made by American astronomers Margaret Geller and John Huchra, working at the Harvard–Smithsonian Center for Astrophysics (CfA), gave an even clearer picture of the universe, showing that galaxies cluster around great voids of empty space. Geller and Huchra's work provided valuable clues to the nature of the very early universe.

A redshift survey uses a wide-angle telescope to select target galaxies, generally millions of light-years away. Astronomers compare the light from each galaxy with benchmark wavelengths to determine the redshift, and thus the distance the light has traveled, allowing them to plot the positions of many galaxies. Huchra started the first redshift survey in 1977; by its completion in 1982, he had mapped 2,200 galaxies.

Margaret Geller

Margaret Geller earned a Ph.D. from Princeton in 1975, and took various fellowships before joining the Harvard–Smithsonian Center for Astrophysics in 1983. She worked there with John Huchra, analyzing the results of his redshift survey. Geller went on to lead the second (CfA2) redshift survey. She is a frequent public speaker and has made several films about the universe, including *Where the Galaxies Are*, which takes viewers on a graphical voyage around the large-scale objects of the observed universe.

See also: Spiral galaxies 156–61 ▪ Beyond the Milky Way 172–77 ▪ Cosmic inflation 272–73 ▪ A digital view of the skies 296

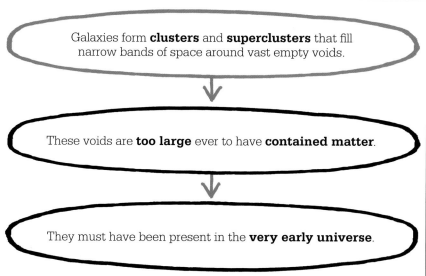

Galaxies form **clusters** and **superclusters** that fill narrow bands of space around vast empty voids.

↓

These voids are **too large** ever to have **contained matter**.

↓

They must have been present in the **very early universe**.

Before Huchra began his survey, it was known that galaxies existed in clusters. For example, the Milky Way is one of at least 54 galaxies in a cluster called the Local Group, which is about 10 million light-years wide. It was assumed that clusters were evenly spread. However, by 1980, Huchra had shown through his redshift survey that dozens of clusters form superclusters hundreds of millions of light-years wide. The Local Group is part of the Laniakea Supercluster, which contains 100,000 other galaxies.

Walls of galaxies

In 1985, Geller began the CfA2 Redshift Survey, taking 10 years to map 15,000 galaxies. Her survey confirmed that superclusters were themselves arranged in sheets and walls enclosing vast voids, like the surface film of a bubble. She found the first "great wall" of galactic superclusters in 1989. The exact size of CfA2 Great Wall is still unclear, but it is estimated at 700 million light-years long, 250 million wide, and 16 million thick. It was the first of several supersized structures now known.

The size of the voids puzzled astronomers. They were too large to have been emptied completely by the gravitational collapse of material that formed the stars and galaxies, which meant that they must have been empty since the beginning of the universe. Cosmologists theorize that the large-scale order of superclusters and voids is the legacy of quantum fluctuations during the inflationary epoch of the universe. Quantum fluctuations are fleeting changes in the amount of energy at points in space. These small but highly significant irregularities were locked into the fabric of the universe in the first fraction of a second of its existence, and remain today. They are now the vast areas of void permeated by a tangled pattern of matter. ■

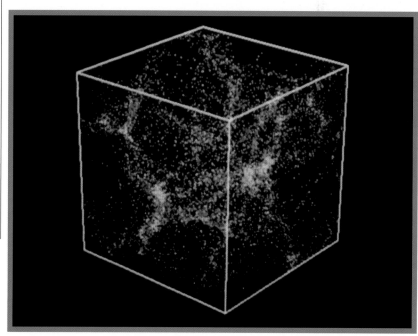

This computer simulation of a portion of the universe shows the distribution of 10,000 galaxies, which cluster in long filaments and "walls," in between vast empty voids.

STARS FORM FROM THE INSIDE OUT

INSIDE GIANT MOLECULAR CLOUDS

IN CONTEXT

KEY ASTRONOMER
Frank Shu (1943–)

BEFORE
1947 Bart Bok observes dark nebulae and suggests that they are sites of star formation.

1966 Frank Shu and Chia-Chiao Lin develop the density wave theory to explain spiral arms in the Milky Way.

AFTER
2003 The Spitzer Space Telescope, an infrared observatory, is launched. It produces the best view yet of stellar nurseries.

2018 First light on the James Webb Space Telescope will allow astronomers to study protostars inside dark Bok globules.

S tars form inside dark globules of dust and gas that are called giant molecular clouds (GMCs). However, the process by which a cloud of gas transforms into an embryonic star, or protostar, has never been observed, partly because the process must take millions of years, and partly because even the most advanced telescopes find it difficult to penetrate the dark density of the cloud.

Without observational evidence, astrophysicists must construct mathematical models for what they think is happening inside those dark globules. The most consistent model of star formation was derived by US mathematician Frank Shu.

The Pillars of Creation are vast clouds of gas and dust where new stars are made. This famous image was captured by the Hubble Space Telescope in 1995.

Shu and his colleagues Fred Adams and Susana Lizano at the University of California at Berkeley presented their model in 1986 after 20 years of work.

The inside-out model

Shu's system is called the "singular isothermal model," or the "inside-out model." It is built from the complex mathematics that define the dynamics of gas clouds, taking into account factors such as temperature, density, electrical charge, and magnetism. Shu's model works by making the process self-similar. A starting condition that causes some of the gas cloud to contract into a dense core will result in the same—or similar—conditions, which cause more gas to join the core, and so on. This process was found to be stable enough to keep the young star together as it grew. Earlier models had failed because they could not find a way to balance the mechanisms that were pulling the gases in and pushing heat out; as a result, these models ended with the young star disintegrating.

GMCs are vast regions of the galaxy filled with hydrogen atoms and molecules mixed with specks of dust and ice. Typically, a GMC contains 100,000 solar masses of material, which is a mixture of primordial gases produced by the Big Bang and the remnants of long-dead stars. GMCs are mostly found in the spiral arms of a galaxy.

In the mid-1960s, Shu and the renowned Chinese–American mathematician Chia-Chiao Lin modeled the rotation of a spiral galaxy, and showed that the arms are located at density waves—"traffic jams" of stars. Such density waves sweep up interstellar material into GMCs, and this triggers the formation of stars. »

Stars are **dense balls** of superhot hydrogen.	They must have formed from **clouds of hydrogen** gas in interstellar space.	Material from near the **middle contracted first**, and then drew in the outer regions.

Stars form from the inside out.

Frank Shu's inside-out model describes the four-stage formation of a star from a giant molecular cloud.

1 Cores form within GMCs as magnetic forces and turbulence calm.

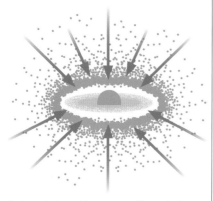

2 A protostar with a surrounding nebular disk forms at the center of a cloud core, collapsing from inside out.

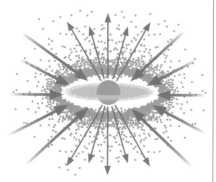

3 A stellar wind breaks out along the rotational axis of the system, creating a bipolar flow.

4 The infall of material ends, revealing a newly formed star with a circumstellar disk.

The impact of a density wave, or something more violent such as a blast from a nearby supernova, creates turbulence inside a GMC. However, highly tangled magnetic fields run through the cloud, and these stop the turbulence from ripping the cloud apart. The magnetism also acts to prevent the cloud from collapsing in on itself under its own gravity.

Cloud cores

Over millions of years, the magnetic pressure and turbulence in the gases dissipate, creating regions of calm, where slowly rotating "cloud cores" form. On closer inspection, GMCs are not uniform, but made up of dark fragments or clumps of denser material, known as Bok globules. Each globule is thought to contain several cloud cores.

Shu's model supposes that the core becomes a single isothermal (equal-temperature) sphere, or something very close to it. This means that the gravity pulling the ball of gas together is balanced by the outward pressure of the moving gas and its magnetic forces. Such a state can never persist for long,

and the contracting gravitational force in the core wins out over the outward pressure.

The inner region of the cloud core contracts to form a dense ball of gas at the center. This is the protostar. Protostars do not form in a rapid process, but take millions of years, and millions more to grow into a full-fledged star. The protostar is also surrounded by a disk of material formed by the system's rotation, and wave upon wave of material is pulled in from the surrounding envelope of gas. With each wave, the mass of the protostar and its more diffuse disk grows, and its gravity grows with it. The increasing gravity steadily pulls in material from farther away, hence the description of the process as an "inside-out collapse."

The star gathers mass

The protostar warms up as it becomes denser, but it is still too small and cold to produce energy by fusing hydrogen in its core. The force of all of the new material landing on its surface also adds to the heat signature given out by the protostar. At this stage, it is giving out only faint infrared and

Frank Shu

Born in Kunming, China, Frank Shu moved to the United States when he was six to join his father, an academic mathematician, who was beginning research at MIT. Frank followed his father to MIT, where he completed a degree in physics in 1963. While there, Shu worked on the density wave theory of spiral arms. He later moved to Harvard to complete his doctorate in astronomy in 1968. Shu worked on his protostar model while at Berkeley and was the head of the astronomy department there by the time he presented the full review of his isothermal sphere model in 1986. Today, Shu holds tenure at Berkeley. In recent years he has used his knowledge of astrophysics to tackle climate change. He often works in collaboration with his graduate students, who are collectively known as the "Shu Factory."

Key work

1981 *The Physical Universe*

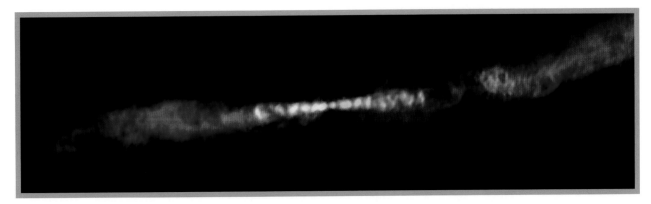

microwave radiation, which makes it hard to see. Eventually, however, the protostar gathers enough mass for fusion to begin, but initially only the deuterium, a heavy isotope of hydrogen, begins to burn. Unlike an "adult" star, a protostar releases its heat entirely by a process of convection. Heat from its core rises up to the surface in the same way that hot water in a pan rolls around as it boils. The convection and rotation of the star create a strong magnetic field, which pushes out from each pole, clearing a narrow hole in the envelope of gas and dust. The growing protostar's heat and a stellar wind of plasma are directed away from the star via these polar jets. These features, explained by Shu's model, have been confirmed by observations.

Becoming a star, almost

A star with the mass of the sun spends about 10 million years as a protostar. As its mass increases, the angle of its polar jets widens, pushing away more of the gas cloud. Eventually, the protostar's stellar wind blasts out from the entire star's surface, and it clears its gas cloud away completely. At this point, the young stellar object is revealed for the first time. Giant stars (above 8 solar masses) have already started burning hydrogen by this point and have become

full-fledged stars destined to live short, bright lives. However, smaller stars—those less than 8 solar masses—have not begun a full fusion process and so are known as pre-main-sequence (PMS) stars.

A PMS star still has a disk of material spinning around it. Some of that will be dispersed by the stellar wind into the wider GMC. What remains, around the smaller stars especially, is likely to form into gas giant planets, and perhaps later, rocky ones as well.

Final ignition

The final phase of star formation is a contraction of the fast-spinning PMS star. Red, orange, and yellow dwarfs (M, K, G, and F type stars)

A protoplanetary disk surrounds the young star HL Tauri in the constellation Taurus. The dark patches are thought to represent the possible positions of newly forming planets.

An infant star sits at the center of two nearly symmetrical jets of dense gas. Known as CARMA-7, the star is about 1,400 light-years from Earth.

form from PMS stars that are less than 2 solar masses. They are considerably wider and less dense than their adult forms, and appear much brighter as they give out light from larger surface areas, frequently punctuated by high-energy outbursts of X-rays. This energy is the product of gravitational contraction, not nuclear fusion. It takes about 100 million years for the PMS star to compress itself enough to begin burning hydrogen, and by that time, it will have lost half to three-quarters of its initial mass. Larger PMS stars (those between 2 and 8 solar masses) take a different route to achieving fusion and form rare blue dwarfs (A and B type stars).

PMS stars are the earliest stage of star formation that have been seen clearly. Infrared space telescopes such as Spitzer and Hubble have given faint glimpses of protostars but mostly they are too heavily shrouded by the dark dust clouds. NASA's new infrared James Webb Space Telescope is designed to be sensitive enough to see through that dust, so perhaps soon the moment when a star is born may at last be observed. ∎

WRINKLES IN TIME

OBSERVING THE CMB

IN CONTEXT

KEY ASTRONOMERS
George Smoot (1945–)
John Mather (1946–)

BEFORE
1964 The cosmic microwave background—an echo of the Big Bang itself—is discovered.

1981 Alan Guth proposes cosmic inflation, a theory in which fluctuations of energy density were locked into space during the Big Bang.

1983 Redshift surveys show that galaxies are clustered around voids of nothingness.

AFTER
2001 Wilkinson Microwave Anisotropy Probe is launched to refine the map of the CMB.

2015 The Planck observatory studies the CMB to refine the age of the universe to 13.813 billion years +/- 38 million years. Combining this with other data, the latest estimate is 13.799 billion years +/- 21 million years.

I always think of space-time as being the real substance of space, and the galaxies and the stars just like the foam on the ocean.
George Smoot

The Cosmic Microwave Background, or CMB, was discovered in 1964. This is the afterglow of the Big Bang and it is as near as scientists can get to observing the event that brought the universe into existence, 13.8 billion years ago. Linking the structures observed in the universe to the features discerned in the CMB remains a key challenge for cosmologists.

Wrinkled time

The first great breakthrough came from the Cosmic Microwave Background Explorer, known as COBE, a NASA satellite launched in 1989. The detectors on COBE, designed and run by George Smoot, John Mather, and Mike Hauser, were able to find the oldest structures in the visible universe, described by Smoot as "wrinkles in time." These wrinkles in otherwise uniform space were once dense regions containing the matter that would form stars and galaxies. They correspond to the large-scale galaxy superclusters and great walls seen in the universe today, and add weight

The Cosmic Microwave Background Explorer (COBE) spent four years in space collecting information about the CMB, scanning the celestial sphere every six months.

to the inflationary model of the early universe proposed by American Alan Guth.

The CMB is a flash of radiation that was released about 380,000 years after the Big Bang, at the time the first atoms formed (pp.196–97). The expanding universe had cooled enough for stable ions (positively charged nuclei) of hydrogen and helium to form, and then, after a little more cooling, the ions began to collect electrons to make neutral atoms. The removal of free electrons from space led to the release of photons (particles of radiation).

Those photons are visible now as the CMB. The CMB comes from the whole sky, without exception. It has redshifted (the wavelengths have stretched), and it now has wavelengths of a few millimeters, while the original radiation's wavelengths would be measured in nanometers (billionths of a meter).

See also: The birth of the universe 168–71 ■ Searching for the Big Bang 222–27 ■ Cosmic inflation 272–73 ■ Redshift surveys 274–75 ■ Tegmark (Directory) 339

One of the key observations of the CMB came in the 1970s, and removed any doubt that it was an echo of the Big Bang. This was the discovery that the thermal spectrum of radiation from the CMB tallied very closely with that of a theoretical black body (p.225).

Black bodies

Black bodies do not really exist—they cannot be made and no object observed in the universe functions as black bodies do in theory. However, the CMB is the closest match that has ever been found.

A black body absorbs all radiation that hits it. Nothing is reflected. However, the absorbed radiation adds to the thermal energy of the object, and this is released as radiation. In 1900, German Max Planck, the founding figure of quantum physics, showed that the spectrum of radiation released by a black body is entirely dependent on temperature.

In an everyday example of radiation varying with temperature, an iron bar glows red when first

The **cosmic microwave background** is a flash of radiation produced 380,000 years after the Big Bang.

The CMB's **wavelength** shows how **hot the universe was** when the CMB was emitted.

The CMB is **not smooth and uniform**, but contains tiny fluctuations in temperature.

These fluctuations, or "wrinkles in time", are the oldest structures ever found and represent the formation of the first stars and galaxies.

heated. Heating it more makes it orange, and eventually the bar will glow "blue hot." Metalworkers learn to roughly judge the temperature of iron by its color. The metal is not particularly close to a black body in the theoretical sense, but stars and other astronomical objects are a much closer match to a black body, and so the color, or wavelengths of their emissions, can be compared to the thermal spectrum of a theoretical black body to give a relatively precise temperature. **»**

George Smoot

After a childhood in Florida and Ohio, Smoot began his career as a particle physicist working at MIT. His interests switched to cosmology and he moved across the country to the Lawrence Berkeley National Laboratory. It was there that Smoot studied the CMB and developed ways of measuring its radiation.

Smoot's early work involved fitting detectors to high-altitude U2 spyplanes, but in the late 1970s, he became involved in the COBE project to take his detector into space. After his success with COBE, Smoot cowrote *Wrinkles in Time* with Keay Davidson to explain the discovery. Smoot won the Nobel Prize in 2006, along with John Mather, for his work on COBE. He reportedly gave his prize money to charity. However, three years later, Smoot won an even greater sum when he bagged the $1 million jackpot on the TV game show *Are You Smarter Than a 5th Grader?*

Key work

1994 *Wrinkles in Time* (with Keay Davidson)

The temperature of the CMB today is a chilly 2.7 K. The thermal spectrum at that temperature contains no visible light, which is why space looks black to human eyes. However, the spectrum has redshifted (stretched) over time as the universe has expanded. Extrapolating back to the moment the CMB was emitted gives an original temperature of about 3,000 K. The color of radiation at this temperature is orange, so the CMB started out as a flash of orange light that shone out from every point in space.

Smooth signal

The early observations of the CMB suggested that it was isotropic, which means that its spectrum is the same everywhere. In cosmology, the terms density, energy, and temperature are somewhat synonymous when discussing the early universe. So the isotropic nature of the CMB suggested that, in those early days, space had a uniform density, or spread of energy. However, this did not tally with the developing theories of the Big Bang, which demanded that matter and energy were not evenly spread through the young universe, but had been concentrated together

in places. These denser areas, or anisotropies, were where the stars and galaxies formed. COBE was sent into space to take a close look at the CMB to see if it could find any anisotropies, to find out whether the CMB changed, however slightly, depending on where it looked.

COBE's mission

A mission to study the CMB from space had been in the planning stages since the mid-1970s. Construction of COBE began in 1981. It was initially designed to enter polar orbit (its orbit passing over both poles). However, the Challenger disaster of 1986 grounded the shuttle fleet, and the COBE team had to look for another launch system. In 1989, the satellite was launched using a Delta rocket, and it was placed in a sun-synchronous geocentric orbit—orbiting in a way that saw it pass over each place on Earth at the same time of day. This worked just as well as a polar orbit in that it allowed COBE to point away from Earth and scan the entire celestial sphere, strip by strip.

The spacecraft carried three instruments, all protected from the sun's heat and light by a cone-shaped shield, and chilled to

[COBE has made] the greatest discovery of the century, if not of all time.
Stephen Hawking

2 K (colder than space itself) using 100 gallons (650 liters) of liquid helium. George Smoot ran the Differential Microwave Radiometer (DMR), which mapped the precise wavelengths of the CMB, while John Mather was in charge of FIRAS, the Far-InfraRed Absolute Spectrophotometer, which collected data on the spectra of the CMB. These two experiments were looking for anisotropies. The third detector on COBE had a slightly different goal. The Diffuse Infrared Background Experiment, run by Mike Hauser, found galaxies that were so ancient and far away that they are only visible by their heat radiation (or infrared).

COBE's instruments created the most accurate map of the CMB to date. However, it was not a simple surveying job. Smoot and Mather were interested in primary anisotropies—that is, the density differences that were present at the time the CMB formed. To find these, they needed to filter out the secondary fluctuations caused by obstacles that lay between COBE

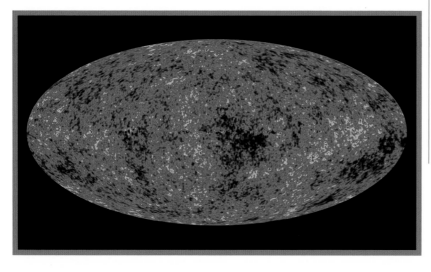

The full-sky map produced by WMAP in 2011 showed many fine details of the isotropy of the CMB. Colder spots are blue, while hotter spots are yellow and red.

In addition to mapping the CMB, WMAP measured the age of the universe as 13.77 billion years, dark matter as 24.0 percent of the universe, and dark energy as 71.4 percent.

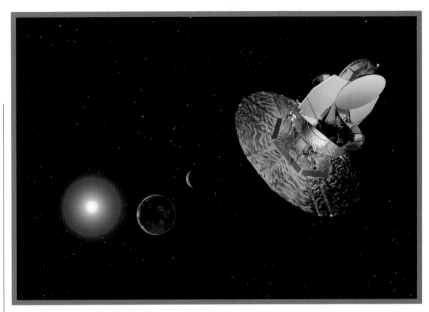

and the edge of the universe. Dust clouds and the effects of gravity had interfered with the radiation on its long journey to Earth. The data from the three instruments were used to detect and correct these so-called secondary anisotropies.

Tiny fluctuations

After 10 months in space, COBE's helium ran out, which limited the function of the two infrared detectors, but the DMR continued working until 1993. By 1992, the COBE team's analysis had shown what they were looking for. The CMB, and thus the early universe, was not a uniform blob of energy. Instead it was riddled with tiny but significant fluctuations. The differences were minute, with density variations of about 0.001 percent. However, the pattern was enough to explain why the contents of the universe are clustered together, while the rest of space is made from vast empty voids.

Since COBE, two subsequent missions have added detail to the picture of the CMB. Between 2001 and 2010, NASA's Wilkinson Microwave Anisotropy Probe (WMAP) mapped the CMB to a higher resolution than COBE. Then, from 2009–2013, the ESA's Planck Observatory produced the most accurate map to date.

Every wrinkle on the map is the seed from which an entire galaxy formed about 13 billion years ago.

However, no known galaxy can be seen forming in the CMB. The CMB radiation detected today has traveled from near the edge of the observable universe over the course of most of the age of the universe. Astronomers can only see 13.8 billion light-years away, but most of the universe now lies farther away than that. The galaxies forming in the CMB are now far beyond what can be observed, and are receding faster than the speed of light. ∎

Improving resolution of the CMB

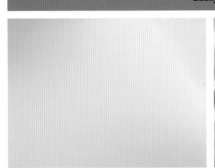

COBE's imaging of the CMB shows slight variations in a 10-sq-degree panel of its all-sky map, proving that the CMB is not uniform.

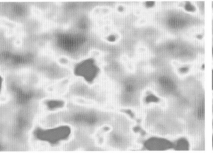

WMAP's map of the CMB shows greater detail within the same panel, revealing smaller-scale features that COBE could not identify.

Planck's resolution is 2.5 times greater than that of WMAP, showing features as small as $1/12$ of a degree. This is the most detailed map of the CMB to date.

THE KUIPER BELT IS REAL

EXPLORING BEYOND NEPTUNE

IN CONTEXT

KEY ASTRONOMERS
David Jewitt (1958–)
Jane Luu (1963–)

BEFORE
1930 American astronomer
Clyde Tombaugh discovers Pluto
orbiting beyond Neptune. It is
initially identified as the ninth
planet but is later reclassified.

1943 Kenneth Edgeworth
suggests that Pluto is just
one of many objects in the
outer solar system.

1950 Fred Whipple describes
the icy nature of comets as
"dirty snowballs."

AFTER
2003 Sedna is discovered
orbiting 76 AU–1,000 AU from
the sun, beyond the outer edge
of the Kuiper belt.

2005 Eris is seen in the disk
beyond the Kuiper belt.

2008 Two Kuiper Belt Objects
are classified as dwarf planets
along with Eris, Pluto, and Ceres.

The outer solar system contains the
leftover material from the formation of the planets.

Some of the material travels
from the **edge of the solar
system** in the form of
long-period comets.

Short-period comets
must come from a
nearer source.

The Kuiper belt, a theoretical reservoir of icy bodies beyond the
orbit of Neptune, could be **the source of short-period comets**.

In 1950, Dutch astronomer Jan
Oort proposed that a spherical
shell of potential comets
surrounds the solar system half a
light-year away. The so-called Oort
cloud was the source of long-period
comets, which took millennia to
orbit the sun. But the source of
the short-period comets that orbit
the sun every few centuries must
be nearer. In 1943, Irish scientist
Kenneth Edgeworth speculated
that the comet reservoir was a

belt beyond Neptune. But the
Dutch–American astronomer
Gerard Kuiper argued in 1951 that,
although there was once such a
belt, it would have been scattered
away by the gravity of the outer
planets. It was a puzzle, and comet
nuclei that far away would be too
faint for even the best telescopes.

In the 1980s, sensitive new CCD
(charge-coupled device) detectors
became available. With these,
astronomers at last had a chance

See also: The Kuiper belt 184 ▪ The Oort cloud 206 ▪ Studying Pluto 314–17

The egg-shaped dwarf planet Haumea hangs in the sky above one of its two moons, Namaka. Haumea, discovered in 2004, is the third-largest dwarf planet.

of spotting small icy objects beyond Neptune. Americans David Jewitt and Jane Luu were among the astronomers who set about the difficult task. After five years of searching, in 1992, Jewitt and Luu discovered an object formally designated 1992 QB1, the first body to be found beyond Neptune since Pluto, and the first evidence that the Kuiper belt was real.

Cubewanos and plutinos

More than 1,000 Kuiper Belt Objects (KBOs) are now known and there are probably thousands more. They are designated as asteroids, but unlike most asteroids, KBOs are typically a mixture of rock and ices. The largest are several hundred miles across and many of them have moons.

1992 QB1 is typical of the KBOs in the most densely populated middle part of the Kuiper belt, about 45 AU from the sun. These KBOs are sometimes called "cubewanos." Closer in, at around 40 AU, the gravity of Neptune has thinned out the Kuiper belt, leaving a family of objects (including Pluto itself) called "plutinos," in orbits that are unaffected by Neptune's gravity. Beyond the main Kuiper belt lies a region called the "scattered disk," which includes the large objects Eris and Sedna. It is now believed that

this region is the source of short-period comets. In 2006, Eris was designated a dwarf planet along with Pluto. Since then, two more cubewanos, Makemake and Haumea (Haumea is orbited by two small moons), have been classed as dwarf planets, with many more KBOs listed as candidate dwarf planets. Scientists believe that these KBOs resemble the primitive bodies that formed the planets. ▪

Gerard Kuiper

Gerard Kuiper was born in the Netherlands in 1905. At a time when few other astronomers were interested in the planets, Kuiper, working mostly at the University of Chicago, made many discoveries that changed the course of space science: he found that the Martian atmosphere was mostly carbon dioxide, that Saturn's rings comprised billions of chunks of ice, and that the moon was covered in a fine rock dust. In 1949, Kuiper's idea that the planets were formed from a cloud of gas and dust that

surrounded the young sun changed scientists' view of the early solar system.

In the 1960s, Kuiper helped identify landing sites on the moon for the Apollo program and cataloged several binary stars. He died of a heart attack in 1973, at age 68. Since 1984, the Kuiper Prize has been awarded annually by the American Astronomical Society to recognize achievement in planetary science, a field of astronomy in which many consider Gerard Kuiper to have been the pioneer.

MOST STARS ARE ORBITED BY PLANETS

EXOPLANETS

IN CONTEXT

KEY ASTRONOMERS
Michel Mayor (1942–)
Didier Queloz (1966–)

BEFORE
1952 US scientist Otto Struve proposes the radial velocity method to find exoplanets.

1992 The first such planet is found, orbiting a pulsar and not a main sequence star.

AFTER
2004 Construction begins on the James Webb Space Telescope, which will be able to image exoplanets.

2005 The Nice model offers a new idea for the evolution of the solar system that places the giant planets closer to the sun.

2014 The construction of the European Extremely Large Telescope begins.

2015 Kepler 442-b, an Earth-sized rocky exoplanet around an orange dwarf, is discovered.

I n 1995, two Swiss astronomers, Michel Mayor and Didier Queloz, researching at the Observatoire de Haute-Provence near Marseille, found a planet orbiting 51 Pegasi, a sunlike star 60 light-years away in the constellation of Pegasus. This was the first confirmed observation of a true extrasolar planet, or exoplanet—a planet beyond the solar system. It was orbiting a main sequence star, and was therefore assumed to have formed by the same process as that which created the solar system.

Mayor and Queloz named the new planet 51 Pegasi b, but it is unofficially known as Bellerophon after the hero who rode Pegasus, the winged horse of ancient Greek myth. Its discovery prompted a major hunt to find more exoplanets. Since 1995, several thousand exoplanets have been found, many in multiple star systems. Astronomers now estimate that there is an average of one planet around every star in the galaxy, although this is probably a very conservative figure. Some stars have no planets, but most, like the sun, have several.

For more than 2,000 years, people have dreamed of finding other habitable worlds.
Michel Mayor

The discovery of 51 Pegasi b marked the final milestone in a process that has forced astronomers to abandon any lingering notion that Earth occupies a privileged place in the universe.

Copernican principle

In the 1950s, the Anglo–Austrian astronomer Hermann Bondi had described a new way for humans to think about themselves, which he called the Copernican principle. According to Bondi, humankind could no longer regard itself as a unique phenomenon of central importance to the universe. On the contrary, humans should now

---

Michel Mayor

Michel Mayor was born in Lausanne, Switzerland, and has spent most of his career working at the University of Geneva. His interest in exoplanets arose from his earlier study of the proper motion of stars in the Milky Way. To measure this motion more accurately, he developed a series of spectrographs, which eventually culminated in ELODIE. The ELODIE project with Didier Queloz was initially intended to search for brown dwarfs—objects that were bigger than planets but not quite large enough to be stars. However, the system was

sensitive enough to spot giant planets as well, and, following their 1995 discovery, Mayor is currently the chief investigator at the HARPS program for the European Southern Observatory in Chile. His team has found about half of all the exoplanets discovered to date. In 2004, Mayor was awarded the Albert Einstein medal.

Key work

1995 *A Jupiter-mass Companion to a Solar-type Star* (with Didier Queloz)

See also: The Copernican model 32–39 ▪ Radio telescopes 210–11 ▪ Studying distant stars 304–05 ▪ Looking farther into space 326–27 ▪ Kumar (Directory) 339

understand that their existence is insignificant in the context of the universe.

The principle is named after Nicolaus Copernicus, who changed the way humankind saw itself by relegating Earth from the center of the solar system to one of several planets that orbited the sun. By the late 20th century, successive discoveries had moved the solar system from the center of the universe to a quiet wing at the edge of a galaxy containing 200 billion other stars. The galaxy was not special either, simply one of at least 100 billion arranged in vast filaments that extended for hundreds of millions of light-years. Nevertheless, planet Earth and the solar system were still regarded as very special—since there was no evidence that any other stars had planets, let alone planets capable of supporting life. Since Mayor's and Queloz's discovery, however, this idea has also succumbed to the Copernican principle.

Wobbling light

Queloz and Mayor found 51 Pegasi b using a system called Doppler spectroscopy. Also known as the radial velocity or "wobble" method, Doppler spectroscopy can detect an exoplanet by its gravitational effects on its host star. The star's gravity is far greater than that of the planet, and this is what keeps the planet in orbit. However, the planet's gravity also has a small effect on the star, making it wobble back and forth as the planet moves around it. The effect is tiny: Jupiter changes the sun's speed by about 12 miles/s (7.4 km/s) over a period of 11 years, while Earth's effect is only 0.1 miles/s (0.16 km/s) each year.

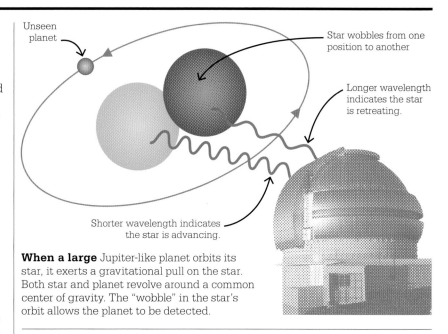

When a large Jupiter-like planet orbits its star, it exerts a gravitational pull on the star. Both star and planet revolve around a common center of gravity. The "wobble" in the star's orbit allows the planet to be detected.

Labels on figure: Unseen planet · Star wobbles from one position to another · Longer wavelength indicates the star is retreating. · Shorter wavelength indicates the star is advancing.

In 1952, US astronomer Otto Struve had suggested that this kind of star wobble could be detected as small fluctuations in a star's spectrum. As the star moved away from Earth, its emissions would be slightly redshifted from the norm. When it wobbled back again toward the observer, the light would be blueshifted. The theory was solid but detecting the wobble required an ultrasensitive detector.

We are getting much closer to seeing solar systems like our own.
Didier Queloz

That detector was a spectrograph named ELODIE developed by Mayor in 1993. ELODIE was about 30 times more sensitive than any previous instrument. Even then, it was only capable of measuring velocity changes of 7 miles/s (11 km/s), which meant it was limited to detecting planets about the size of Jupiter.

Improving the search

In 1998, an even more sensitive spectrograph, named CORALIE, was installed at La Silla Observatory in Chile, which again was searching for planets using the radial velocity technique. In 2002, Michel Mayor began overseeing HARPS (High Accuracy Radial velocity Planet Searcher) at the same site, using a spectrograph capable of detecting exoplanets about the size of Earth. The wobble method of detection was very slow, so new techniques of spotting exoplanets were developed. »

The most successful method was the transit method, which looked for periodic changes in the brightness of a star. These changes were very small and happened when a planet transited the star, passing between the star and the observer, and causing it to dim very slightly. The best place to look for exoplanets by the transit method was out in space and so, in 2009, the Kepler observatory, named after the man who first described planetary orbits (pp.50–55), was launched to do just that.

Staring at one place

Kepler was placed in a heliocentric orbit, trailing behind Earth as it circled the sun. The craft was designed to keep its aperture firmly fixed on a single patch of space, called the Kepler field. This made up only about 0.25 percent of the whole sky, but the spacecraft could see 150,000 stars in that area. To find exoplanets, Kelper would have to concentrate on this single field of view for years on end. It was

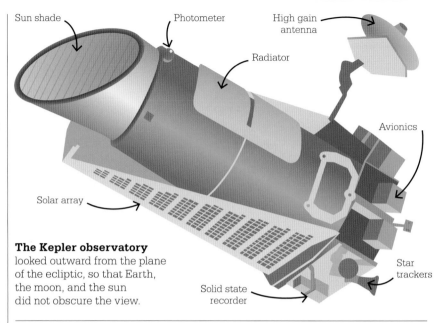

Sun shade Photometer High gain antenna Radiator Avionics Solar array Star trackers Solid state recorder

The Kepler observatory looked outward from the plane of the ecliptic, so that Earth, the moon, and the sun did not obscure the view.

unable to see individual exoplanets, but could identify stars that were likely to have them.

Kepler could only detect the transits of exoplanets with orbital paths that crossed the spacecraft's line of sight. Many exoplanets would be orbiting at the wrong angle for

that. Those that were correctly oriented would only transit their star once every orbital period (the planet's year), so Kepler's method was better at finding planets that orbited close to their star, taking a few years and months (or even weeks and days) to complete each revolution.

Candidate stars

By the start of 2013, Kepler had identified about 4,300 candidate stars that might have extrasolar planetary systems. Unfortunately, the guidance system used to keep Kepler locked on target then failed, bringing its planet hunt to an end about three years sooner than expected. However, the data it had collected was enough to keep researchers busy for years to come. Kepler's candidate stars could only be confirmed as planetary systems using radial velocity measurements from ground-based observatories, such as HARPS in Chile and the Keck Telescope in Hawaii. (Radial velocity is the velocity of the star in the direction of Earth.) So far, about a tenth of Kepler's candidate stars

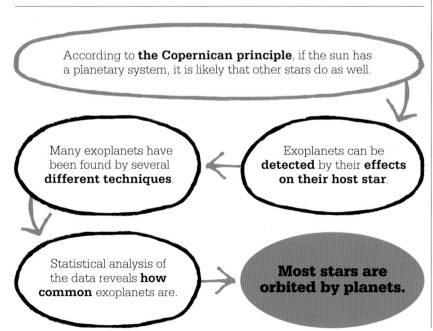

According to **the Copernican principle**, if the sun has a planetary system, it is likely that other stars do as well.

Many exoplanets have been found by several **different techniques**.

Exoplanets can be **detected** by their **effects on their host star**.

Statistical analysis of the data reveals **how common** exoplanets are.

Most stars are orbited by planets.

have proved to be false positives but, after three years of analysis, the program had identified 1,284 exoplanets, with more than 3,000 stars left to examine. The statistics for the exoplanets in the Kepler field are striking—most stars are part of a planetary system. This means that the number of planets in the universe is likely to exceed the number of stars.

The amount of dimming during a transit gives an indication of how big an exoplanet might be, but the study of an exoplanet's size and characteristics is still in its early stages. The light reflected from a planet is about 10 billion times fainter than the star it orbits. Astronomers are waiting for the James Webb Space Telescope in 2018 and the European Extremely Large Telescope in 2024 to image this light directly and analyze the chemistry of exoplanets. Until then, they have to speculate using very little data: an approximate mass of the planet, its radius, the orbital

> We were not expecting to find a planet with a 4-day [orbital] period. No one was expecting this.
> **Michel Mayor**

distance, and the temperature of the star. This tells them what the planet is probably made of and allows them to conjecture what the surface conditions are likely to be.

Hot and super Jupiters

The exoplanets discovered so far have added a host of weird worlds to the neat family portrait that is the sun's planetary system. For

example, 51 Pegasi b was the first of many "hot Jupiters." These have a mass similar to Jupiter's and a large size that shows that they are mostly made of gas. 51 Pegasi b is half as massive as Jupiter, but is slightly larger. This gas giant orbits its sunlike star every four days. That means it is much closer to its star than Mercury is to the sun. Such proximity means it is tidally locked to the star—one side always faces the scorching stellar surface, and the other always faces away. Many hot Jupiters have been found. They have confounded scientists, who are trying to understand how gas planets can exist so close to a star without evaporating. Some exoplanets are dozens of times more massive than Jupiter, and are known as "super-Jupiters." »

The "super-Jupiter" Kappa Andromedae b, shown here in an artist's render, has a mass of 13 times Jupiter's. It glows a reddish color, and may yet be reclassed as a brown dwarf.

> Red dwarfs with rocky planets could be ubiquitous in the universe.
> **Phil Muirhead**
> *Professor of Astronomy*
> *Boston University*

These super-Jupiter planets do not appear to grow in size as their mass increases. For instance, Corot-3b is a super-Jupiter that is 22 times as heavy as Jupiter but more or less the same size, due to its gravity holding its gaseous contents together. Astronomers have calculated that the density of Corot-3b is greater than that of gold and even osmium, the densest element on Earth.

Brown dwarfs and rogues

When a super-Jupiter reaches 60 Jupiter masses, it is no longer regarded as a planet, but as a brown dwarf. A brown dwarf is essentially a failed star—a ball of gas that is too small to burn brightly through nuclear fusion. The brown dwarf and its star are seen as a binary star system, not a planetary one. Some super-Jupiters and small brown dwarfs have broken free of their star to become free-floating rogue planets. One, named MOA-

Kepler 10b in the Draco constellation is shown transiting its star in an artist's impression. Its extremely hot surface temperature and dizzying orbit mean life there is improbable.

2011-BLG-262, is thought to have a satellite, and could be the first exoplanet found with an exomoon.

Another class of planet are called the super-Earths. These have a mass 10 times that of Earth but less than that of an ice giant like Neptune. Super-Earths are not rocky but made from gas and ice: alternative names for them are mini-Neptunes or gas dwarfs.

Living planets

Earth's solar system has terrestrial planets (planets with a rocky surface), of which Earth is the largest. So far, exoplanet searches have struggled to find many terrestrial planets, because they are generally small and beyond the sensitivity of the planet detectors. The first confirmed terrestrial exoplanet was Kepler-10b, which is three times the mass of Earth and is so close to its star

that it orbits once an Earth day and has a surface temperature that would melt iron. Life seems highly unlikely there, but the hunt continues for rocky planets that might be more hospitable.

Astrobiologists—scientists who search for alien life—focus on the particular conditions that all life needs. When choosing likely places to look, they assume that alien life-forms will require liquid water and carbon-based chemicals, just like life on Earth. Living planets would also need an atmosphere to shield the surface from damaging cosmic rays and to act as a blanket that retains some of the planet's heat during the night.

The region around a star where the temperatures would allow planets to have liquid water, carbon chemistry, and an atmosphere, is known as its habitable zone, also called the "Goldilocks zone"—

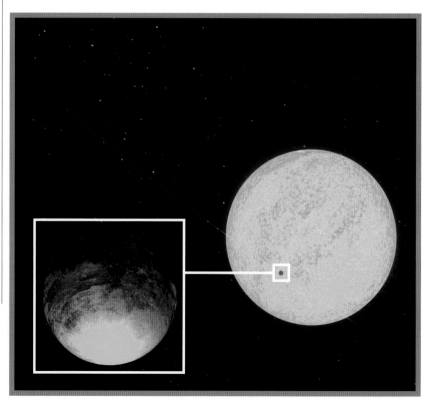

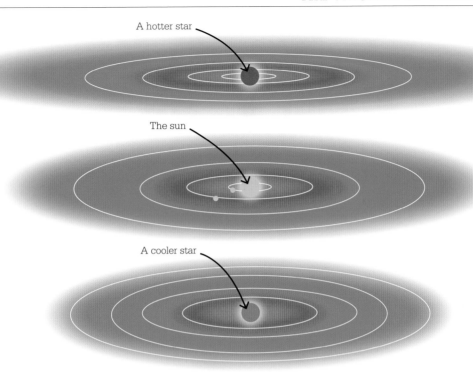

The size of the habitable zone (green) depends on the size of the star. The red zone is too hot, while the blue zone is too cold. The habitable zone is closer to cooler stars than it is to hotter stars. The size of a planet, the shape of its orbit, and the speed of its rotation between night and day also affect its habitability.

like Baby Bear's porridge in the fairy tale, "not too hot, not too cold." The size and locations of habitable zones depend on the activity of the host star. For example, if Earth were orbiting a K-type star, an orange dwarf that is considerably cooler than the sun (the sun is a G-type, or yellow dwarf), it would need to orbit at about one-third its current distance to receive the same amount of warmth.

Of the thousands of exoplanets that have been identified, only a tiny proportion are candidates orbiting in their star's habitable zones, with Earth-like conditions for life—rocky surface with liquid water. Typically, they are larger than Earth, and very few have good prospects for being Earth-like. If and when Earth-like planets are found, astrobiologists will look at the atmospheric chemistry for signs of life, such as the presence of elevated levels of oxygen, produced by photosynthesizing life-forms. How life evolved from nonliving material on Earth is still a mystery but the study of Earth-like planets

If we keep working as well and we keep being as enthusiastic ... the issue about life on other planets will be solved.
Didier Queloz

may throw light on that process. Even if life is found, it is likely that most extraterrestrial natural histories will not have moved beyond microorganisms. As every step toward evolving more complex life-forms becomes ever more unlikely, so alien civilizations that match humankind's will be a lot less common. However, if only G-type stars, like the sun, are counted, there are about 50 billion in the galaxy. It is estimated that 22 percent of them have an Earth-like planet in their habitable zones, which equals 11 billion possible Earths. Adding in other types of stars such as orange and red dwarfs, that number rises to 40 billion. Even if the probability of civilizations evolving is one in a billion, the chances are that humankind is not alone. ∎

THE MOST AMBITIOUS MAP OF THE UNIVERSE EVER
A DIGITAL VIEW OF THE SKIES

IN CONTEXT

KEY ASTRONOMER
Donald York (1944–)

BEFORE
1929 Edwin Hubble proves that the universe is expanding.

1963 Maarten Schmidt discovers quasi-stellar objects, or quasars, which turn out to be young galaxies.

1999 Saul Perlmutter, Brian Schmidt, and Adam Riess show that the expansion of the universe is increasing due to the mysterious effects of "dark energy."

AFTER
2004 Construction begins on the James Webb Space Telescope, which will use infrared to see the first stars that formed after the Big Bang.

2014 Approval is given for the European Extremely Large Telescope, which will have a 128-ft (39-m) segmented main mirror, making it the most sensitive optical telescope ever.

Set up to produce "a field guide to the heavens," the Sloan Digital Sky Survey (SDSS) began operating in 1998. The ambitious goal was to make a map of the universe on an immense scale—not just a survey of objects on a celestial sphere, but a three-dimensional model of a large portion of deep space. The project was initially headed by US astronomer Donald York, but is now a collaboration between 300 astronomers from 25 institutions. SDSS uses an 8-ft 3-in (2.5-m) telescope at Apache Point, New Mexico. The telescope's wide-angle camera has digitized objects visible from the northern hemisphere.

From the 500 million objects visible, the brightest 800,000 galaxies and 100,000 quasars were selected, and their sizes and positions in the sky accurately transposed as holes drilled into hundreds of aluminum disks. When fitted to the telescope, a disk blocks unwanted light, and feeds the light from each target galaxy into its own dedicated optical fiber and

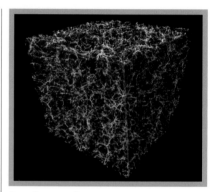

A cube section of the SDSS sky map shows the intricate distribution of matter in space. The tangles of light are interconnected galaxies.

onto a spectroscope. From these accurate galactic spectra, the astronomers can figure out how far away each galaxy is. Data collection began in 2000 and is expected to continue until 2020. The information gathered so far has revealed galaxies in clusters and superclusters, and even galactic "walls"—immense structures containing millions of galaxies, forming a tangled cosmic web with vast voids in between. ∎

See also: Beyond the Milky Way 172–77 ▪ Quasars and black holes 218–21 ▪ Studying distant stars 304–05 ▪ Looking farther into space 326–27

OUR GALAXY HARBORS A MASSIVE CENTRAL BLACK HOLE
THE HEART OF THE MILKY WAY

In 1935, Karl Jansky found a source of radio waves named Sagittarius A (Sgr A) at the center of the Milky Way. Hidden from light telescopes by cosmic dust, the radio waves emanated from several sources. In 1974, radio telescopes pinpointed the most intense source, named Sagittarius A* (Sgr A*). It was small and produced intense X-rays, suggesting that matter at the heart of the galaxy was being ripped apart by a gigantic black hole, emitting X-rays in the process. However, this remained hypothetical until Andrea Ghez, an astronomer at UCLA, used a method for observing stars through the dust using infrared.

In 1980, Hawaii's Keck Observatory began measuring the speed of stars orbiting close to the galactic center. This data made it possible to calculate the mass of the invisible object inside Sgr A*. Ghez's team found that the stars closest to Sgr A* were orbiting at a quarter of the speed of light. Such speed indicated an immense gravitational presence: a black hole 4 million times heavier than the sun, which must have swallowed up stars and other black holes when the galaxy was young. ■

An X-ray flare shoots from the black hole at the heart of the Milky Way. The discovery suggests all galaxies may have black holes at their hearts.

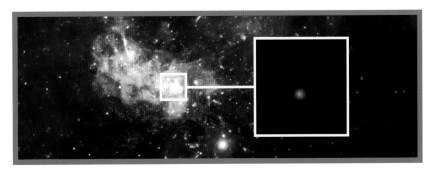

See also: Radio astronomy 179 ▪ Discovering black holes 254 ▪ Studying distant stars 304–05

COSMIC EXPANSION IS ACCELERATING

DARK ENERGY

IN CONTEXT

KEY ASTRONOMERS
Saul Perlmutter (1959–)
Brian Schmidt (1967–)
Adam Riess (1969–)

BEFORE
1917 Albert Einstein adds a cosmological constant to his field calculations to keep the universe static.

1927 Georges Lemaître suggests that the universe could be dynamic, not static.

1928 Edwin Hubble finds evidence of cosmic expansion.

1948 Fred Hoyle, Hermann Bondi, and Thomas Gold propose the steady-state theory of the expanding universe.

AFTER
2013 The Dark Energy Survey begins to map the universe.

2016 The Hubble Space Telescope shows that cosmic acceleration is 9 percent faster than originally measured.

The Big Bang theory has at its heart a simple idea—the universe started out very small and then expanded. In 1998, two teams of scientists discovered that the expansion of the universe is itself speeding up. This discovery revealed that what astronomers can directly detect makes up just 5 percent of the total mass and energy in the universe. Invisible dark matter makes up another 24 percent, while the rest is a mysterious phenomenon, known simply as dark energy. In 2011, three Americans, Saul Perlmutter, Brian Schmidt, and Adam Riess, won the Nobel Prize for Physics for this discovery.

Expanding space
The year after Georges Lemaître's paper hypothesized the Big Bang, Edwin Hubble found proof of the expanding universe when he showed that galaxies were moving away from Earth—and the ones that were farther away were moving faster. These were not simply objects blasting away from each other through space; this was space itself growing in size and moving the matter with it. The

It's everywhere, really. It's between the galaxies. It is in this room. We believe that everywhere that you have space, empty space, you cannot avoid having some of this dark energy.
Adam Riess

galaxies are not only moving away from Earth; they are expanding away from everywhere all at once.

Better picture
Subsequent observations helped to tell the history of the expanding universe. The 1964 discovery of the cosmic microwave background (CMB), a cold glow left over from the Big Bang, showed that the universe has been expanding for approximately 13.8 billion years. Surveys of

---

The **expansion** of the universe is assumed to be **slowing down** due to the force of gravity.

→

Measuring this deceleration should reveal the **ultimate fate** of the universe.

However, when measured, cosmic expansion is found to be accelerating.

This acceleration must be due to a previously unknown **force that works against gravity**, called **dark energy**.

the large-scale structure of the universe have since revealed that billions of galaxies are clustered together around vast empty voids (p.296). This structure corresponds to minute ripples in the CMB that show how observable matter—the stars and galaxies—emerged in anomalous regions in otherwise empty space. However, the future of the universe was uncertain. It was unknown whether it would expand forever or one day collapse under its own gravity.

Decelerating universe

Throughout the 20th century, cosmologists assumed that the rate of expansion was slowing down. Following a rapid initial expansion, gravity would start decelerating. It seemed there were two main possibilities. If the universe was heavy enough, its gravity would eventually slow the expansion to a stop and begin to pull matter back together in a cataclysmic Big Crunch, a kind of Big Bang in reverse. The second possibility was that the universe was too light to stop the expansion, which would therefore continue forever, gradually slowing down. This process would result in heat death, where the material of the universe had broken up, become infinitely dispersed, and ceased to interact in any way at all. A measurement of the deceleration of the universe's expansion would tell cosmologists which possible future the universe was heading for.

The Chandra X-ray Observatory took this image of the remnant of type 1a supernova SN 1572 in Cassiopeia. It is also known as Tycho's nova, as it was observed by Tycho Brahe.

By the mid 1990s, two programs were under way to measure the rate of expansion of the universe. The Supernova Cosmology Project was headed by Saul Perlmutter at Lawrence Berkeley National Laboratory, while Brian Schmidt, based at the Australian National University, led the High-Z Supernova Search Team. Adam Riess, of the Space Telescope Science Institute, was the lead author for the latter project. The project leaders considered merging, but had different ideas about how to proceed, and so opted instead for a healthy rivalry.

Both projects were using a discovery made by the Calán/Solodo Supernova Survey, carried out in Chile between 1989 and

If you're puzzled by what dark energy is, you're in good company.
Saul Perlmutter

1995. The survey found that type 1a supernovae could be used as standard candles, or objects that can be used to measure distances across space. A standard candle is an object of known brightness, »

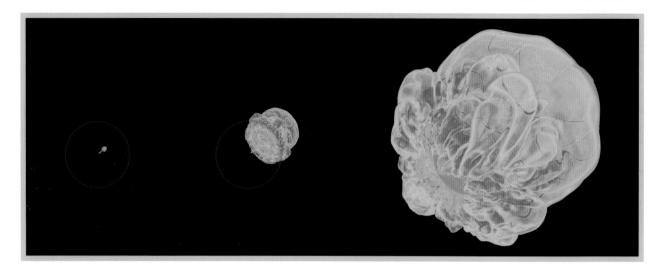

and so its apparent magnitude (brightness as seen from Earth) shows how far away it is.

A type 1a supernova is a little different from a standard supernova, which forms when large stars run out of fuel and explode. A type 1a forms in a binary star system, in which a pair of stars orbit each other. One is a giant star, the other is a white dwarf. The white dwarf's gravitational pull hauls stellar material over from the giant. The material accretes on the surface of the white dwarf until it has grown to 1.38 solar masses. At this point, the temperature and pressure are such that a runaway nuclear fusion explosion ignites the star, creating an object billions of times brighter than the sun.

Distance and motion

Both surveys used the Cerro Tololo Inter-American Observatory in Chile to find type 1a supernovae. The plan was not simply to plot the positions of the supernovae. They used the Keck Telescope in Hawaii to take spectra of each explosion, giving its redshift (the lengthening the spectra have undergone).

A computer simulation shows a white dwarf star exploding in a type 1a supernova. A flame bubble forms inside the star (left), rises above the surface (center), and envelops the star (right).

The brightness, or magnitude, of each star gave the distance—often billions of light-years—while its redshift indicated its speed relative to Earth, caused by the expansion of the universe. The teams were aiming to measure the rate at which the expansion was changing. The rate of expansion, as indicated by more distant objects, was expected to be tailing off. Exactly how fast it was doing this would show if the universe was "heavy" or "light." However, when the teams looked beyond about 5 billion light-years (meaning that they were looking 5 billion years into the past), they found that the opposite was happening—the expansion of the universe was not slowing down but speeding up.

Dark Energy Survey

In 2013, the Dark Energy Survey began a five-year project to map the expansion of the universe in detail. The project uses the Dark Energy Camera (left) at Cerro Tololo Inter-American Observatory, Chile. The camera has one of the widest fields of view in the world. In addition to searching for type 1a supernovae, the project is looking for baryon acoustic oscillations. These are regular ripples in the distribution of normal matter about 490 million light-years apart, which can be used as a "standard ruler" to show up cosmic expansion.

Dark energy

This result was first thought to be an error, but successive checks showed it was not—and both teams found the same thing. In 1998, Perlmutter and Schmidt went public with their findings. The results shook the scientific world. Using Einstein's field equations for general relativity, Adam Riess had found that the results appeared to give the universe a negative mass. In other words, it appeared that a kind of antigravity force was pushing matter apart. This source of energy was named dark energy, because it was a complete mystery.

In 2016, new observations were used to calculate a more accurate, and slightly faster, figure for the acceleration of the universe's expansion. If dark energy continues to push the universe apart (it may not, no one really knows), it will disperse the galaxies so that eventually they would all be too far away to be seen from Earth (which itself will no longer exist). Eventually, it may scatter the stars within the Milky Way until the sky goes dark. The sun and the planets in the solar system would be pulled apart, and finally the particles in atoms will also be scattered, resulting in a form of heat death dubbed the Big Rip.

Reviving Einstein's mistake

Dark energy may indicate that the universe is not as homogenous as cosmologists think it is, and that the apparent acceleration seen is due to the fact that it is inside a region with less matter in it than elsewhere. It may also be showing that Einstein's theory of gravity is incorrect on the largest scales. On the other hand, dark energy might also be explained by a mathematical device Einstein created in 1917 called the cosmological constant. Einstein used this as a value that would counteract the pull of gravity and make the universe a static, unchanging place. However, when Lemaître used Einstein's own equations to show that the universe could only be dynamic—

> This discovery has led us to believe that there is some unknown form of energy that is ripping the universe apart.
> **Brian Schmidt**

expanding or contracting— Einstein dropped the constant from his theories, calling it a mistake.

The value of Einstein's cosmological constant is set to match the energy contained in a vacuum—in empty space. This was assumed to be zero. However, according to quantum theory, even a vacuum contains "virtual"

particles, which exist for a Planck time (10^{-43} seconds, the smallest possible amount of time) and then disappear again. Dark energy may match this idea—a form of energy arising from these virtual particles, which creates a negative pressure that pulls space apart, and represents a nonzero value for the cosmological constant.

The expansion was not always accelerating. There was a time when gravity and other forces pulled matter together and was more powerful than dark energy. However, once the universe became big and empty enough, the effects of dark energy appear to have become dominant. It may be that a different force takes over in the future, or dark energy's effects may continue to grow. One suggestion is that a Big Rip would be so powerful that dark energy will tear apart spacetime itself, creating a singularity—the next Big Bang. ∎

Four possible futures

  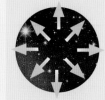

If the average density of the universe is above a certain critical value, it should be closed, and end with a Big Crunch. The critical value is estimated to be the equivalent of five protons per cubic meter.

If the density is equal to the critical density, the universe's geometry will be flat, and the universe ought to continue into the future, neither expanding nor contracting.

If the density is below the critical value, the universe should be open and expand forever, to end eventually in a heat death.

Observations suggest that the universe's expansion is accelerating due to mysterious "dark energy." The measured density is very close to the critical density, but dark energy is accelerating expansion.

PEERING BACK OVER 13.5 BILLION YEARS

STUDYING DISTANT STARS

IN CONTEXT

KEY DEVELOPMENT
James Webb Space Telescope (2002–)

BEFORE
1935 Karl Jansky shows that radiation other than light can be used to view the universe.

1946 Lyman Spitzer Jr suggests placing telescopes in space to avoid atmospheric interference.

1998 The Sloan Digital Sky Survey begins to make a 3D map of the galaxies.

AFTER
2003 The Spitzer Space Telescope, an infrared observatory, is launched.

2014 The European Extremely Large Telescope project is approved, with a primary mirror 39 m (128 ft) in diameter.

2016 LIGO announces the discovery of gravitational waves, suggesting a possible means for looking even farther than the JWST.

An artist's impression of the JWST in space shows the layered stack of the sunshield unfolded beneath the telescope. The beryllium mirror is coated in gold for optimal reflection.

The James Webb Space Telescope (JWST) is designed to be the most powerful astronomical tool in space, able to see farther than even the Hubble Space Telescope. Named in 2002 after the NASA director who oversaw the Apollo program, the JWST is an infrared telescope equipped with a 21-ft (6.5-m)-wide gold-plated mirror. This will allow it to see more than 13.5 billion light-years into the distance—to the time when the universe's first stars were forming.

Conceived in 1995 as the successor to Hubble, the JWST has had a long road to completion, encountering multiple technical hurdles. When launched in 2018, it will take up a tight orbit around L2 (Lagrange point 2), a location 1 million miles (1.5 million kilometers) beyond Earth's orbit, away from the sun.

See also: Radio astronomy 179 ▪ Space telescopes 188–95 ▪ A digital view of the skies 296 ▪
Gravitational waves 328–31 ▪ Lagrange (Directory) 336

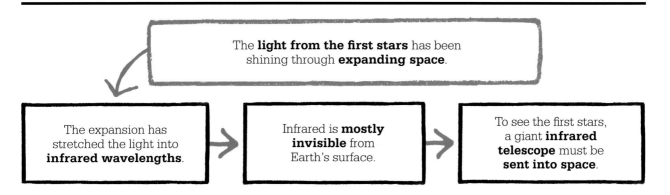

The **light from the first stars** has been shining through **expanding space**.

The expansion has stretched the light into **infrared wavelengths**.

Infrared is **mostly invisible** from Earth's surface.

To see the first stars, a giant **infrared telescope** must be **sent into space**.

L2 is a place in space where the gravity of the sun and Earth work together to pull an orbiting object around the sun at the same rate as Earth, making one orbit every year. This means the JWST will be largely in the shadow of Earth, blocking out any heat pollution from the sun and allowing the telescope to detect very faint infrared sources in deep space. NASA claims that the telescope could detect the heat of a bumblebee on the moon.

Heat seeker
The JWST's vast primary mirror is seven times the area of Hubble's and, instead of polished glass, the mirror contains 18 hexagonal units made from beryllium for maximum reflection. The 270-sq-ft (25-m²) mirror is too large to be launched flat, so it is designed to unfold once in orbit.

To pick up the faint heat signatures of the most distant stars, the telescope's detectors must always be extremely cold—never more than −370°F (−223°C). To accomplish this, the JWST has a heat shield the size of a tennis court. Again, this is folded away for launch. The shield is made from five layers of shiny plastic that reflect most of the light and heat. Any heat

that penetrates the top layer is then radiated sideways by successive inner layers so that almost nothing reaches the telescope itself.

First light
The light waves from the first stars to form have been stretched as they shine through the expanding universe, changing them from visible light to infrared, so they

are a prime target of observation for the JWST. At the same time, this ultra-sensitive eye on the infrared sky has three other main goals. It will investigate how galaxies have been built over billions of years, study the birth of stars and planets, and provide data about extrasolar planets. NASA hopes that the telescope will be in operation for at least 10 years. ■

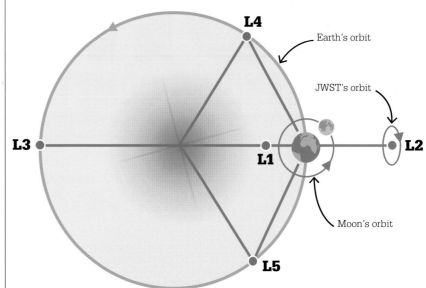

JSWT will not be exactly at the L2 point, but will circle around it in a halo orbit. Lagrange points are positions in the orbit of two large bodies where a smaller object can keep a stable position relative to those two large bodies. There are five L points in the orbital plane of Earth and the sun.

OUR MISSION
IS TO LAND ON A
COMET

UNDERSTANDING COMETS

IN CONTEXT

KEY DEVELOPMENT
ESA—Rosetta (2004)

BEFORE
1986 The Halley Armada of eight spacecraft, led by ESA's Giotto, make observations of Halley's comet.

2005 The Deep Impact mission fires a probe at comet Tempel 1 to create a crater in the surface, and analyzes what is underneath.

2006 The Stardust mission collects a capsule of cometary dust from the tail of comet Wild 2 and returns to Earth.

AFTER
2015 New Horizons flies by Pluto and begins an exploration of the Kuiper belt.

2016 NASA's OSIRIS-REx spacecraft is launched with a mission to collect and return a sample from the asteroid 101955 Bennu.

By studying comets, astronomers hope to shed new light on various questions about the early solar system, the formation of Earth, and even the origins of life.

Earth is the only planet known to have a surface ocean of liquid water. The origin of this water is one of the enduring mysteries of Earth science. A leading theory is that the hot, young planet sweated out the water from its rocks, releasing water vapor into the atmosphere. Once the planet had cooled sufficiently, this vapor condensed and fell as a deluge of rain that filled the oceans. Another theory argues that at least some of the water arrived from space, specifically in the hundreds of thousands of icy comets that rained down on Earth during the first half billion years of its existence and vaporized on impact.

A flyby of Halley's comet in 1986 by a flotilla of spacecraft led by ESA's Giotto got the first close-up look at a comet's core, or nucleus. The Halley encounter provided conclusive proof that comets are largely made from water ice mixed with organic dust and chunks of

Giotto ignited the planetary science community in Europe.
Gerhard Schwehm
Giotto Project scientist

rock. This revived the theory that this is where Earth's oceans came from. One theory concerning the origin of life was that the complex chemical building blocks necessary for life, such as amino acids and nucleic acids, arrived on Earth from space. Perhaps these organic compounds were also delivered to Earth by comets. The only way to find out was to send a spacecraft to meet up with a comet and land on its surface. In 2004, the ESA-led Rosetta mission blasted off on a 10-year journey to do just that.

Fresh target
Rosetta's intended target was Comet 67P/Churyumov–Gerasimenko, or 67P for short. In 1959, this comet had been captured by the gravity of Jupiter, which had pulled it into a shorter six-year orbit of the sun. Before that, 67P had been circling the sun much farther away. This excited the Rosetta scientists because the tail of a comet—its most familiar feature—is caused by solar radiation heating the surface of the nucleus, which

In 2005, the Deep Impact impactor collided with comet Tempel 1, releasing debris from the comet's interior. Analysis showed the comet to be less icy than expected.

creates streams of dust, gas, and plasma hundreds of millions of miles long. The material in the tail is lost from the comet forever. 67P had only been close to the sun a handful of times in its existence. That meant it was still "fresh" with its primordial composition intact.

All aboard

Rosetta was launched by an Ariane 5 rocket from ESA's space center in French Guiana. The spacecraft weighed just under 3 tons, with

We didn't just land once—
maybe we landed twice!
Stephan Ulamec
Philae landing manager

An artist's impression shows Rosetta releasing the Philae lander above comet 67P. The lander bounced on landing, flying up from one lobe of the comet to land again on the other.

a central body about as large as a small van. A folded solar array unfurled to provide 690 sq ft (64 m^2) of photovoltaic cells, which would power the craft throughout the mission.

Most of Rosetta's instruments were designed to study the comet while in orbit. They included various spectroscopes and microwave radars for studying the composition of the comet surface and the dust and gases that would be released when 67P neared the sun and began to heat up. One of the most important instruments on board was CONSERT (Comet Nucleus Sounding Experiment by Radiowave Transmission), which would blast a beam of radio waves through the comet to find out what lay inside. CONSERT would operate with the help of the lander Philae. Once on

Comets are the leftovers from the **formation of the planets**.

Earth's **water** and the **chemicals needed for life** may have come from comets.

To find out, we need to land on a comet.

First indications are that Earth's water and organic chemicals **did not come from comets**.

the surface, Philae would pick up signals from CONSERT, sent out while Rosetta was orbiting on the far side. Philae was equipped with solar panels and rechargeable batteries and was designed to work on the comet's surface in order to analyze its chemistry.

Both the names Rosetta and Philae referred to ancient Egyptian artifacts. The Rosetta Stone has a carved inscription in three languages: Hieroglyphs, Demotic Egyptian, and ancient Greek. »

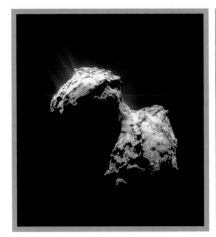

Rosetta captured this image of comet 67P/Churyumov–Gerasimenko on July 14, 2015, from a distance of 96 miles (154 km), as the comet neared its closest point to the sun.

In the early 19th century, it allowed scholars to decipher the hieroglyphic writing system, thus unlocking the meaning of many ancient Egyptian writings. Philae refers to an obelisk with multiple inscriptions that was used in a similar way. Comets are remnants left over from the formation of the solar system, so these names were chosen because the Rosetta and Philae missions at comet 67P were intended as a way to unlock new knowledge of the primordial material that formed the planets.

Comet cruise

Rosetta took a circuitous route to the comet, using three flybys of Earth and one of Mars (a risky maneuver, skimming its atmosphere only 150 miles [250 km] up) to boost speed through gravity assists. This process took five years, after which Rosetta had enough speed to fly through the asteroid belt (getting a very close look at some asteroids), and out beyond the orbit of Jupiter. There, it began to swing back around,

and was soon bearing down on 67P at great speed. For the journey to deep space, Rosetta had been powered down to save energy, but it powered back up and contacted Earth on schedule as it neared the comet in August 2014. Rosetta's controllers then began a series of thruster burns to make Rosetta zigzag through space and slow from 2,540 ft/s (775 m/s) to 26 ft/s (7.9 m/s). On September 10, the spacecraft went into orbit around 67P, offering the first look at the target world.

Bumpy landing

Comet 67P is about 2.5 miles (4 km) long and turned out to be more irregular in shape than expected. From some viewpoints, the comet looks like a vast rubber duck, with its two lobes, one larger than the other, connected by a narrow neck. (It is assumed the comet was formed from two smaller objects making a low-speed impact.) The surface was riddled with boulder fields and ridges, and the Rosetta team struggled to find a clear location to set down the Philae lander.

> We are on the surface of the comet! Whatever we do has never been done before. The data we get there is unique.
> **Matt Taylor**
> *Rosetta Project scientist*

A landing zone on the "head" of the comet was selected, and at 8:35 GMT on November 12, 2014, Philae was released from Rosetta. It took almost eight hours to confirm that Philae was on the surface, much longer than expected. The lander was designed to touch down at a slow speed—slower than an object dropped from shoulder height on Earth—and attach itself to the ground using harpoons fired from the tips of its legs. However, something had gone wrong. It is

Rosetta received gravity assists from both Earth and Mars en route to Comet 67P. As it swung around the planets, their gravitational fields threw the spacecraft forward at greatly increased speed.

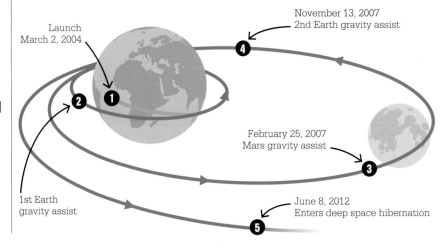

Launch
March 2, 2004

November 13, 2007
2nd Earth gravity assist

February 25, 2007
Mars gravity assist

1st Earth
gravity assist

June 8, 2012
Enters deep space hibernation

On July 16, 2016, Rosetta was just 8 miles (12.8 km) from the center of comet 67P. This image covers an area about 1,500 ft (450 m) across. It shows a dust-covered rocky surface.

thought that the lander landed awkwardly and hit a boulder, and the very low gravity of the comet meant Philae bounced right off again. It was later calculated that Philae bounced up about 3,000 ft (1 km) from the surface before falling back again, tumbling to a resting place on the edge of the target landing zone. Unfortunately, the lander ended up in the shadow of a cliff and appeared to be at an angle. Without sunlight to recharge its batteries, Philae had only about 48 hours of power to perform its primary science missions, returning data on the chemical composition of the dust and ice, and performing scans with the CONSERT instrument on Rosetta. A last-ditch plan to push the lander out into the sunlight using the harpoons (which had not fired on landing) failed, and Philae shut down into safe mode.

Approaching the sun

Despite this setback, the perilous Philae landing was deemed a success. The hope was that Philae's shaded location would become sunnier as the comet approached the sun. The comet would reach its perihelion, or closest point, in August 2015. On the approach, comet 67P began to heat up and its surface erupted with jets of dust and plasma. Rosetta was sent on a complex orbital path so that it could fly low over the comet and pass through the denser regions of the coma, or cloud of material, that was forming around 67P. Its path also took it farther out, providing a more complete picture

of the ways in which the comet was changing as it entered the warmer part of the solar system.

In mid-June 2015, Philae received enough sunlight to wake up, and began intermittent communication with Rosetta, allowing further CONSERT scans. In early July, however, it fell silent again. Fortunately, it was spotted by the OSIRIS camera on September 2, 2016, as it approached within 1.7 miles (2.7 km) of the comet. Knowing Philae's precise landing spot allows scientists to place the information it sent back a year earlier into context.

After comet 67P passed perihelion in August 2015, the solar power available to Rosetta fell rapidly. In September 2016, Rosetta was instructed to get slowly nearer to the comet. It ended its mission on

September 30 by making a controlled crash-landing, returning data right up to the moment of impact.

Alien water

The amount of deuterium ("heavy hydrogen") found in 67P's water is much greater than in the water found on Earth, evidence against the idea that Earth's water is of extraterrestrial origin. The mission has found many carbon-based compounds, but only one amino acid (the building block of proteins) and no nucleic acids (the ingredient of DNA) have been detected in the data.

Rosetta's results will allow astronomers to better understand comets and whether 67P is a typical body. Combined with discoveries from the Kuiper belt, this is hoped to reveal what the solar system was made of as the sun formed. ∎

THE VIOLENT BIRTH OF THE SOLAR SYSTEM
THE NICE MODEL

IN CONTEXT

KEY ASTRONOMERS
Rodney Gomes (1954–)
Hal Levison (1959–)
Alessandro Morbidelli (1966–)
Kleomenis Tsiganis (1974–)

BEFORE
1943 Kenneth Edgeworth suggests that Pluto is just one of many objects in the outer solar system.

1950 Jan Oort suggests that long-period comets come from a distant cloud surrounding the solar system.

1951 Gerard Kuiper proposes that a comet belt existed beyond Pluto in the early stages of the solar system.

1993 American planetary scientist Renu Malhotra suggests that planet migration took place in the solar system.

1998 The Kuiper belt is confirmed to exist.

AFTER
2015 Spacecraft New Horizons reaches the Kuiper belt.

B y the start of the 21st century, the solar system was known to contain many kinds of object. In addition to the planets and the asteroid belt, there were cometlike bodies called centaurs located in between the giant planets, trojan asteroids sharing the orbits of many planets, and the outer Kuiper belt had also just been discovered.

The **solar system** is filled with many kinds of object, all orbiting the sun.

The arrangement of these objects formed as **the outermost planets** Saturn, Uranus, and Neptune **migrated** out from the sun.

The outermost planets **swept away a vast disk of material**, leaving the system seen today.

Surrounding all these bodies was a distant sphere of comet material, called the Oort cloud.

It was difficult to explain how a system like this had evolved from a proto-solar cloud of dust and gas. Evidence from extra-solar systems showed that giant planets were often much closer to their star than was previously thought possible. It was at least feasible, therefore, that the giant planets of Earth's solar system had formed closer to the sun.

Planetary migration
In 2005, four astronomers in Nice, France, used computer simulations to develop a theory to explain the evolution of the solar system. This is now known as the Nice model. They suggested that the solar system's three outer planets, Saturn, Uranus, and Neptune, were once much closer to the sun than they are now. Jupiter was slightly farther away than it is now at 5.5 astronomical units (AU), but Neptune was much closer, at 17 AU (it now orbits at 30 AU). From Neptune's orbit, a vast disk of smaller objects called planetesimals spread to 35 AU. The giant planets pulled these

See also: The discovery of Ceres 94–99 ▪ The Kuiper belt 184 ▪ The Oort cloud 206 ▪ Investigating craters 212 ▪ Exploring beyond Neptune 286–87

Rodney Gomes

Brazilian scientist Rodney Gomes is a member of the Nice model quartet of scientists that came to prominence in 2005. It also includes American Hal Levison, Italian Alessandro Morbidelli, and Greek Kleomenis Tsiganis. Gomes, who has worked at Brazil's national observatory in Rio de Janeiro since the 1980s, is a leading expert in the gravitational modeling of the solar system, and has applied techniques similar to those used in building the Nice model to understand the motion of several Kuiper belt objects (KBOs) that appear to be following unusual orbits. In 2012, he shook up the accepted view of the solar system yet again. Gomes proposes that a Neptune-sized planet (four times as heavy as Earth) is orbiting 140 billion miles (225 billion km) from Earth (at 1,500 AU) and that this mysterious planet is distorting the orbits of the KBOs. The search is now on to locate this "Planet X."

The Nice model changed the whole community's perspective on how the planets formed and how they moved in these violent events.
Hal Levison

planetesmials inward and, in return, Saturn, Uranus, and Neptune started slowly edging farther away from the sun. Planetesimals encountering Jupiter's powerful gravity were fired out to the edge of the solar system to form the Oort cloud, and this had the effect of shifting Jupiter inward (its current orbital distance is 5.2 AU).

Resonant orbit

Eventually, Saturn shifted to a resonant 1:2 orbit with Jupiter, which meant that Saturn orbited once for every two orbits of Jupiter. The gravitational effects of this resonant orbit swung Saturn, then Uranus and Neptune into more eccentric orbits (on more stretched ellipses). The ice giants swept through the remaining planetesimal disk, scattering most of it, to create what is known as the Late Heavy Bombardment, which occurred about 4 billion years ago. Tens

of thousands of meteorites were punched from the outer disk and rained down on the inner planets.

Much of the planetesimal disk became the Kuiper belt, tied to Neptune's orbit at 40 AU. Some planetesimals were captured by the planets to become moons, others filled stable orbits as trojans, and some may have entered the asteroid belt. Planetesimals were also scattered farther out, including the dwarf planets Sedna and Eris, discovered in 2003 and 2005.

The Nice model works well for many starting scenarios for the solar system. There is even one in which Uranus is the outermost planet, only to swap places with Neptune 3.5 billion years ago. ▪

During the Late Heavy Bombardment, the moon would have glowed as it was struck by meteorites. Most of the early Earth's surface was volcanic.

A CLOSE-UP VIEW OF AN ODDBALL OF THE SOLAR SYSTEM

STUDYING PLUTO

IN CONTEXT

KEY ASTRONOMER
Alan Stern (1957–)

BEFORE
1930 Clyde Tombaugh discovers Pluto, which is named as the ninth planet.

1992 Pluto is found to be one of many Kuiper Belt Objects orbiting the sun beyond Neptune.

2005 Another Pluto-sized object is found beyond the orbit of Neptune. It is called Eris.

AFTER
2006 Pluto, Eris, and several other objects are reclassified as dwarf planets.

2016 A skewing of the orbits of Kuiper Belt Objects suggests that there is a Neptune-sized planet much farther out in space, orbiting the sun every 15,000 years. The search is now on for this object.

I n January 2006, NASA's New Horizons spacecraft lifted off from Cape Canaveral on a voyage to the planet Pluto and beyond. The moment was testament to the perseverance of the principal investigator for New Horizons, Alan Stern.

Planetary demotion

At the time, nobody knew what Pluto actually looked like. It was small and far away on the inner rim of the Kuiper belt, and even the mighty Hubble Space Telescope could only render it as a pixelated ball of light and dark patches. Plans to explore Pluto close-up were thwarted during the 1990s as NASA budgets were

See also: The Oort cloud 206 ▪ The composition of comets 207 ▪
Exploring the solar system 260–67 ▪ Exploring beyond Neptune 286–87

Just as a Chihuahua is still
a dog, these ice dwarfs
are still planetary bodies.
Alan Stern

squeezed. By 2000, the plans had
been shelved, but Stern made the
case for sending a mission to Pluto,
the smallest, most distant planet,
which had been discovered by US
astronomer Clyde Tombaugh in 1930.

In 2003, Stern's New Horizons
proposal was given the green
light, and the 2006 launch set the
spacecraft on a nine-year flight to
Pluto. It occurred not a moment too
soon. In August 2006, prompted
by the discovery of a possible
tenth planet beyond the orbit of
Pluto, the general assembly of the
International Astronomical Union
(IAU) gathered in Prague to discuss
issues raised by the new discovery.
The first question was whether it

was a planet at all. The IAU agreed
that the new body, to be named
Eris, was not a planet. Its gravity
was too weak to clear other bodies
from its orbit. The planets from
Mercury to Neptune are big enough
to do this, but the bodies of the
asteroid belt manifestly are not—
and nor was Pluto. However,
Pluto and Eris were not like most
asteroids. They were massive
enough to be spherical rather than
irregular chunks of rock and ice.
So the IAU created a new class of
object: that of dwarf planet. Pluto,
Eris, and several large Kuiper Belt
Objects (KBOs) were given dwarf
planet status, as was Ceres, the
largest body in the asteroid belt.
For most of these objects, this was
a promotion in the hierarchy of
the solar system, but not for Pluto.
If Pluto had been declassified as
a planet prior to New Horizons'
launch, it is uncertain whether
the mission would have happened.

Long journey
Although Pluto's orbit does bring
it closer to the sun than Neptune
for some of its 248-year revolution
around the sun, the New Horizons
probe had the longest journey to the
most distant target in the history »

Alan Stern

Born Sol Alan Stern in New
Orleans, Louisiana, in 1957,
Stern's fascination with
Pluto began in 1989 when
he worked with the Voyager
program. While he was there,
Stern witnessed the final
encounter of Voyager 2
as it flew past Neptune and its
moon Triton. Triton appeared
as a ball of ice, and looked
very much like the Pluto
Stern and other scientists
had imagined. (Triton is
thought to be a Kuiper Belt
Object that has been captured
by Neptune.)

In the 1990s, Stern trained
as a Space Shuttle payload
specialist (technical expert),
but he never got the chance
to fly into space. Instead, he
returned to the study of Pluto,
the Kuiper belt, and the Oort
cloud. In addition to leading
the New Horizons mission
as principal investigator,
Stern is active in developing
new instruments for space
exploration and more cost-
effective ways of putting
astronauts into orbit.

Key work

2005 *Pluto and Charon: Ice
Worlds on the Ragged Edge
of the solar system*

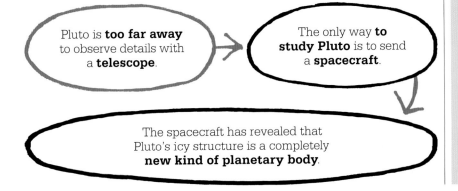

Pluto is **too far away**
to observe details with
a **telescope**.

The only way **to
study Pluto** is to send
a **spacecraft**.

The spacecraft has revealed that
Pluto's icy structure is a completely
new kind of planetary body.

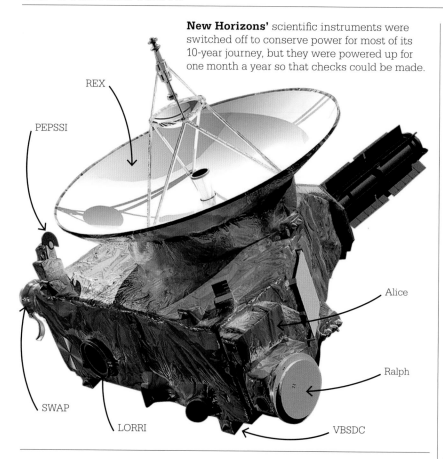

New Horizons' scientific instruments were switched off to conserve power for most of its 10-year journey, but they were powered up for one month a year so that checks could be made.

REX

PEPSSI

Alice

Ralph

SWAP

LORRI

VBSDC

Range Reconnaissance Imager), a telescopic camera, would produce the highest resolution pictures of the Pluto system; SWAP (Solar Wind Around Pluto) would, as the name suggests, observe Pluto's interaction with the solar wind, while PEPSSI (Pluto Energetic Particle Spectrometer Science Investigation) detected the plasma given off by Pluto. This would help in understanding the way the dwarf planet's atmosphere is formed by sublimation (the change of a solid directly into a gas) from the icy surface during Pluto's "summer" as it nears the sun (and then freezes again in winter). Finally the SDC (Student Dust Counter) was an instrument operated by university students throughout the mission. This experiment was renamed VBSDC for Venetia Burney, the British girl who had proposed the name Pluto.

Destination reached

New Horizons began its approach in January 2015. One of the first things it did was to make an accurate measurement of Pluto's size. This had always been a tricky problem to solve. When it was first discovered,

of space exploration—30 AU, or 2.7 billion miles (4.4 billion km) from Earth. To achieve this, the spacecraft was given the fastest launch ever, blasting off with an escape velocity of 36,373 mph (58,536 km/h). A year after launch, the spacecraft reached Jupiter. In addition to making some observations of the Jovian system, New Horizons used Jupiter's gravity to gain a 20 percent speed boost. This cut the flight time to Pluto from over 12 years to 9.5 years.

Instruments on board

The precision of the trajectory from Jupiter was crucial to the success of New Horizons. If it was just slightly off, the craft would miss Pluto altogether. The main observation window was about 12 hours long, after which New Horizons would

be leaving Pluto behind. It takes radio signals from Pluto 4.5 hours to reach Earth, plus the same again to send a return message. Therefore, it would take at least nine hours to make even tiny course corrections, by which time the primary mission would be almost over.

New Horizons carried seven instruments. They included two imaging spectrometers, built to work together and named after the characters in the 1950s US sitcom *The Honeymooners*. Ralph was the visible and infrared spectrometer used to make maps of Pluto's surface, while Alice was sensitive to ultraviolet and was tasked with studying Pluto's thin atmosphere. REX (Radio science EXperiment) would take the temperature of Pluto and its moons; LORRI (Long

It used to be that Pluto was a misfit. Now it turns out that Earth is the misfit. Most planets in the solar system look like Pluto, and not like the terrestrial planets.
Alan Stern

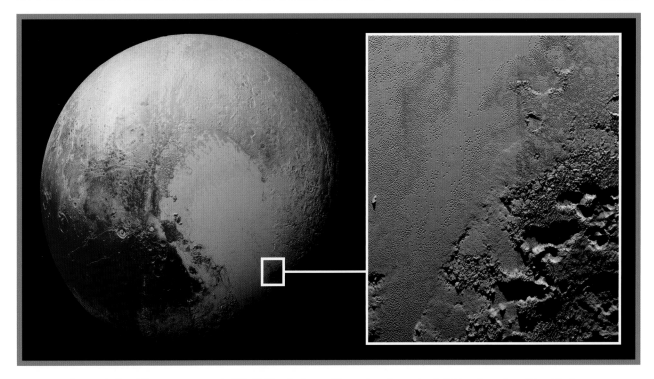

it was estimated that Pluto was seven times the size of Earth. By 1978, it was clear that Pluto was smaller than Earth's Moon. However, it also had a huge satellite, named Charon (the boatman of the dead in Greek mythology), which was about a third of the size of Pluto, and the two bodies moved around each other as a binary system. At launch, planners also took into account two more tiny moons, Nix and Hydra, but by 2012, with New Horizons already well on the way, it was found that there were two more—Kerberos and Styx—which could potentially disrupt the mission.

Measuring Pluto

In the end, these worries were misplaced, and LORRI was able to obtain a measurement for all of these bodies. Pluto is 1,470 miles (2,370 kilometers) wide, meaning that it is larger than Eris (although Eris is heavier). On July 14, 2015, New Horizons flew 7,750 miles

(12,472 km) above Pluto, its closest approach. Its instruments were collecting huge quantities of data to be fed back to Earth. The close-up view of Pluto showed it to be a world of pale ice plains and dark highlands. The ice is largely frozen nitrogen, which makes Pluto a very bright object for its size. The highlands are also ice (although mixed with tar-like hydrocarbons). The ice is thrust into lumpy peaks that tower 2 miles (3 km) above the plains. Quite how such huge features arose on such a cold and small body is one of the mysteries of the New Horizons mission. In addition, craterlike structures have been identified as possible ice volcanoes.

Naming the landmarks

Pluto's surface features have been given unofficial names by NASA scientists. Cthulhu Regio is a large whale-shaped dark patch in the southern hemisphere. Other regions

This view from New Horizons zooms in on the southeastern portion of Pluto's great ice plains, where the plains border rugged, dark highlands.

are named after past missions: Voyager, Venera, and Pioneer. Two main mountain ranges have been imaged clearly: Norgay Montes and Hillary Montes, named after the first two climbers to reach the summit of Mount Everest. However, the central feature of New Horizons' partial map of Pluto is Tombaugh Regio, a heart-shaped plain. Half of this area is made up of Sputnik Planitia, a vast ice floe riddled with cracks and troughs, but with no craters, which suggests it is a young feature that is carving out new surface features like glaciers on Earth.

Now past Pluto, the craft is on course to meet up with other KBOs. Its nuclear power source should last until around 2030, and the mission should make many more discoveries. ∎

A LABORATORY ON MARS

EXPLORING MARS

IN CONTEXT

KEY ORGANIZATION
NASA—Mars exploration

BEFORE
1970 The Soviets' Lunokhod 1 becomes the first vehicle to be used on another body when it touches down on the moon.

1971 The Lunar Roving Vehicle is driven on the moon for the first time during NASA's Apollo 15 mission.

1977 NASA's Sojourner is the first rover to reach Mars.

AFTER
2014 Opportunity breaks the distance record for a rover on an extraterrestrial body.

2020 The NASA Mars 2020 rover is set to be launched as a replacement for Curiosity.

2020/21 The ExoMars rover is due to be deployed by the European Space Agency in Oxia Planum, a depression filled with clay-bearing rocks.

In August 2012, the Mars Science Laboratory Rover, better known as Curiosity, landed on Mars. This 2,000-lb (900-kg) wheeled vehicle, which is still roaming the Martian surface, is a mobile laboratory equipped to conduct geological experiments aimed at figuring out the natural history of the red planet. It is the latest robot explorer to reach Mars, and the largest and most advanced in a long line of rovers sent to explore other worlds.

Wanderers

The potential of rovers in space was clear as far back as 1971, when Apollo 15 carried a four-wheel Lunar Roving Vehicle to the moon. This agile two-seater widened the scope of lunar exploration for the last three Apollo missions. For instance, during the first moon landing in 1969, Neil Armstrong and Buzz Aldrin spent just two and a half hours moonwalking, and the farthest they moved from their lunar module was 200 ft (60 m). By contrast, however, in the final Apollo moon mission, Apollo 17, in 1972, the crew of two—Eugene Cernan and Harrison Schmitt—

Mars has been flown by, orbited, smacked into, radar-examined, and rocketed onto, as well as bounced upon, rolled over, shoveled, drilled into, baked, and even blasted. Still to come: Mars being stepped on.
Buzz Aldrin

spent more than 22 hours outside. In their rover, they covered 22 miles (36 km) in total, with one drive taking the pair 4.7 miles (7.6 km) from their spacecraft. The Lunar Roving Vehicle, or moon buggy, was used to collect rocks. The six Apollo missions returned to Earth with 840 lb (381 kg) of them.

Analysis of these rocks revealed much about the history of the moon. The oldest were about 4.6 billion years old, and their chemical composition clearly showed a common ancestry with rocks on Earth. Tests revealed no evidence of organic compounds, indicating that the moon has always been a dry and lifeless world.

Lunokhod 1

The Soviet lunar program, which began in the early 1960s, relied on unmanned probes to explore the moon. Three of the Soviet Luna

Geologist–astronaut Harrison Schmitt collects samples from the lunar surface during the 1972 Apollo 17 mission. He spent many hours exploring the surface on the moon buggy.

See also: The Space Race 242–49 ▪ Exploring the solar system 260–67 ▪ Understanding comets 306–11 ▪ Studying Pluto 314–17

The Soviet Lunokhod 1 rover, seen here in tests on Earth, was the first rover ever to land on an alien world—its predecessor, Lunokhod 0, was launched in 1969 but never reached orbit.

probes returned with a total of 11.5 oz (326 g) of rock. Then, in November 1970, the Soviet lander Luna 17 arrived at a large lunar plain called the Sea of Rains (many lunar areas are named after the weather conditions they were once thought to influence on Earth). Luna 17 carried the remote-controlled rover Lunokhod 1 (Lunokhod means "moonwalker"). This was the first wheeled vehicle to traverse an extraterrestrial world, arriving about eight months before the first Apollo buggy. The concept behind it was simple—instead of sending moon rocks to Earth, the rover would do the analysis there.

Remote-controlled explorer

The Lunokhod rover was 7½-ft (2.3-m) long and resembled a motorized bathtub. The wheels were independently powered so that they could retain traction on the rough lunar terrain. Lunokhod was equipped with video cameras that sent back TV footage of the

moon. An X-ray spectrometer was used to analyze the chemical composition of rocks, and a device called a penetrometer was pushed into the lunar regolith (soil) to measure its density.

Lunokhod was powered by batteries that were charged by day using an array of solar panels that folded out from the top of the rover. At night, a source of radioactive polonium inside the machine acted

Over time you could terraform Mars to look like Earth … So it's a fixer-upper of a planet.
Elon Musk
Canadian space entrepreneur

as a heater to keep the machinery working. The rover received commands from controllers on Earth about where to go and when to perform experiments. A human might have done a better job, but rovers could stay in space for months on end, and did not require food and water from Earth.

Lunokhod 1 was designed to work for three months, but lasted almost 11. In January 1973, Lunokhod 2 landed in the Le Monnier Crater on the edge of the Sea of Serenity. By June, Lunokhod 2 had traveled a total of 24 miles (39 km), a record that would stand for more than three decades.

Martian walker

As Lunokhod 1 was exploring the moon, the Soviet space program was eyeing an even greater prize: a rover on Mars. In December 1971, two Soviet spacecraft, code-named Mars 2 and Mars 3, sent modules to land on the red planet. Mars 2 crashed, but Mars 3 made a successful touchdown—the first-ever landing on Mars. However, it lost all communications just 14.5 seconds later, probably due to damage from an intense dust storm. Scientists never found out what happened to Mars 3's cargo: a Prop-M rover, a tiny 10-lb (4.5-kg) vehicle designed to walk on two ski-shaped feet. It was powered through a 50-ft (15-m) umbilical cord, and once on the surface was designed to take readings of the Martian soil. It is unlikely that » the Prop-M ever carried out its mission, but it was programmed to operate without input from Earth. A radio signal between the moon and Earth travels in less than 2 seconds, but a signal to or from Mars takes between 3 and 21 minutes to arrive, varying with the planet's distance from Earth. For a Martian rover to be a successful explorer, it needed to work autonomously.

Bounce down

In 1976, NASA's two Viking landers sent back the first pictures of Mars. Following this success, many more rovers were planned, but most of these projects never reached their destination, succumbing to what the press dubbed the "Martian Curse."

NASA eventually had a success with its 1997 Mars Pathfinder mission. In July of that year, the Pathfinder spacecraft entered the Martian atmosphere. Slowed first by the friction of a heat shield and then by a large parachute, the spacecraft jettisoned its outer shielding, and the lander inside was lowered on a 65-ft (20-m) tether. As it neared the surface,

During its 83 days of operation, the tiny Sojourner rover explored around 2,691 sq ft (250 sq m) of the planet's surface and recorded 550 images.

We landed in a nice flat spot. Beautiful, really beautiful.
Adam Steltzner
Lead landing engineer, Curiosity

huge protective airbags inflated around the lander, and retrorockets on the spacecraft holding the tether fired to slow the speed of descent. The tether was then cut, and the lander bounced across the Martian surface until it rolled to a stop. Fortunately, once the airbags had deflated, the lander was the right side up. The three upper sides or "petals" of the tetrahedral lander folded outward, revealing the 24-lb (11-kg) rover.

During development, the rover was called MFEX, short for Microrover Flight Experiment. However, it was known to the public as Sojourner, meaning "traveler" and chosen for its link to Sojourner Truth, a 19th-century US abolitionist and rights activist.

Rolling on Mars

Sojourner was the first rover to take a tour of the Martian surface. However, the Pathfinder mission was really a test for the innovative landing system and the technology that would power larger rovers in the future. The minuscule vehicle

traveled just 300 ft (100 m) during its 83-day mission, and never ventured farther than 40 ft (12 m) from the lander. Now named the Carl Sagan Memorial Station, the lander was used to relay data from the rover back to Earth. Most of the rover's power came from small solar panels on the top. One of the goals of the mission was to see how these panels stood up to extreme temperatures and what power could be generated in the faint Martian sunlight.

The rover's activities were run from NASA's Jet Propulsion Laboratory (JPL) in California, and JPL has remained the lead agency in developing Martian rovers. With the time delays inherent in communicating with Mars, it is not possible to drive a rover in real time, so every leg of a journey must be preprogrammed. To achieve this, cameras on the lander were used to create a virtual model of the surface around Sojourner. Human controllers could view the area in 3-D from any angle before mapping a route for the rover.

Spirit and Opportunity

Despite its limitations in terms of size and power, Sojourner's mission was a great success, and NASA

Whatever the reason you're on Mars is, I'm glad you're there. And I wish I was with you.
Carl Sagan
in a message for future explorers

An artist's impression portrays a NASA Mars Exploration Rover. Rovers Opportunity and Spirit were launched a few weeks apart in 2003 and landed in January 2004 at two sites on Mars.

pressed ahead with two Mars Exploration Rovers (MERs). In June 2003, MER A, named Spirit, and MER B, Opportunity, were ready for launch. They were about the same size as a Lunokhod rover, but were much lighter, at around 400 lb (180 kg). By the end of January the following year, both were traveling across the Martian deserts, hills, and plains, photographing surface features and chemically analyzing rock samples and atmosphere. They sent back the most glorious vistas of the Martian landscape ever seen, enabling geologists to examine the large-scale structures of the planet.

Spirit and Opportunity had landed using the same airbag-and-tether system as Sojourner. Like Sojourner, both relied on solar panels, but the new rovers were built as self-contained units, able to wander far from their landers. Each vehicle's six wheels were attached to a rocking mechanism, which made it possible for the rovers to keep at least two wheels on the ground as they crossed rugged terrain. The software offered a degree of autonomy so that the rovers could respond to unpredictable events, such as a sudden dust storm, without needing to wait for instructions from Earth.

Low expectations

Nevertheless, expectations for these rovers were low. JPL expected that they would cover about 2,000 ft (600 m) and last for 90 Martian sols (equivalent to about 90 Earth days). During the Martian winter, however, the team did not know whether the solar-powered rovers would retain adequate power to keep working. Of all the solar system's rocky planets, the seasons of Mars are the most Earth-like, due to the similar tilts of the planets' rotational axes. Martin winters are dark and bitterly cold, with surface temperatures falling to as low as −225°F (−143°C) near the polar ice caps.

As predicted, Martian winds blew fine dust onto the solar arrays, cutting their generating power; but the wind also blew the panels clean from time to time. As winter drew nearer, the JPL team searched for suitable locations in which the rovers could safely hibernate. To do this, they used a 3-D viewer built from the images taken from the rover's stereoscopic cameras. They chose steep slopes that faced the rising sun in order to maximize electricity generation and to top off the batteries. All nonessential equipment was shut down so that power could be diverted to heaters that kept the rovers' internal temperature above −40°F (−40°C).

Continuing mission

The hibernation worked, and incredibly, JPL has managed to extend the rover missions from a few days to several years. More than five years into its mission, however, Spirit became bogged down in soft soil; all attempts to free it by remote control from Earth failed, and unable to move to a winter refuge, Spirit finally lost power 10 months later. It had traveled 4.8 miles (7.73 km). Opportunity, meanwhile, has »

avoided mishap and continues to operate. In 2014, it beat Lunokhod 2's distance record, and by August 2015 it had completed the marathon distance of 26.4 miles (42.45 km). This was no mean feat on a planet located some hundreds of millions of miles from Earth.

Curiosity needed

Spirit and Opportunity were equipped with the latest detectors; including a microscope for imaging mineral structures and a grinding tool for accessing samples from the interiors of rocks.

However, Curiosity, the next rover to arrive on the planet in August 2012, carried instruments that not only studied the geology of Mars but also looked for biosignatures—the organic substances that would indicate whether Mars once harbored life. These included the SAM or Sample Analysis at Mars device, which

vaporized samples of ground rock to reveal their chemicals. In addition, the rover monitored radiation levels to see whether the planet would be safe for future human colonization.

Considerably larger than previous rovers, Curiosity was delivered to Mars in an unusual way. During the landing phase of the mission, the radio delay (caused

The Seven Minutes of Terror has turned into the Seven Minutes of Triumph.
John Grunsfeld
NASA associate administrator

In the "Kimberley" formation on Mars, photographed by Curiosity, strata indicate a flow of water. In the distance is Mount Sharp, named after US geologist Robert P. Sharp in 2012.

by the sheer distance from Earth) was 14 minutes, and the journey through the atmosphere to the surface would take just seven—all on autopilot (not remotely controlled from Earth). This created "seven minutes of terror": the engineers on Earth knew that by the time a signal arrived informing them that Curiosity had entered the Martian atmosphere, the rover would already have been on the ground for seven minutes—and would be operational or smashed to pieces.

Safe landing

As Curiosity's landing craft moved through the upper atmosphere, its heat shield glowed with heat, while rockets adjusted the descent speed to reach the Gale Crater, an ancient crater caused by a massive meteorite impact. A parachute slowed the craft to about 200 mph (320 km/h), but this was still too fast for a landing. It continued to slow its descent over a flat region of the crater, avoiding the 20,000-ft (6,000-m) mountain at its center. The craft reached about 60 ft (20 m) above the surface and then had to hover, since going too low would create a dust cloud that might wreck its instruments. The rover was finally delivered to the surface via a rocket-powered hovering platform called a sky crane. The sky crane then had to be detached and blasted clear of the area so that its eventual impact did not upset any future exploration.

Having survived the landing, Curiosity signaled to Earth that it had arrived safely. Curiosity's power supply is expected to last at least 14 years, and the initial two-year mission has now been extended indefinitely. So far, it has measured radiation levels, revealing that it may be possible for humans to survive on Mars; discovered an ancient stream bed, suggesting a past presence of water and perhaps even life; and found many of the key elements for life, including nitrogen, oxygen, hydrogen, and carbon. ∎

ExoMars

In 2020, the European Space Agency, in collaboration with the Russian space agency, Roscosmos, will launch its first Mars rover, ExoMars (Exobiology on Mars), with the goal of landing on Mars the following year. In addition to looking for signs of alien life, the solar-powered rover will carry a ground-penetrating radar that will look deep into Martian rocks to find groundwater. The ExoMars rover will communicate with Earth via the ExoMars Trace Gas Orbiter, which was launched in 2016. This system will limit data transfer to twice a day. The rover is designed to drive by itself; its control software will build a virtual model of the terrain and navigate through that. The rover software was taught how to drive in Stevenage, England, at a mockup of the Martian surface called the Mars Yard (above).

The ExoMars rover is expected to operate for at least seven months and to travel 2.5 miles (4 km) across the Martian surface. It will be delivered to the surface by a robotic platform that will then remain in place to study the area around the landing site.

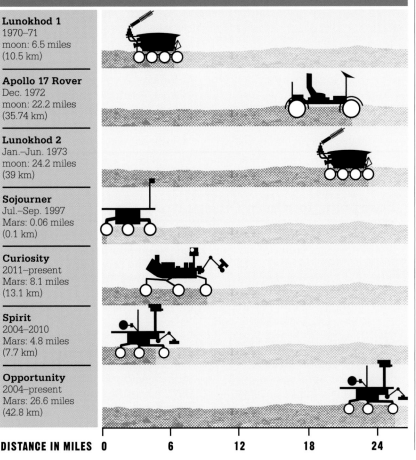

Distances traveled by extraterrestrial rovers

Lunokhod 1
1970–71
moon: 6.5 miles
(10.5 km)

Apollo 17 Rover
Dec. 1972
moon: 22.2 miles
(35.74 km)

Lunokhod 2
Jan.–Jun. 1973
moon: 24.2 miles
(39 km)

Sojourner
Jul.–Sep. 1997
Mars: 0.06 miles
(0.1 km)

Curiosity
2011–present
Mars: 8.1 miles
(13.1 km)

Spirit
2004–2010
Mars: 4.8 miles
(7.7 km)

Opportunity
2004–present
Mars: 26.6 miles
(42.8 km)

DISTANCE IN MILES   0   6   12   18   24

THE BIGGEST EYE ON THE SKY
LOOKING FARTHER INTO SPACE

IN CONTEXT

KEY DEVELOPMENT
**European Extremely
Large Telescope** (2014–)

BEFORE
1610 Galileo Galilei makes
the first recorded astronomical
observations using a telescope.

1668 Isaac Newton makes the
first usable reflecting telescope.

1946 Lyman Spitzer Jr.
suggests putting telescopes
in space to avoid Earth's
atmospheric interference.

1990 The Hubble Space
Telescope is launched.

AFTER
2015 Construction begins in
Chile on the US-led 72-ft (22-m)
Giant Magellan Telescope.

2016 LIGO detects the
gravitational waves of objects
in space.

2018 The James Webb
Space Telescope will become
the largest telescope ever
launched into space.

Despite its name, the ESO, or European Southern Observatory, is located in northern Chile, a region of dry desert and high mountains ideal for ground-based astronomy. This collaborative organization of 15 European countries, along with Brazil and Chile, has been pushing the limits of astronomy for more than 50 years.

Big telescopes

The ESO uses literal names for its telescopes. In 1989, it began operating the New Technology Telescope, the new technology being adaptive optics that reduce the blurring effect on images caused by the turbulence of the atmosphere. In 1999, it opened its Very Large Telescope, which comprises four 27-ft (8.2-m) reflecting telescopes that can be used together. The Atacama Large Millimeter Array, a vast radio telescope with 66 antennae, then became operational in 2013. This is the largest ESO program to date, and the largest ground-based astronomical project of all time. In 2014, however, the ESO received

European Southern Observatory

Formed in 1962, the ESO today has 17 member countries: Austria, Belgium, the Czech Republic, Denmark, Finland, France, Germany, Italy, the Netherlands, Poland, Portugal, Spain, Sweden, Switzerland, and the UK, with Chile and Brazil. It is located in the Atacama Desert of Chile, chosen for its clear, moisture-free skies and the absence of light pollution. The ESO's headquarters is near Munich, Germany, but its working base is the Paranal Observatory, an ultra-modern science center in the remote desert. The observatory's subterranean living quarters were used as a Bond villain's lair in the 2008 movie *Quantum of Solace*. The site's new Extremely Large Telescope is costing $1.1 billion (€1 billion) to build. The ESO opted for this project after rejecting the far costlier OWL (Overwhelmingly Large Telescope), the proposed design of which had a 330-ft (100-m) wide primary mirror.

See also: Galileo's telescope 56–63 ▪ Gravitational theory 66–73 ▪ Space telescopes 188–95 ▪ Studying distant stars 304–05

The E-ELT's dome is shown opening as the sun sets over the desert in this artist's impression. The completed structure will be 256 ft (78 m) high.

funding for the European Extremely Large Telescope (E-ELT). When completed in 2024, this will be the largest optical telescope ever built, with a resolution 15 times sharper than the Hubble Space Telescope (pp.172–77).

Giant mirror

The E-ELT has an unusual five-mirror design housed inside a dome half the size of a football stadium. The primary mirror (M1), which collects the visible light (and near infrared) is built from 798 hexagonal segments that are 4 ft 10 in (1.45 m) wide. Together they will make a mirror that is 129 ft (39.3 m) across. In contrast, the Hubble's primary mirror is just 7 ft 11 in (2.4 m) wide; even the E-ELT's secondary mirror (M2) is larger than that, at 13 ft 10 in (4.2 m).

The shape of M1 can be fine-tuned to account for distortions caused by temperature changes, and by the gravitational effect as the telescope swings into different positions. M2 directs the light from M1 through a hole in the fourth mirror (M4) onto the third mirror (M3). From there light is reflected back onto M4, the adaptive optics mirror, which greatly reduces atmospheric blurring of the image. M4 follows the twinkling of an

artificial "star," which is created by firing a laser into the sky. M4 can alter its shape 1,000 times a second using 8,000 pistons housed underneath. In other words, the 798 segments of this astonishing mirror

can ripple and warp in real time to counteract any atmospheric distortions. Finally, M5 directs the image into the camera.

The E-ELT will pick up a narrower band of the spectrum than space telescopes, but it can do so on a much larger scale. As a result, the E-ELT will be able to see exoplanets, protoplanetary discs (including their chemistry), black holes, and the first galaxies in greater detail than ever before. ▪

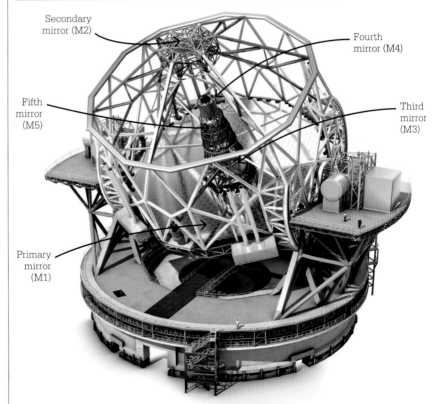

Secondary mirror (M2)

Fourth mirror (M4)

Fifth mirror (M5)

Third mirror (M3)

Primary mirror (M1)

At the heart of the E-ELT's complex arrangement of mirrors is the huge dish of the primary mirror. It will gather 13 times more light than the largest existing optical telescopes, and will be aided by six laser guide star units.

RIPPLES THROUGH SPACETIME

GRAVITATIONAL WAVES

IN CONTEXT

KEY ORGANISATION
LIGO (2016)

BEFORE
1687 Isaac Newton formulates the universal law of gravitation, which sees gravity as a force between masses.

1915 Albert Einstein presents the general theory of relativity, which explains gravity as the distortion of spacetime by mass, predicting the existence of gravitational waves.

1960 American physicist Joseph Weber attempts to measure gravitational waves.

1984 Rai Weiss and Kip Thorne found LIGO.

AFTER
2034 eLISA is scheduled to search for gravitational waves using three spacecraft in heliocentric orbits, between which lasers will be fired.

I n 1916, as he worked on his theory of relativity, Albert Einstein predicted that, as a mass moved, its gravity would cause ripples in the fabric of spacetime. Every mass would do this, although larger masses would make bigger waves, in the same way that a pebble dropped in a pond makes an ever-increasing circle of ripples, while a meteor impacting the ocean creates tsunami-sized waves.

In 2016, 100 years after Einstein's predictions, a collaboration between scientists working under the name LIGO announced that they had discovered these ripples, or gravitational waves. Their

See also: Gravitational theory 66–73 ▪ The theory of relativity 146–53

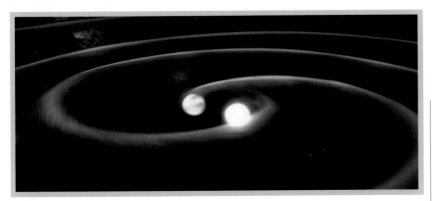

Within a period of 20 milliseconds, the two black holes LIGO had detected increased their orbital speed from 30 times a second to 250 times a second before colliding.

decades-long search had revealed the gravitational equivalent of tsunamis created by two black holes spiraling around each other and then colliding.

It is hoped that the discovery of gravitational waves will provide a new way of observing the universe. Instead of using light or other electromagnetic radiation, astronomers are hoping to map the universe by the gravitational effects of its contents. While radiation is obscured in many ways, including by the opaque plasma of the early universe up to 380,000 years after the Big Bang, gravitational waves pass through everything. This means that gravitational astronomy could see back to the very beginning of time, a trillionth of a second after the Big Bang.

Wave behaviors

LIGO stands for Laser Interferometer Gravitational-Wave Observatory. It is a remarkable set of instruments for measuring expansions and contractions in space itself. This is no easy task. A ruler cannot do it because, as space changes in size, so does the ruler, so the observer measures no change at all. LIGO succeeded using the benchmark that remains constant whatever

space is doing: the speed of light. Light behaves like a wave, but it does not require a medium through which to travel. Instead, light (and any kind of electromagnetic radiation) is an oscillation of an electromagnetic field: in other words, light is a disturbance in a field permeating all of space.

Gravitational waves can be understood as disturbances in the gravitational field that permeates the universe. Einstein described how these disturbances are caused

by the mass of objects curving the space around them. What is understood as the "pull of gravity" is a small mass appearing to alter its motion and "fall" toward a larger mass as it encounters a region of warped space.

All masses are in motion— planets, stars, even galaxies—and as they move, they leave a trail of gravitational disturbances in their wake. Gravity waves propagate in a comparable way to sound waves, by distorting the medium through which they travel. In the case of sound waves, that medium is made of molecules, which are made to oscillate. In the case of gravity, the medium is spacetime, the very fabric of the universe. Einstein predicted that the speed of gravity »

Relativity reveals that gravity is **the warping of spacetime by mass**.

↓

Moving objects create ripples through spacetime, or gravitational waves.

↓

Gravitational waves can be detected by measuring the **expansion and compression** of spacetime.

→

Gravitational waves let astronomers **see farther into space** than ever before.

would be the same as the speed of light, and that the ripples in spacetime would move outward in all directions. The intensity of these ripples diminishes rapidly with distance (by a square of the distance), so detecting a distinct gravity wave from a known object far out in space would require a very powerful source of waves and a very sensitive instrument.

Laser-guided

As its name suggests, LIGO employs a technique called laser interferometry. This makes use of a property of waves called interference. When two waves meet, they interfere with one another to create a single wave. How they do this depends on their phase—the relative timing of their oscillations. If the waves are exactly in phase—rising and falling perfectly in sync—they will interfere constructively, merging to create a wave with double the intensity. By contrast, if the waves are exactly out of phase—one rising as the other falls—the interference will be destructive. The two waves will merge and cancel one another

With no gravitational waves, LIGO's light waves cancel one another out when they are recombined. Gravitational waves stretch one tube while compressing the other, so that the waves are no longer perfectly aligned, and a signal is produced.

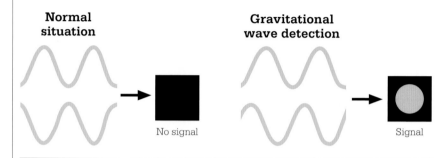

Normal situation → No signal

Gravitational wave detection → Signal

out, disappearing completely. LIGO's source of waves is a laser, which is a light beam that contains a single color, or wavelength, of light. In addition, the light in a laser beam is coherent, which means that its oscillations are all perfectly in time. Such beams can be made to interfere with one another in very precise ways.

The laser beam is split in two and the resulting beams are sent off perpendicular to one another. They both hit a mirror and bounce straight back to the starting point. The distance traveled by each beam is very precisely controlled so that one has to travel exactly

half a wavelength farther than the other (a difference of a few hundred billionths of a meter). When the beams meet each other again, they are exactly out of phase as they interfere, and promptly disappear—unless a gravitational wave has passed through space while the beams were traveling. If present, a gravitational wave would stretch one of the laser tracks and compress the other, so the beams would end up traveling slightly altered distances.

Noise filter

The laser beams are split and sent on a 695-mile (1,120-km) journey up and down LIGO's 2.5-mile (4-km) long arms before being recombined. This gives LIGO the sensitivity to detect minute perturbations in space that add up to a few thousandths of the width of a proton. With the distances put very slightly out of sync, the interfering beams would no longer cancel each other out. Instead, they would create a flickering pattern of light, perhaps indicating a gravity wave passing through LIGO's corner of space.

The difficulty was that such a sensitive detector was prone to distortions from the frequent seismic waves that run through Earth's surface. To be sure that a laser flicker

Rai Weiss and Kip Thorne

LIGO is a collaboration between Caltech and MIT, and also shares its data with a similar experiment called Virgo, which is running in France and Poland. Hundreds of researchers have contributed to the discovery of gravitational waves. However, there are two people, both Americans, who stand out among them all: Rainer "Rai" Weiss (1932–) and Kip Thorne (1940–). In 1967, while at MIT, Weiss developed the laser interferometry technique used by LIGO, working from the initial ideas of Joseph Weber, one of the inventors of the laser. In 1984, Weiss cofounded LIGO with Thorne, a counterpart at Caltech, who is a leading expert on the theory of relativity. LIGO is the most expensive science project ever funded by the US government, with a current cost of $1.1 billion. After 32 years of trying, in 2016 Weiss and Thorne announced their discovery of gravitational waves at a news conference in Washington, D.C.

LIGO's precision instruments must be kept completely clean. Maintaining the purity of the laser beams is one of the project's biggest challenges.

was not an earth tremor, two identical detectors were built at opposite ends of the United States: one in Louisiana, the other in Washington state. Only signals registering on both detectors were gravitational waves (the signals are in fact 10 milliseconds apart—the time it takes for light, and gravitational waves, to travel from Louisiana to Washington). Ligo operated from 2002–2010 with no success, then started up again in 2015 with enhanced sensitivity.

Colliding black holes

On September 14, 2015, at 9:50:45 GMT, two black holes a billion light-years away collided and unleashed huge warps in the fabric of space. In fact, this event occurred a billion years ago but it had taken that long for the ripples they had released to reach Earth—where they were detected by both LIGO detectors. The researchers took another few months to check their result and went public in February 2016.

The search is now on for more gravitational waves, and the best place to do it is from space. In December 2015, the spacecraft LISA Pathfinder was launched. It is headed to an orbit at L1, which is a gravitationally stable position between the sun and Earth. There, the spacecraft will test laser interferometry instruments in space, in the hope that they can be used in an ambitious experiment called eLISA (evolved Laser Interferometer Space Antenna). Provisionally scheduled for 2034, eLISA will use three spacecraft triangulated around the sun. Lasers will be fired between the spacecraft, making a laser track 2 million miles (3 million km) long that is many times more sensitive to gravitational waves than LIGO.

The discovery of gravitational waves has the potential to transform astronomers' view of the universe. The patterns in the fluctuations in the light signals from LIGO and future projects will produce new information, providing a detailed map of mass across the universe. ∎

Gravitational waves will bring us exquisitely accurate maps of black holes—maps of their spacetime.
Kip Thorne

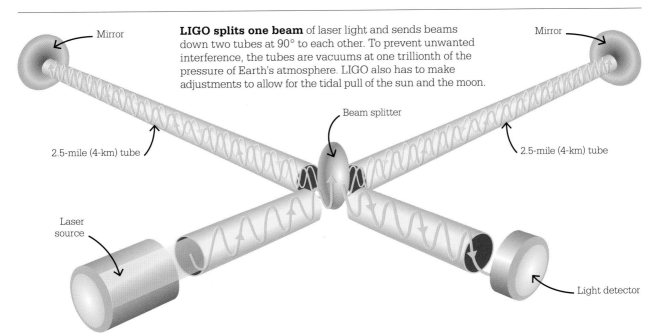

LIGO splits one beam of laser light and sends beams down two tubes at 90° to each other. To prevent unwanted interference, the tubes are vacuums at one trillionth of the pressure of Earth's atmosphere. LIGO also has to make adjustments to allow for the tidal pull of the sun and the moon.

Mirror

Mirror

Beam splitter

2.5-mile (4-km) tube

2.5-mile (4-km) tube

Laser source

Light detector

DIRECTO

RY

DIRECTORY

For a field of enquiry as broad as astronomy, there has not been room to include every significant scientist as a main entry in this book. The following pages list astronomers who have also made important contributions across time, from the 7th century BCE to the present day. In its early stages, astronomy usually involved individuals or small groups making observations and calculations. Modern high-tech "big astronomy," meanwhile, often requires large-scale collaborations of hundreds or thousands of scientists. Whether they are booking time for an experiment at a particle accelerator or requesting that a space telescope be pointed in a particular direction, today's astronomers form part of a huge community developing the big ideas of tomorrow.

ANAXIMANDER OF MILETUS
610–546 BCE

Greek philosopher Anaximander provided one of the earliest attempts at a rational explanation of the universe. He speculated that the celestial bodies made full circles around Earth, which led him to the conclusion that Earth must float freely and unsupported in space. He also stated that celestial bodies lie behind one another, meaning that there was depth to the universe—the first recorded conception of the idea of "space." Anaximander placed the celestial bodies in the wrong order, however, believing that the stars were nearest to Earth, followed by the moon, and then the sun.
See also: The geocentric model 20

ERATOSTHENES
c.276–c.194 BCE

The third chief librarian at the famous Library of Alexandria, Greek scholar Eratosthenes made major contributions to the field of geography. He measured

the circumference of Earth by comparing the angle of the noon shadow at midsummer in Alexandria with that in Syene (present-day Aswan). He knew the distance between the two locations, and his measurement allowed him to figure out the proportion of the entire circumference that this represented. He also produced an accurate measurement of Earth's axial tilt, measured the distances to the sun and the moon, introduced the leap day to correct the length of the year, and produced one of the first ever maps of the world.
See also: Consolidating knowledge 24–25

ZU CHONGZHI
429–500 CE

Tasked with producing a new calendar by the Emperor Xiaowu, Chinese mathematician Zu Chongzhi made highly accurate measurements of the lengths of the sidereal year (Earth's rotation period measured relative to the background stars), the tropical year (the period between successive

vernal equinoxes), and the lunar month. He used his calculations to predict four solar eclipses correctly. Zu measured the length of Jupiter's year as 11.858 Earth years, which is less than 0.1 percent away from the current accepted figure.
See also: The solar year 28–29

AL-BATTANI
c.858–929

Arab astronomer and mathematician Al-Battani made accurate observations to refine the figures for the length of the year, the inclination of the ecliptic, and the precession of the equinoxes. He developed trigonometric methods to improve on Ptolemy's calculations, and showed that the distance of the sun from Earth varies over time. Al-Battani's most influential work was a compilation of astronomical tables, which was translated into Latin in the 12th century and was a major influence on Copernicus.
See also: Consolidating knowledge 24–25 ▪ The Copernican model 32–39

IBN AL-HAYTHAM
c.965–1040

Also known by his Latinized name Alhazen, Ibn al-Haytham worked at the court of the Fatimid Caliphate in Cairo. A pioneer of the scientific method, whereby hypotheses are tested by experiment, al-Haytham wrote a work popularizing Ptolemy's *Almagest* and, later, a book casting doubts on aspects of Ptolemy's system.
See also: Consolidating knowledge 24–25

ROBERT GROSSETESTE
c.1175–1253

English bishop Robert Grosseteste wrote treatises concerning optics, mathematics, and astronomy. He translated Greek and Arabic texts into Latin, introducing the ideas of Aristotle and Ptolemy into medieval European thought. In his work *De luce* (*On light*), Grosseteste made an early attempt to describe the entire universe using a single set of mathematical laws. He called light the first form of existence, which, he said, allowed the universe to spread out in all directions, in a description reminiscent of the Big Bang theory.
See also: The geocentric model 20 ▪ Consolidating knowledge 24–25

JOHANNES HEVELIUS
1611–1687
ELISABETHA HEVELIUS
1647–1693

From an observatory he built on top of his house, Polish astronomer Johannes Hevelius made detailed maps of the surface of the moon.

Although he made and used telescopes, he preferred to map star positions with just a sextant and the naked eye, making him the last major astronomer to do so. Hevelius's second wife, Elisabetha, whom he married in 1663, helped him to compile a catalog of more than 1,500 stars, which she completed and published following his death. A tireless and skilled observer in her own right, Elisabetha was one of the first notable female astronomers.
See also: The Tychonic model 44–47

CHRISTIAAN HUYGENS
1629–1695

Dutch mathematician and astronomer Christiaan Huygens was fascinated by Saturn and the strange "handles" that telescopes revealed to protrude from either side of it. With his brother Constantijn, he constructed a powerful telescope with improved lenses through which to study the planet. Huygens was the first to describe the true shape of Saturn's rings, explaining that they were thin and flat, and tilted at an angle of 20 degrees to the plane of the planet's orbit. He published his findings in 1659 in the book *Systema Saturnium*. Four years earlier, he had discovered Titan, Saturn's largest moon.
See also: Observing Saturn's rings 65

OLE RØMER
1644–1710

Working at the Paris Observatory, Danish astronomer Ole Rømer demonstrated that light has a finite speed. Rømer was working on a project to calculate the time of day using the eclipses of the moons of Jupiter, a method first proposed by Galileo to solve the problem of measuring longitude at sea. Over a number of years, Rømer carefully timed the eclipses of the moon Io and found that their duration varied depending on whether Earth was moving toward Jupiter or away from it. He reasoned that this variation was due to a difference in the time it took the light from Io to reach Earth, and estimated that light takes 22 minutes to travel a distance equal to the diameter of Earth's orbit of the sun. This gave the speed of light as 140,000 miles/s (220,000 km/s), about 75 percent of its true value. Rømer's finding that light has a finite speed was confirmed in 1726, when James Bradley explained the phenomenon of stellar aberration in terms of light speed.
See also: Stellar aberration 78

JOHN MICHELL
1724–1793

English clergyman John Michell studied a wide range of scientific fields, including seismology, magnetism, and gravity. He designed the torsion balance, which his friend Henry Cavendish later used to measure the strength of gravity. Michell was also the first person to propose that an object might be so massive that light would be unable to escape its gravitational pull. He calculated that a star 500 times the size of the sun would be such an object, which he called a "dark star." Michell's idea was largely forgotten until the 20th century, when astronomers started to take the concept of black holes seriously.
See also: Curves in spacetime 154–55 ▪ Hawking radiation 255

JOSEPH-LOUIS LAGRANGE
1736–1813

French–Italian mathematician and astronomer Joseph-Louis Lagrange studied celestial mechanics and the effects of gravity. He explored mathematically the ways in which the gravitational pulls within a system of three bodies, such as the sun, Earth, and the moon, combine with one another. His work led to the discovery of positions with stable orbits for a small body orbiting two larger ones, now called Lagrange points. Space telescopes are often placed near Lagrange points for their orbits around Earth and the sun.
See also: Gravitational theory 66–73 ▪ Studying distant stars 304–05

JEAN BAPTISTE JOSEPH DELAMBRE
1736–1813

A leading figure in scientific circles during the French Revolution, in 1792 Delambre was tasked with measuring the length of the arc of the meridian from Dunkirk to Barcelona. This was to refine the new metric system, which defined the meter as 1/10,000,000 of the distance from the North Pole to the equator. He completed the task in 1798. From 1804, Delambre served as the director of the prestigious Paris Observatory. His astronomical work included the production of accurate tables showing the positions of Jupiter's moons. In 1809, he estimated that light from the sun takes 8 minutes 12 seconds to reach Earth (the figure is now measured at 8 minutes 20 seconds).
See also: Gravitational disturbances 92–93

BENJAMIN APTHORP GOULD
1824–1896

A child prodigy, American Benjamin Apthorp Gould graduated early from Harvard University before moving to Germany to study under the renowned mathematician Friedrich Gauss in 1845. In Europe, he earned a Ph.D. in astronomy—the first American to receive a doctorate in the subject. He returned to the US in 1849 determined to raise the profile of American astronomy. To this end, he founded *The Astronomical Journal* to publish research from the United States; the journal continues to this day. Between 1868 and 1885, Gould worked in Argentina, where he founded the National Observatory in Córdoba. He also helped to set up the Argentine National Weather Service. Gould produced a comprehensive catalog of the bright stars visible from the southern hemisphere, which he published in 1879 as the *Uranometria Argentina*.

RICHARD CARRINGTON
1826–1875

British amateur astronomer Richard Carrington carried out careful observations of the sun over the course of many years. In 1859, he was the first person to observe a solar flare—a magnetic explosion on the surface of the sun that causes a surge of visible light. The flare was followed by disruption to worldwide telegraph systems, and Carrington suggested that such solar activity might have an electrical effect on Earth. In 1863, through his records of the movements of sunspots, he demonstrated that different parts of the sun were rotating at different speeds.
See also: Galileo's telescope 56–63 ▪ The surface of the sun 103

ISAAC ROBERTS
1829–1904

In the 1880s, British amateur astronomer Isaac Roberts made important advances in the field of astrophotography, enabling photographs of the night sky to reveal structures invisible to the naked eye for the first time. Roberts developed an instrument that allowed very long exposure times, and thus the collection of more light. He kept the telescope pointing at exactly the same point in the sky by adjusting it to compensate for the rotation of Earth. Roberts' most famous image is an 1888 photograph of the Andromeda nebula, now known to be a galaxy, which revealed its spiral structure in unprecedented detail.
See also: Astrophotography 118–19

HENRY DRAPER
1837–1882

A pioneer of astrophotography, medical doctor Henry Draper resigned as dean of medicine at New York University in 1873 to devote himself to astronomy. With the assistance of his wife, Anna Mary, Draper photographed the transit of Venus in 1874, was the first to capture the Orion nebula on camera in 1880, and was also the first to take a wide-angled photograph of a comet's tail in 1881. He developed new techniques for astrophotography, but died of pleurisy in 1882, a few years before photography began to be taken seriously by astronomers as a means of discovery. After his death, his wife created a foundation in his name, which funded the Henry Draper Catalogue, a huge photographic survey of the stars carried out by Edward C. Pickering and his team of female astronomers.

See also: The star catalog 120–21 ▪ The characteristics of the stars 122–27

JACOBUS KAPTEYN
1851–1922

Using photographic plates supplied to him from South Africa by David Gill, Dutch astronomer Jacobus Kapteyn cataloged more than 450,000 southern stars. After grouping stars in different parts of the galaxy and measuring their magnitudes, radial velocities, and proper motions, Kapteyn carried out vast statistical analyzes that revealed the phenomenon of star streaming—which shows how the motions of stars are not random, but grouped together in two opposite directions. This was the first definitive evidence that the Milky Way galaxy is rotating.

See also: Astrophotography 118–19

EDWARD WALTER MAUNDER
1851–1928

ANNIE SCOTT DILL MAUNDER
1868–1947

British husband and wife team Edward Walter Maunder and Annie Maunder (née Scott Dill) collaborated at the Greenwich Royal Observatory in the study of the sun. Their investigations of

sunspots uncovered a correlation between their number and Earth's climate. This led them to discover a period of reduced solar activity between 1645 and 1715, now called the Maunder Minimum, which coincided with lower-than-average temperatures in Europe. When the ban on women at the society was lifted in 1916, Annie Maunder was elected a Fellow of the Royal Astronomical Society, after which her observations were published under her own name. Prior to that, much of her work had appeared in papers under her husband's name.

See also: The surface of the sun 103 ▪ The properties of sunspots 129

E. E. BARNARD
1857–1923

US astronomer Edward Emerson Barnard was a renowned observer, who discovered about 30 new comets and numerous nebulae. In 1892, Barnard discovered a fifth moon around Jupiter, called Amalthea, which was to be the last moon to be discovered through visual observation rather than through the study of photographic plates. Himself a pioneer of astrophotography, Barnard produced a series of stunning long-exposure photographs of the Milky Way, which was published posthumously in 1927 as the *Atlas of Selected Regions of the Milky Way*. Barnard's star is named after him; in 1916, he discovered that this faint red dwarf has the largest known proper motion (rate at which a star changes its position on the celestial sphere) of all known stars.

See also: Galileo's telescope 56–63 ▪ Astrophotography 118–19

HEBER D. CURTIS
1872–1942

American classics professor Heber Doust Curtis switched to astronomy in 1900 when he became a volunteer observer for the Lick Observatory in California. After receiving his Ph.D. in astronomy in 1902, Curtis enjoyed a long association with the Lick Observatory, carrying out a detailed survey of the known nebulae, which he completed in 1918. In 1920, he took part in the "Great Debate" with fellow astronomer Harlow Shapley at the Smithsonian museum. Curtis argued that distant nebulae were separate galaxies far from the Milky Way, while Shapley asserted that they lay within it.

See also: Spiral galaxies 156–61 ▪ Beyond the Milky Way 172–77

JAMES JEANS
1877–1946

British mathematician James Jeans worked on a variety of theoretical problems relating to astrophysics. In 1902, he calculated the conditions under which a cloud of interstellar gas becomes unstable and collapses to form a new star. In developing his theory of gases in 1916, he explained how gas atoms can gradually escape from a planet's atmosphere over time. In later life, Jeans devoted his time to writing and became well-known for his nine popular books, including *Through Space and Time* and *The Stars in Their Courses*. He promoted an idealist philosophy that saw both mind and matter as central to understanding the universe, which he described as "nearer to a great thought than to a great machine."

See also: Inside giant molecular clouds 276–79

ERNST ÖPIK
1893–1985

Estonian astrophysicist Ernst Öpik obtained his doctorate at the University of Tartu, Estonia, where he worked from 1921 to 1944, specializing in the study of minor objects such as asteroids, comets, and meteors. In 1922, he estimated the distance of the Andromeda galaxy using a new method based on the galaxy's speed of rotation. This method is still used today. Öpik also suggested that comets originated from a cloud beyond Pluto, now known commonly as the Oort cloud, but sometimes referred to as the Öpik–Oort cloud. As the Red Army approached Estonia in 1944, Öpik fled into exile, eventually settling in Northern Ireland, where he took a position at Armagh Observatory.
See also: The Oort cloud 206

CLYDE TOMBAUGH
1906–1997

In the late 1920s, the Lowell Observatory in Arizona embarked upon a systematic search for a planet believed to be causing perturbations to the orbit of Uranus. To carry out the work, the director Vesto Slipher hired the young amateur astronomer Clyde Tombaugh, who had impressed him with drawings of Jupiter and Mars made using a homemade telescope. After 10 months examining photographs, on February 18, 1930, Tombaugh discovered an object orbiting the sun beyond Neptune. Named Pluto after the Roman god of the underworld, it was initially classified as the ninth planet, but has since been demoted to the status of dwarf planet. Following his discovery, Tombaugh earned a degree and pursued a career as a professional astronomer.
See also: Spiral galaxies 156–61 ▪ Studying Pluto 314–17

VICTOR AMBARTSUMIAN
1908–1996

Soviet–Armenian astronomer Victor Ambartsumian was a founding figure in the field of theoretical astrophysics, contributing to theories of star formation and galactic evolution. He was one of the first people to suggest that young stars formed from protostars. In 1946, he organized the construction of the Byurakan Observatory in Armenia, where he was the director until 1988. A popular lecturer with a colorful and engaging style, Ambartsumian served as the president of the International Astronomical Union from 1961–64, and hosted several conferences on the search for extraterrestrial life.
See also: Dense molecular clouds 200–01 ▪ Inside giant molecular clouds 276–79

GROTE REBER
1911–2002

In 1937, American radio engineer Grote Reber built his own radio telescope in his backyard after hearing of Karl Jansky's discovery of galactic radio waves. Over the next few years, Reber was effectively the only radio astronomer in the world, conducting the first radio survey of the sky and publishing his results in astronomy and engineering journals. Reber's work was to form the basis for the development of radio astronomy after the end of World War II. To conduct further radio investigations in clear atmospheric conditions, in 1954 Reber moved to Tasmania, where he remained for the rest of his life.
See also: Radio astronomy 179

IOSIF SHKLOVSKY
1916–1985

In 1962, Soviet astrophysicist Iosif Shklovsky wrote a popular book examining the possibility of extraterrestrial life, which was republished four years later in an expanded edition, co-authored by Carl Sagan, as *Intelligent Life in the Universe*. In this later edition, paragraphs by the two authors are alternated with one another, as Sagan provides a commentary and expansion on Shklovsky's original points. Many of the latter's ideas were highly speculative, including a suggestion that an observed acceleration of Mars's moon Phobos was due to the fact that it was a hollow artificial structure, a monument to a long-gone Martian civilization.
See also: Life on other planets 228–35

MARTIN RYLE
1918–1984

Like many pioneer radio astronomers, Briton Martin Ryle started his career developing radar technology during World War II. Subsequently, he joined the Cavendish Radio Astronomy Group in Cambridge, where he worked alongside Antony Hewish and Jocelyn Bell Burnell, developing new techniques in radio astronomy and producing a number of catalogs of radio sources. Deeply affected by his experiences of war, Ryle devoted his final years to the promotion of the peaceful use of science, warning against the dangers of nuclear weapons and power, and advocating research into alternative energy.

See also: Radio astronomy 179 ▪ Quasars and pulsars 236–39

HALTON ARP
1927–2013

A staff astronomer at the Mount Wilson Observatory in California for nearly 30 years, Halton Arp gained a reputation as a skilled observer. In 1966, he produced his *Atlas of Peculiar Galaxies*, which cataloged, for the first time, hundreds of odd structures that had been seen in nearby galaxies. Today it is known that many of these features are the result of galaxies colliding. Later in his career, Arp found himself professionally marginalized when he cast doubt on the Big Bang theory. He contended that objects with very different degrees of redshift were close to one another and not at vastly different distances.

See also: Beyond the Milky Way 172–77

ROGER PENROSE
1931–

In the 1960s, British mathematician and physicist Roger Penrose figured out much of the complex mathematics relating to the curvature of spacetime around a black hole. In collaboration with Stephen Hawking, he showed how matter within a black hole collapses into a singularity. More recently,

Penrose has proposed a theory of a cyclic cosmology, in which the heat death (end state) of one universe produces the conditions for the Big Bang of another universe. Penrose has also produced a series of popular science books in which he explains the physics of the universe and suggests novel explanations for the origins of consciousness.

See also: Curves in spacetime 154–55 ▪ Hawking radiation 255

SHIV S. KUMAR
1939–

Indian-born astronomer Shiv S. Kumar earned a doctorate in astronomy at the University of Michigan and has made his career in the United States, working on theoretical problems concerning matters including the origin of the solar system, the development of life in the universe, and exoplanets. In 1962, Kumar predicted the existence of low-mass stars that would be too small to sustain nuclear fusion. Later named brown dwarfs by Jill Tarter, their existence was confirmed in 1995.

See also: Exoplanets 288–95

BRANDON CARTER
1942–

In 1974, Australian physicist Brandon Carter formulated the anthropic principle, which states that the universe must necessarily have certain characteristics for humankind to exist. That is to say that the physical properties of the universe, such as the strength of the fundamental forces, must fall within very narrow limits for sunlike stars capable of sustaining

life to develop. Since 1986, Carter has been the director of research at the Paris–Meudon Observatory. He has also made contributions to understanding the properties of black holes.

See also: Life on other planets 228–35 ▪ Hawking radiation 255

JILL TARTER
1944–

As director of the Center for SETI Research in California, Jill Tarter was a leading figure in the search for extra-terrestrial life for more than 30 years, lecturing widely on the subject before her retirement in 2012. In 1975, she coined the term "brown dwarf" for the type of star, discovered by Shiv S. Kumar, that is not massive enough to sustain nuclear fusion. Carl Sagan based the protagonist in his novel and film *Contact* on Tarter.

See also: Life on other planets 228–35

MAX TEGMARK
1967–

Swedish cosmologist Max Tegmark's research at MIT has focused on developing methods to analyze the vast amounts of data produced by surveys of the cosmic microwave background. Tegmark is a leading proponent of the idea that the results of quantum mechanics are best explained by the existence of a multiverse. He has developed the mathematical universe hypothesis, which proposes that the universe is best understood as a purely mathematical structure.

See also: Observing the CMB 280–85

GLOSSARY

Absolute magnitude A measure of the intrinsic brightness of a star. It is defined as the apparent magnitude of the star from a distance of 10 parsecs (32.6 light-years).

Accretion The process by which smaller particles or bodies collide and join together to form larger bodies.

Aphelion The point on its elliptical orbit around the sun at which a planet, asteroid, or comet is farthest from the sun.

Apparent magnitude A measure of the brightness of a star as seen from Earth. The fainter the object, the higher the value of its apparent magnitude. The faintest stars visible to the naked eye are of a magnitude 6.

Armillary sphere An instrument that models the celestial sphere. At its center is Earth or the sun, around which is a framework of rings representing lines of celestial longitude and latitude.

Asteroid A small body that orbits the sun independently. Asteroids are found throughout the solar system, with the greatest concentration in the asteroid belt between the orbits of Mars and Jupiter. Their diameters range from a few yards to 600 miles (1,000 km).

Astronomical unit (AU) A distance equal to the average distance between Earth and the sun. 1 AU = 92,956,000 miles (149,598,000 km).

Big Bang The event with which the universe is thought to have begun, at a particular time in the past, from a hot, dense initial state.

Black body A theoretical, idealized body that absorbs all the radiation that falls on it, reflecting nothing. A black body would emit a spectrum of radiation with a peak at a particular wavelength, depending on its temperature.

Black hole A region of spacetime surrounding a mass that is so dense that its gravitational pull allows no mass or radiation to escape from it.

Blueshift A shift in a spectrum of light or other radiation toward shorter wavelengths that occurs when the source of the light is moving toward the observer.

Bok globule Small, dark clouds of cold gas and dust, within which it is thought that new stars are forming.

Brown dwarf A starlike ball of gas that is not massive enough to sustain nuclear fusion in its core.

Celestial sphere An imaginary sphere surrounding Earth. The positions of stars and other celestial bodies can be defined by their places on this sphere if they were imagined to be attached to it.

Cepheid variable A pulsating star whose brightness increases and decreases over a regular period. The more luminous it is, the longer the period of its variation.

Comet A small, icy body in orbit around the sun. When a comet approaches the sun, gas and dust evaporate from its nucleus (solid core) to produce a cloud called a coma and one or more tails.

Constellation One of 88 named regions on the celestial sphere, containing an identifiable pattern of naked-eye stars.

Cosmic Microwave Background (CMB) Faint microwave radiation that is detectable from all directions. The CMB is the oldest radiation in the universe, emitted when the universe was 380,000 years old. Its existence was predicted by the Big Bang theory, and it was first detected in 1964.

Cosmic rays Highly energetic particles, such as electrons and protons, that travel through space at close to the speed of light.

Cosmological constant A term that Albert Einstein added to his general relativity equations, which may correspond to the dark energy that is accelerating the expansion of the universe.

Dark energy A little-understood form of energy that exerts a repulsive force, causing the expansion of the universe to accelerate.

Dark matter A form of matter that does not emit radiation or interact with other matter in any way other than through the effect of its gravity. It comprises 85 percent of all mass in the universe.

Degeneracy pressure An outward pressure within a concentrated ball of gas, such as a collapsed star, that is exerted due to the principle that no two particles with mass can exist in the same quantum state.

Doppler effect The change in frequency of radiation experienced by an observer in relative motion to the source of the radiation.

Dwarf planet An object in orbit around a star that is large enough to have formed a spherical shape but that has not cleared its orbital path of other material. Examples in the solar system include Pluto and Ceres.

Dwarf star Also called a main sequence star, a star that shines by converting hydrogen to helium. About 90 percent of stars are dwarf stars.

Eclipse The blocking of light from one celestial body, caused by another body passing between it and an observer, or it and a light source that it reflects.

Ecliptic The apparent path along which the sun travels across the celestial sphere. It is equivalent to the plane of Earth's orbit.

Electromagnetic radiation Waves that carry energy through space in the form of oscillating electric and magnetic disturbances. The electromagnetic spectrum ranges from short, high-energy gamma rays to long, low-energy radio waves, and includes the visible spectrum.

Electron A subatomic particle with negative charge. In an atom, a cloud of electrons orbits a central, positively charged nucleus.

Equinox A twice-yearly occasion when the sun is directly overhead at a planet's equator, meaning that day and night are of roughly equal duration across the entire planet.

Escape velocity The minimum velocity an object needs to be traveling at to escape the gravitational pull of a larger body such as a planet.

Event horizon A boundary around a black hole beyond which no mass or light can escape its gravity. At this point, the escape velocity of the black hole equals the speed of light.

Exoplanet A planet that orbits a star other than the sun.

Fraunhofer lines Dark absorption lines found in the spectrum of the sun, first identified by German Joseph von Fraunhofer in the 19th century.

Galaxy A large collection of stars and clouds of gas and dust that is held together by gravity.

Galilean moon One of the four biggest moons of Jupiter, first discovered in 1610 by Galileo.

General theory of relativity A theory that describes gravity as a warping of spacetime by the presence of mass. Formulated by Albert Einstein in 1916, many of its predictions, such as gravitational waves, have now been confirmed experimentally.

Geocentric Of a system or an orbit, treated as having Earth at the center.

Gnomon The part of a sundial that casts a shadow.

Gravitational wave A distortion of space that travels at the speed of light, generated by the acceleration of mass.

Harvard Spectral Classification A system first devised by the Harvard Observatory in the late 19th century to classify stars by the appearance of their spectra.

Heliocentric Of a system or an orbit, treated as having the sun at the center.

Hertzsprung–Russell diagram A scatter diagram on which stars are plotted according to their luminosity and surface temperature.

Hubble's law The observed relationship between the redshifts and distances of galaxies, which shows galaxies receding with a velocity proportional to their distance. The number that quantifies the relationship is called Hubble's constant (H_0).

Inflation A short period of rapid expansion that the universe is thought to have undergone moments after the Big Bang.

Ionization The process by which an atom or molecule gains or loses electrons to gain a positive or negative charge. The resultant charged particles are called ions.

Kepler's laws of planetary motion Three laws devised by Johannes Kepler to describe the shapes and speeds of the orbits of the planets around the sun.

Kuiper belt A region of space beyond Neptune in which a large number of comets orbit the sun. It is the source of short-period comets.

Light-year (ly) A unit of distance that is the distance traveled by light in one year, equal to 5,878 million miles (9,460 billion km).

Main sequence *See* dwarf star.

Messier object One of the nebulae first cataloged by Charles Messier in 1781.

Meteorite A lump of rock or metal that falls from space and reaches the surface of Earth in one piece or in many fragments.

Nebula A cloud of gas and dust in interstellar space. Before the 20th century, any diffuse object in the sky was known as a nebula; many of these are now known to be galaxies.

Neutrino A subatomic particle with very low mass and zero electric charge, which travels at close to the speed of light.

Neutron A subatomic particle made of three quarks with zero electric charge.

Neutron star A very dense, compact star composed almost entirely of densely packed neutrons. Neutron stars form when the core of a high-mass star collapses in a supernova explosion.

Nova A star that suddenly becomes thousands of times brighter before returning to its original brightness over a period of weeks or months.

Nuclear fusion A process whereby atomic nuclei join together to form heavier nuclei, releasing energy. Inside stars like the sun, this process involves the fusion of hydrogen atoms to make helium.

Oort cloud Also known as the Oort–Öpik cloud. A spherical region at the edge of the solar system containing planetesimals and comets. It is the origin of long-period comets.

Orbit The path of a body around another, more massive, body.

Parallax The apparent shift in position of an object due to the movement of an observer to a different place.

Perihelion The point on its elliptical orbit around the sun at which a planet, asteroid, or comet is closest to the sun.

Perturbation A change in the orbit of a body, caused by the gravitational influence of other orbiting bodies. Observed perturbations in the orbit of the planet Uranus led to the discovery of Neptune.

Planet A non-luminous body that orbits a star such as the sun, is large enough to be spherical in shape, and has cleared its neighborhood of smaller objects.

Planetesimal A small body of rock or ice. The planets formed from planetesimals that joined together by the process of accretion.

Precession A change in the orientation of a rotating body's axis of rotation, caused by the gravitational influence of neighboring bodies.

Proper motion The rate at which a star changes its position on the celestial sphere. This change is caused by the star's motion relative to the motion of other stars.

Proton A subatomic particle with a positive charge, made of three quarks. The nucleus of the element hydrogen contains a single proton.

Protostar A star in the early stages of its formation, comprising a collapsing cloud that is accreting matter but in which nuclear fusion has not yet begun.

Pulsar A rapidly rotating neutron star. Pulsars are detected on Earth by their rapid, regular pulses of radio waves.

Quadrant An instrument for measuring angles of up to 90°. Ancient astronomers used quadrants to measure a star's position on the celestial sphere.

Quark A fundamental subatomic particle. Neutrons and protons are made of three quarks.

Quasar Short for "quasi-stellar radio source," a compact but powerful source of radiation that is believed to be an active galactic nucleus.

Radial velocity The part of the velocity of a star or other body that is along the line of sight directly toward or directly away from an observer.

Radio astronomy The branch of astronomy that studies radiation in the long radio wavelength, first discovered to be coming from space in the 1930s.

Red dwarf A cool, red, low-luminosity star.

Red giant A large, highly luminous star. A main sequence star becomes a red giant near the end of its life.

Redshift A shift in a spectrum of light or other radiation toward longer wavelengths that occurs when the source of the light is moving away from an observer.

Reflecting telescope A telescope in which an image is formed by reflecting light on a curved mirror.

Refracting telescope A telescope that creates an image by bending light through a converging lens.

Relativity Theories developed by Albert Einstein to describe the nature of space and time. *See also* general theory of relativity.

Satellite A small body that orbits a larger one.

Schwarzschild radius The distance from the center of a black hole to its event horizon.

SETI Short for Search for Extra-Terrestrial Intelligence, the scientific search for alien life.

Seyfert galaxy A spiral galaxy with a bright, compact nucleus.

Sidereal Relating to the stars. A sidereal day corresponds to Earth's rotation period measured relative to the background stars.

Singularity A point of infinite density at which the known laws of physics appear to break down. It is theorized that there is a singularity at the center of a black hole.

Solar wind A stream of fast-moving, charged particles emanating from the sun that flows out through the solar system. It consists mostly of electrons and protons.

Spacetime The four-dimensional combination of the three dimensions of space and one of time. According to the theory of relativity, space and time do not exist as separate entities. Rather, they are intimately linked as one continuum.

Spectrum The range of the wavelengths of electromagnetic radiation. The full spectrum ranges from gamma rays, with wavelengths shorter than an atom, to radio waves, whose wavelength may be many feet long.

Spectroscopy The study of the spectra of objects. The spectrum of a star contains information about many of its physical properties.

Spiral galaxy A galaxy that takes the shape of a central bulge or bar surrounded by a flattened disk of stars in a pattern of spiral arms.

Standard candle A celestial body that has a known luminosity, such as a Cepheid variable star. These allow astronomers to measure distances that are too large to measure using stellar parallax.

Star A luminous body of hot gas that generates energy through nuclear fusion.

Steady State theory A theory proposing that matter is constantly created. The theory was an attempt to explain the universe's expansion without the need for a "Big Bang."

Stellar aberration The apparent motion of a star caused by movement of an observer in a direction perpendicular to the direction to the star.

Stellar parallax *See* parallax.

Subatomic particle One of the many kinds of particle that are smaller than atoms. These include electrons, neutrinos, and quarks.

sunspot An area on the surface of the sun that appears dark because it is cooler than its surroundings. sunspots are found in areas of concentrated magnetic field.

Supernova The result of the collapse of a star, which causes an explosion that may be many billions of times brighter than the sun.

Time dilation The phenomenon whereby two objects moving relative to each other, or in different gravitational fields, experience a different rate of flow of time.

TNO Short for Trans-Neptunian Object. Any minor planet (dwarf planet, asteroid, or comet) that orbits the sun at a greater average distance than Neptune (30 AU).

Transit The passage of a celestial body across the face of a larger body.

Wavelength The distance between two successive peaks or troughs in a wave.

White dwarf A star with low luminosity but high surface temperature, compressed by gravity to a diameter close to that of Earth.

Zodiac A band around the celestial sphere, extending 9° on either side of the ecliptic, through which the sun, moon, and planets appear to travel. The zodiac crosses the constellations that correspond to the "signs of the zodiac."

INDEX

M

N

ACKNOWLEDGMENTS

Dorling Kindersley would like to thank Allie Collins, Sam Kennedy, and Kate Taylor for additional editorial assistance, Alexandra Beeden for proofreading, and Helen Peters for the index.

PICTURE CREDITS

The publisher would like to thank the following for their kind permission to reproduce their photographs:

(Key: A-Above; B-Below/Bottom; C-Center; F-Far; L-Left; R-Right; T-Top)

24 Wikipedia (bc). **25 Wikipedia** (tr). **27 ESO:** Dave Jones/http://creativecommons.org/licenses/by/3.0/ (bl). **28 Dreamstime.com:** Yang Zhang (bc). **29 Alamy Stock Photo:** JTB Media Creation, Inc. (bl). **31 Dreamstime.com:** Eranicle (bl). **34 Dreamstime.com:** Nicku (bl). **36 Getty Images:** Bettmann (bl). **39 Tunc Tezel** (t). **45 Alamy Stock Photo:** Heritage Image Partnership Ltd (tr). **46 Alamy Stock Photo:** Heritage Image Partnership Ltd (bl). **47 Wellcome Images:** http://creativecommons.org/licenses/by/4.0/ (bl). **49 NASA:** M. Karovska/CXC/M.Weiss (tl). **52 Getty Images:** Bettmann (tr). **53 Wellcome Images:** http://creativecommons.org/licenses/by/4.0/ (tr). **55 Getty Images:** Print Collector (tr). **59 Dreamstime.com:** Brian Kushner (br). **Getty Images:** UniversalImagesGroup (tl). **61 Dreamstime.com:** Joseph Mercier (tr). **62-63 NASA:** DLR (t). **63 Dreamstime.com:** Nicku (bl). **64 NASA:** SDO/AIA (cr). **65 NASA:** ESA/E. Karkoschka (br). **68 Wellcome Images:** http://creativecommons.org/licenses/by/4.0/ (bl). **69 Science Photo Library:** Science Source (tr). **70 Dreamstime.com:** Zaclurs (bl). **71 NASA:** CXC/U.Texas/S. Park et al/ROSAT (bc). **72 Rice Digital Scholarship Archive:** http://creativecommons.org/licenses/by/3.0/ (bl). **75 Dreamstime.com:** Georgios Kollidas (tr). **Wikipedia** (tl). **77 NASA:** W. Liller (tr). **85 Dreamstime.com:** Georgios Kollidas (bl). **Wikipedia** (cr). **87 Adam Evans:** http://creativecommons.org/licenses/by/2.0/ (b). **88 Dreamstime.com:** Dennis Van De Water (c). **90 Science Photo Library:** Edward Kinsman (br). **91 Getty Images:** UniversalImagesGroup (tr). **93 Wellcome Images:** http://creativecommons.org/licenses/by/4.0/ (bl). **96 Wellcome Images:** http://creativecommons.org/licenses/by/4.0/ (bl). **97 NASA:** UCLA/MPS/DLR/IDA99 (tr). **98 Getty Images:** Science & Society Picture Library (bl). **99 NASA:** UCAL/MPS/DLR/IDA (bc). **100 Dreamstime.com:** Dennis Van De Water (bc). **101 Wellcome Images:** http://creativecommons.org/licenses/by/4.0/ (tr).

103 NASA: SDO (br). **105 Wellcome Images:** http://creativecommons.org/licenses/by/4.0/(tr, bl). **107 Science Photo Library:** Royal Astronomical Society (tr). **115 NASA** (tl); **Wellcome Images:** http://creativecommons.org/licenses/by/4.0/ (cr). **116 Dreamstime.com:** Aarstudio (cr). **117 Wikipedia** (bc). **119 Getty Images:** Gallo Images (tc). **Wikipedia:** J E Mayall (bl). **121 Harvard College Observatory** (tr, bl). **124-125 Science Photo Library:** Christian Darkin (b). **127 Library of Congress, Washington, D.C.** (tr). **NASA** (bl). **129 NASA:** SDO (bc). **135 Dreamstime.com:** Kirsty Pargeter (tl). **136 NASA:** ESA/Hubble Heritage Team (tr). **139 Wikipedia** (bl). **140 NASA:** ESA/J. Hester/A. Loll (bc). **150 Wikipedia** (bl). **152 NASA:** Johns Hopkins University Applied Physics Laboratory/Carnegie Institution of Washington (bl). **155 Alamy Stock Photo:** Mary Evans Picture Library (tr). **158 Alamy Stock Photo:** Brian Green (bl). **160 Lowell Observatory Archives** (bl). **NASA** (tl). **161 NASA:** ESA/Z. Levay/R. van der Marel/STScI/T. Hallas and A. Mellinger (tr). **163 Alamy Stock Photo:** PF-(bygone1) (tr). **164 Wikipedia:** Nick Risinger (cr). **165 ESA** (bl). **167 Library of Congress, Washington, D.C.** (bl). **NASA:** SDO (tl). **169 Getty Images:** Bettmann (tr). **174 Getty Images:** New York Times Co. (bl). **175 Getty Images:** Margaret Bourke-White (tl). **177 ESA:** D. Ducros (t). **179 NRAO:** AUI/NSF/http://creativecommons.org/licenses/by/3.0/ (cr). **181 Getty Images:** Bettmann (tr). **NASA** (bl). **183 Getty Images:** Ralph Morse (tr). **185 NASA:** ESA/A. van der Hoeven (cr). **186 NASA** (br). **190 Princeton Plasma Physics Laboratory:** (bl). **192 ESO:** Y. Beletsky/http://creativecommons.org/licenses/by/3.0/ (bl). **193 ESA** (br). **NASA** (tl). **194 NASA** (tl). **195 NASA** (tr). **199 Getty Images:** Express Newspapers (tr). **200 NASA:** ESA/N. Smith/STScI/AURA (bc). **201 Getty Images:** Jerry Cooke (tr). **207 ESA** (tr). **208 Getty Images:** Keystone-France (cr). **209 Getty Images:** Detlev van Ravenswaay (bc). **Wikipedia** (tr). **211 Dreamstime.com:** Mark Williamson (tr). **215 Getty Images:** Handout (tr). **216 NASA:** CXC/NGST (tl); GSFC/JAXA (bc). **217 ESA:** XMM-Newton/Gunther Hasinger, Nico Cappelluti, and the XMM-COSMOS collaboration (br). **219 NASA:** ESA/M. Mechtley, R. Windhorst, Arizona State University (tl). **220 ESO:** M. Kornmesser/http://creativecommons.org/licenses/by/3.0/ (tl). **221 California Institute of Technology** (bl). **NASA:** L. Ferrarese (Johns Hopkins University) (tc). **225 Science Photo Library:** Emilio Segre Visual Archives/American Institute of Physics (tr). **226 Getty Images:** Ted Thai (bl). **227 Science Photo Library:** Carlos Clarivan (tr); Emilio Segre Visual Archives/American Institute of Physics (bl). **230 Getty Images:** Bettmann (bl). **231 NASA:** Don Davis (tl).

232 NASA AMES Research Centre (bl). **233 Science Photo Library** (tr). **234 NASA** (tr). **234-235 NASA:** Colby Gutierrez-Kraybill/https://creativecommons.org/licenses/by/2.0/ (b). **235 NASA** (tr). **237 NASA** (br). **239 Getty Images:** Daily Herald Archive (tr). **241 NASA:** ESA/Z. Levay/STScI (br). **244 NASA** (bl). **245 NASA:** NASA Archive (tl). **246 NASA** (tr, bl). **247 NASA** (b). **248 NASA** (tl). **249 NASA** (br). **253 Brookhaven National Laboratory** (tr). **254 NASA:** CXC/M.Weiss (br). **262 NASA** (tr). **263 NASA** (bl). **264 NASA** (tr). **265 Science Photo Library:** NASA/Detlev van Ravenswaay (br). **266 NASA** (tl). **267 NASA** (tl). **271 NASA:** ESA/HST (bl). **Science Photo Library:** Detlev van Ravenswaay (tr). **273 Getty Images:** Mike Pont (tr). **274 Massimo Ramella** (bc). **275 Science Photo Library:** Prof. Vincent Icke (br). **277 NASA:** ESA/Hubble Heritage Team (tr). **279 ALMA Observatory:** ESO/NAOJ/NRAO (bc). **ESO:** A. Plunkett/http://creativecommons.org/licenses/by/3.0/ (t). **282 NASA:** COBE Science Team (tr). **283 Michael Hoefner:** http://creativecommons.org/licenses/by/3.0/ (tr). **284 NASA** (bl). **285 NASA** (tr). **287 Getty Images:** Bettmann (bl). **Science Photo Library:** John R. Foster (tr). **290 Alamy Stock Photo:** EPA European Pressphoto Agency b.v. (bl). **291 Dreamstime.com:** Photoblueice (tl). **293 NASA Goddard Space Flight Center:** S. Wiessinger (b). **294 NASA:** Kepler Mission/Dana Berry (bc); Kepler Mission/Dana Berry (br). **296 NASA:** ESA/E. Hallman (cr). **297 NASA:** CXC/Stanford/I. Zhuravleva et al. (br). **301 NASA** (br). **302 Science Photo Library:** Fermi National Accelerator Laboratory/US Department of Energy (bl); **Lawrence Berkeley National Laboratory** (t). **303 Dreamstime.com:** Dmitriy Karelin (br). **304 ESA/Hubble:** C. Carreau (cr). **308 NASA:** UMD (bl). **309 ESA:** C. Carreau/ATG Medialab (tl). **310 Science Photo Library:** ESA/Rosetta/NAVCAM (tl). **311 ESA:** Rosetta/MPS for OSIRIS Team/UPD/LAM/IAA/SSO/INTA/UPM/DASP/IDA (tr). **313 Science Photo Library:** Chris Butler (br). **315 Southwest Research Institute** (tr). **316 NASA:** Johns Hopkins University Applied Physics Laboratory/Southwest Research Institute (tl). **317 NASA:** JHUAPL/SwRI (tl); JHUAPL/SwRI (tr). **320 NASA** (bl). **321 Getty Images:** Sovfoto (tl). **322 Science Photo Library:** NASA (bl). **323 NASA** (tr). **324 NASA:** MSSS (tl). **325 Airbus Defence and Space** (tr). **327 ESO:** http://creativecommons.org/licenses/by/3.0/ (br); L. Calçada/http://creativecommons.org/licenses/by/3.0/ (tl). **329 NASA** (tl). **331 Laser Interferometer Gravitational Wave Observatory (LIGO)** (tl).

All other images © Dorling Kindersley. For more information see: www.dkimages.com